*Intermediate algebra*

**(Intermediate**

# algebra) (Melvin F. Janowitz)

UNIVERSITY OF MASSACHUSETTS

Prentice-Hall, Inc.
Englewood Cliffs, New Jersey

10   9   8   7   6   5   4   3   2   1

Library of Congress Cataloging in Publication Data

Janowitz, M      F
  Intermediate algebra.
  Includes index.
  1.  Algebra.    I.  Title.
QA152.J33           512.9           75-29258
ISBN 0-13-469528-3

PRENTICE-HALL INTERNATIONAL, INC., London

PRENTICE-HALL OF AUSTRALIA, PTY. LTD., Sydney

PRENTICE-HALL OF CANADA, LTD., Toronto

PRENTICE-HALL OF INDIA PRIVATE LIMITED, New Delhi

PRENTICE-HALL OF JAPAN, INC., Tokyo

PRENTICE-HALL OF SOUTHEAST ASIA (PTE.) LTD., Singapore

*On the cover is a reproduction of a woodcut entitled "Circle Limit IV, 1960" by Maurits Escher. It shows an interlocking design of angels and devils covering a fixed area (the circle). The original print is in the collection of the Escher Foundation, Haags Gemeentemuseum, which gave permission for its use.*

# Preface

This book has been prepared for use in introductory algebra courses at the college level. It is designed to acquaint the student with the fundamental language and concepts of algebra, thus providing the necessary background for more advanced mathematics courses, for courses in the sciences, indeed, for any endeavor that requires a working knowledge of algebra.

In writing this text, I have been guided by my experience in teaching mathematics for the past fifteen years. The formal, "theorem-proof" format usually frightens and confuses most students at this level. For that reason I have tried to use a conversational tone to explain concepts and procedures as simply and clearly as possible. Informality notwithstanding, I have made every effort to state results in a mathematically precise way. Plausibility arguments, too, are given to show why various results might be true.

Algebra students benefit from illustrative examples, and this book supplies a wealth of them, in the text and in the exercises. New ideas are introduced with examples, then stated formally and explained and, when necessary, illustrated again. Moreover, many exercises are completely solved for the student in sample solutions sections following each exercise set. In these sections, the idea is to learn how to break complex procedures down into sequences of simpler steps; the student can verify results and methodology.

It would be difficult to thank all the people who have helped me to realize the objectives embodied in this book; so I shall limit my published acknowledgements to Frank C. Denny, Chabot College, Michael Gartenberg, City University of New York, Albert W. Liberi, Westchester Community College, and David Russell, Prince George's Community College. Similar contributions and special assistance were received from Frank Deane, Berkshire Community College and Joseph Will, who helped check for mathematical accuracy.

(∘)

**Note to the instructor** ·

Several aspects of this book's organization should be mentioned. Set theory is introduced in Chapter 1, but it is discussed informally and is used mainly to facilitate the set notation that is needed in the book. Chapter 1 also reviews basic arithmetic and the order properties of the real number system. It does not, however, include an axiomatic treatment of the real number system. To present the axioms for a complete ordered field and then to prove things like $a \cdot 0 = 0$ seems to obscure rather than clarify any understanding of the arithmetic of the real numbers. Of course, there are some proofs, but they are presented in the spirit of explaining why various laws really work. Some instructors will want to cover all the material of the chapter, whereas others may choose to discuss the high points, or even to omit it entirely.

Chapters 2 and 3 review polynomials and fractions. First-degree equations are presented quite early, in Chapter 4. Word problems first appear in this chapter, arranged in various categories. The system of solution is explained for each type of problem, then the student is guided so that he can apply the general system to any problem within the category. Systems of linear equations are discussed in Chapter 9, but the instructor who wishes to cover them earlier may easily do so.

Relations are introduced in Chapter 7, and functions are defined as a certain kind of relation. From that point on, functions are used wherever possible. An extensive treatment of first- and second-degree functions and their graphs is given in Chapter 7. The explanation of inverse functions in Chapter 8 has a strong geometric orientation. Logarithmic functions are defined to be inverses of exponential functions, and in Chapter 10, sequences are treated as functions. The equations that result from word problems are often interpreted as functions whose domain has been restricted.

A few exercises have complicated answers, and some require a fair amount of computation. This is not an accident. Practical problems do not always have nice integral solutions. Students should learn how to solve the hard problems as well as the easy ones.

The text was written for use in a three-credit, one-semester course. If a shorter course is desired, Sections 8.8, 9.2, and 9.5 may safely be omitted since they are self-contained. Chapter 10 is another logical candidate for omission.

A teacher's manual and a student guide supplement the text. The teacher's manual contains suggestions for course sequences; some discussion of the philosophy of, and prerequisites for, each chapter; sample tests and answers; and answers to the text's even-numbered problems. (All odd-numbered answers can be found in the text—either in sample solutions sections or in the backmatter.)

The student guide has numerous additional exercises to supplement the ones in the text. It is designed to help the student discover his own weak points and correct them.

(∘)

**Note to the student**

One of the best ways to learn how to solve problems of all kinds is to examine similar problems that have already been worked out. In each exercise set in this text, you will find special problems that are solved step by step in the sample solutions section that follows the exercise set. Your attention is called to these special problems in two ways. To the left of any line that contains one or more of these problems you will find a device that looks like this (∘); and the special problems themselves have green, boldface problem numbers, whereas all other problem numbers are printed in dark blue.

The green color is used in another way in this textbook. In explanatory sections, important numbers, terms, or mathematical ideas are often isolated by printing them in green so that it will be easy for you to "follow the progress" of such algebraic expressions from the start of an operation to its end. For example, on page 19 the numerals 6 and 12 printed in green can help you actually to "see" the associative nature of the multiplication operation. In other places, green is used for emphasis. On page 33, the exponents printed in green call attention to the results of operations involving exponents. Don't think of the green print as just an attractive touch! It is a tool that can help you to understand the subject of algebra better.

The backmatter of this book includes several different kinds of useful material: A list of important algebraic statements and the text pages on which they are introduced; formulas from geometry; a table of squares and square roots for the positive integers from 1 to 99; a table of common logarithms; the answers to odd-numbered problems other than those that are worked out in sample solutions sections; and the index.

A workbook entitled *Study Guide for Janowitz's Intermediate Algebra* by Frank Deane is available to help you learn the material for this course. The study guide is particularly useful in showing how to work problems and in giving you practice at that.

# Contents

## (1) (Sets and numbers)

1.1 Sets, 1 ◦ 1.2 Unions and intersections, 6 ◦ 1.3 Number lines, 9 ◦ 1.4 Addition and subtraction, 14 ◦ 1.5 Multiplication and division, 19 ◦ 1.6 The distributive law and the order of operations, 22 ◦ 1.7 Order properties of real numbers, 26

## (2) (Polynomials)

2.1 Positive integral exponents, 32 ◦ 2.2 Some basic definitions, 36 ◦ 2.3 Addition and subtraction, 40 2.4 Multiplication, 45 ◦ 2.5 Monomial factors, 50 2.6 Factoring quadratics, 53 ◦ 2.7 More factoring, 57

## (3) (Fractions)

3.1 Basic definitions, 61 ◦ 3.2 Least common multiples. 66 3.3 Using the fundamental principle, 69 ◦ 3.4 Addition and subtraction, 75 ◦ 3.5 Multiplication and division, 79 3.6 Division by $x - a$, 85 ◦ 3.7 Complex fractions, 90

# (4) (First-degree equations and inequalities)

4.1 Equations, 99 ∘ 4.2 Solving equations, 102 ∘ 4.3 Equations in more than one variable, 106 ∘ 4.4 Word problems 109 4.5 More word problems, 117 ∘ 4.6 Inequalities, 126 4.7 Absolute value, 133

# (5) (Exponents and radicals)

5.1 Integer exponents, 138 ∘ 5.2 Rational exponents, 144 5.3 Radical notation, 148 ∘ 5.4 More on radical notation, 152 5.5 Arithmetic of radicals, 158 ∘ 5.6 Sums and differences of complex numbers, 162 ∘ 5.7 Products and quotients of complex numbers, 165

# (6) (Quadratic equations and inequalities)

6.1 Solution by factoring, 169 ∘ 6.2 Extraction of roots and completing the square, 173 ∘ 6.3 The quadratic formula, 178 6.4 Equations quadratic in form 182 ∘ 6.5 Equations involving square roots, 185 ∘ 6.6 More inequalities, 188 ∘ 6.7 Word problems, 193

# (7) (Functions and relations)

7.1 Graphing in two dimensions, 201 ∘ 7.2 Relations and functions, 207 ∘ 7.3 The distance formula, 214 7.4 Equations for straight lines, 218 ∘ 7.5 Graphing quadratic functions, 229 ∘ 7.6 Sketching parabolas, 237 ∘ 7.7 Circles, ellipses, and hyperbolas, 249 ∘ 7.8 Direct and inverse variation, 257

# (8) (Exponential and logarithmic functions)

8.1 Scientific notation, 263 ∘ 8.2 Inverses of functions, 267 8.3 Exponential functions, 275 ∘ 8.4 Logarithmic functions, 284 8.5 Properties of logarithms, 289 ∘ 8.6 Common logarithms, 294 ∘ 8.7 Computations with logarithms, 300 8.8 Exponential equations, 310

## (9) (Systems of equations)

9.1 Systems of linear equations in two variables, 317
9.2 Systems in two variables—solution by determinants, 330
9.3 Systems of linear equations in three variables, 336
9.4 Systems in three variables—solution by determinants, 346
9.5 Systems involving second-degree equations, 352

## (10) (Sequences and series)

10.1 Basic terminology, 364   ∘   10.2 Arithmetic sequences, 368
10.3 Geometric sequences, 374   ∘   10.4 Binomial expansions, 383

**Important algebraic statements,** *390*

**Formulas from geometry,** *394*

**Tables of powers and roots,** *396*

**Common logarithms,** *397*

**Answers to selected problems,** *399*

**Index,** *434*

*Intermediate algebra*

# (1)

## Sets and numbers

### (1.1) (Sets)

People need to communicate, and they create elaborate systems for that purpose. The communication system that comes most quickly to mind is language, both spoken and written. Language, which is made up of words that have fairly reliable meanings, is governed by rules of grammar that tell us how to make meaningful word combinations.

Not everyone in the world uses the same language, of course. A person who speaks only English will have trouble communicating with someone who speaks only Swahili. These two people may have many common interests, be potentially the best of friends, yet never get to know each other because they do not share a language.

Mathematics, too, is a system of communication. It enables you to communicate with others by means of numbers and symbols rather than words—but all the same it is a way of using meaningful combinations to express thoughts and to find out what others are thinking. In order for you to understand the ideas of mathematics, it is necessary for you to learn its language.

Let's begin our study of the language of math with the notion
of *set*. There is nothing complicated or mysterious about sets. A
set is a collection of some kind. The only restriction placed on
this collection is that there must be a definite way to decide whether
a particular object belongs to the collection. The objects that belong
to a set are called the *elements,* or *members,* of the set.

*set*

*elements, members*

Sets are often symbolized by capital letters—*A, B, C,* and so
on—and members of sets by small letters—*a, b, c,* and so on. Another
symbolism for sets uses braces, { }, to enclose either a list of elements
of the set or some phrase that describes the elements. Thus, both
{1, 2, 3} and {first three natural numbers} represent the set whose
elements are the numbers 1, 2, and 3.

If you can count the elements of a set one by one until you
reach a final element, that set is called *finite.* When you *can't* count
the elements of a set, you are dealing with an *infinite set.* If you
start listing the even numbers—2, 4, 6, 8,...—you have started on
an infinite set; you can never reach a "last" element. A set that
contains no members is called the *empty set* or *null set.* It is denoted
by the symbol Ø and is considered a finite set.

*finite set*

*infinite set*

*empty set, null set*

*Ø*

The sets that will concern us consist mainly of numbers and
sometimes of letter symbols. But numbers are a bit more complicated
than they look. The number that we call "two" is really an abstract
idea associated with the collection of all sets containing a pair
of objects, and the symbol 2, called a *numeral,* is used to represent
this abstract idea in writing. Once you understand the distinction
between the abstract idea embodied by a number and the symbol
(numeral) that stands for this idea, you are permitted to abbreviate
a bit. There can be no confusion if you speak of "the number two"
or write "2" when you really mean "the abstract idea that is
represented by the word two or the numeral 2."

*numeral*

You will be expected to be familiar with the following sets of
numbers:

*N, natural numbers*

**1** The set *N* of *natural numbers,* consisting of the counting numbers 1,
2, 3,..., and so on.

*I, integers*

**2** The set *I* of *integers,* which contains the natural numbers, their negatives,
and zero.

*Q, rational numbers*

**3** The set *Q* of *rational numbers.* A typical element of *Q* can be expressed
in the form $a/b$, with $a$ and $b$ integers and $b$ not zero. Some examples
of rational numbers are $1/2$, $7/8$, $-2/3$, and 8, which is rational since
it can be expressed in the form $8/1$.

*P, irrational numbers*

**4** The set *P* of *irrational numbers.* The members of this set are those real
numbers that are *not* rational. Examples are such numbers as $\sqrt{2}$, $-\sqrt{3}$,
and $\pi$.

**5** The set $R$ of *real numbers*.

Before you can hope to do very much with these sets, however, you will need some additional terminology. Two sets are called *equal* if they have the same members. Thus $\{1, 2, 3\}$ = {first three natural numbers} since both sets comprise the numbers 1, 2, and 3.

The set $A$ is called a *subset* of the set $B$ if every element of $A$ is also an element of $B$. The symbol for a subset is $\subset$, and $A \subset B$ is read "$A$ is a subset of $B$" or "$A$ is contained in $B$." Thus, $\{1, 2, 4\} \subset \{1, 2, 3, 4\}$ and $\{1, 2, 4\} \subset \{1, 2, 4\}$ are each valid assertions. The empty set is taken to be a subset of every set.

The fact that the object $x$ is an element of the set $A$ is denoted by writing $x \in A$. To indicate that $x$ is *not* an element of $A$, one writes $x \notin A$. The slant-bar symbol, $/$, can be used consistently to negate symbols. Thus, $\neq$ means "is not equal to" and $\not\subset$ means "is not a subset of."

Of the sets mentioned earlier in this section, we have that $N \subset I$, $I \subset Q$, $Q \subset R$, and $P \subset R$. (Can you add to this list?) In the following exercise set, you will get some practice in manipulating the symbols that have been introduced.

## Exercise set 1.1

*The left-hand column contains six sets described in words, and the right-hand column contains seven sets described in symbols. For each set in the left-hand column, find a set in the right-hand column to which it is equal. Of course, you will have one set left unmatched.*

$(\circ)$

| | | | |
|---|---|---|---|
| 1 | {Natural numbers less than 5} | **a** | $\{-3, +3\}$ |
| 2 | {First three natural numbers} | **b** | $\{4, 6, 8,...\}$ |
| 3 | {Even integers between 3 and 9} | **c** | $\{1, 2, 3\}$ |
| 4 | {Natural-number multiples of 4} | **d** | $\{4, 8, 12,...\}$ |
| 5 | {Real numbers whose square is 9} | **e** | $\{1, 2, 3, 4\}$ |
| 6 | {First three even natural numbers} | **f** | $\{4, 6, 8\}$ |
| | | **g** | $\{2, 4, 6\}$ |

*Specify each set by enclosing a list of its members within braces.*

$(\circ)$

7 {Natural-number multiples of 3}

8 {First four natural numbers}

   9   {Even integers between $-3$ and $+5$}

10   {Odd integers between 4 and 6}

11   {Natural numbers whose square is 4}

12   {Integers whose square is 4}

13   {Natural-number multiples of 9}

14   {First four even natural numbers}

*The following sets are described in symbols. Describe each of them in words.*

   15   {1, 3, 5}              16   {2, 4, 6}

17   {5, 10, 15,...}          18   {6, 7, 8, 9, 10}

19   {1, 2, 3, 4, 5, 6}      20   {4}

21   {$-4$, $+4$}            22   {6, 12, 18}

*In problems 23 to 30, determine whether the given set is finite or infinite.*

   23   {Natural numbers whose last digit is 5}

24   {First billion natural numbers}

   25   {Natural numbers whose numerals contain three digits}

26   {Real numbers between 0 and 1}

27   {Natural numbers whose square does not exceed 100}

28   {Real numbers whose square does not exceed 100}

29   {Rational numbers between 0 and 1}

30   {Integers not exceeding 4}

*In problems 31 to 38, let A = {1, 2, 3, 4}, B = {1, 2}, C = {1}, and D = {2, 3}. Fill in each blank line with the symbol $\subset$ or the symbol $\not\subset$, so that the resulting statement is true.*

   31   $A$ _____ $B$           32   $B$ _____ $A$

   33   $C$ _____ $B$           34   $B$ _____ $C$

35   $B$ _____ $D$           36   $D$ _____ $B$

37   $C$ _____ $D$           38   $A$ _____ $A$

*In problems 39 to 46, fill in the blank spaces with the symbol $\in$ or the symbol $\notin$, so that the resulting statement is true.*

   39   3 _____ {3, 4}          40   $-2$ _____ {1, 2, 3, 4}

$( \circ )$    41   $\sqrt{2}$ ____ $Q$             42   $\dfrac{3}{2}$ ____ $I$

43   $-2$ ____ $N$                                44   $341$ ____ $N$

45   $3$ ____ {even integers}                      46   $2$ ____ {even integers}

In problems 47 to 54, fill in each blank space with the symbol $\in$ or the symbol $\subset$, so that the resulting statement is true.

$( \circ )$    47   $2$ ____ $\{2, 3\}$            48   $\{2\}$ ____ $\{2, 3\}$
$( \circ )$    49   $\emptyset$ ____ $\{1, 2\}$    50   $0$ ____ $I$

51   $\{0\}$ ____ $I$                              52   $3$ ____ $N$

53   $2$ ____ $\{2\}$                              54   $N$ ____ $I$

Suppose you are asked to list the subsets of $\{a, b\}$. The empty set is a subset with no elements, the one-element subsets are $\{a\}$ and $\{b\}$, and the only two-element subset is $\{a, b\}$. Thus the subsets of $\{a, b\}$ are $\emptyset$, $\{a\}$, $\{b\}$, and $\{a, b\}$.

List all subsets of each of the following sets.

$( \circ )$    55   $\{0, 1\}$                     56   $\{1, 2\}$

57   $\{1, 2, 3\}$                                 58   $\{0, 1, 2, 3\}$

$( \circ )$

*Sample solutions*
*for exercise set 1.1*

1   The natural numbers less than 5 are 1, 2, 3, and 4. Hence, item **(1)** matches up with item **(e)**.

7   $\{3, 6, 9,...\}$. The notation "..." here is really ambiguous. Anything could go in there. Yet for simple sets like the natural-number multiples of 3, we can agree to let the notation mean that the elements of the set continue indefinitely in the relationship we have established.

9   The elements of the set of even integers are 0, +2, −2, +4, −4, +6, −6,.... Those between −3 and +5 are −2, 0, 2, and 4. The correct answer is therefore $\{-2, 0, 2, 4\}$.

15   Each of the following descriptions is correct, and you should be able to provide one or two more: {First three odd natural numbers}, {Odd integers between 0 and 6}, and {Odd natural numbers not exceeding 5}.

23   The set $\{5, 15, 25, 35,...\}$ is clearly infinite.

25   The elements of this set are 100, 101, 102,...,999. Hence, it is finite.

31   $A \not\subset B$. This statement is true because 3 and 4 are elements of $A$ but not of $B$.

**33** Every element of $C$ is also an element of $B$; so $C \subset B$.

**39** $3 \in \{3, 4\}$

**41** $\sqrt{2} \notin Q$ since $\sqrt{2}$ is not a rational number.

**47** $2 \in \{2, 3\}$

**49** The empty set is a subset of every set; hence, $\emptyset \subset \{1, 2\}$.

**55** Every subset of $\{0, 1\}$ has zero, one, or two elements. The subset with zero elements is the empty set, $\emptyset$. The one-element subsets are $\{0\}$ and $\{1\}$. A single subset has two elements, namely $\{0, 1\}$. Don't forget this one, and don't confuse $\emptyset$ and $\{0\}$! The correct answer is $\emptyset$, $\{0\}$, $\{1\}$, and $\{0, 1\}$.

## (1.2) (Unions and intersections)

We've just considered some introductory notions about sets. In this section we develop these ideas further by introducing the operations of *set union* and *set intersection*.

*union,* ∪

Suppose $A$ and $B$ are sets. The *union* of $A$ and $B$, denoted $A \cup B$, is defined to be the set of all objects that are in either $A$, or $B$, or both $A$ and $B$. The symbol $A \cup B$ is read "$A$ union $B$," and to say that $x \in A \cup B$ is to say that $x$ is a member of $A$, or $x$ is a member of $B$, or $x$ is a member of both $A$ and $B$. You can visualize this concept in Figure 1.1. The idea is to think of the region enclosed by each closed curve as representing the elements of the given set. Thus curve $A$ encloses the elements of set $A$, curve $B$ encloses the elements of set $B$, and $A \cup B$ is represented by the shaded region. It is assumed here that both $A$ and $B$ are subsets of some general set $U$.

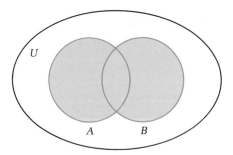

Fig. 1.1

Another notation for $A \cup B$ is

$$A \cup B = \{x \mid x \in A \text{ or } x \in B\}.$$

The word *or* is used here in the inclusive sense of *and/or*. This
notation is called *set-builder notation* and is read "the set of all
$x$ such that $x$ is an element of $A$ or $x$ is an element of $B$." The
letter $x$ is called a *variable* because it denotes an unspecified element of $A$ or $B$. Thus, in set-builder notation we have an expression
of the form

*set-builder notation*

*variable*

$$\{ \qquad \mid \qquad \qquad \}$$

The set     such
of all      that

The vertical line is read "such that." To its left a variable appears
and to its right, a phrase or formula that describes the possible
values of the variable.

*intersection*

∩

Another notation is that of *intersection*. The intersection of $A$
and $B$ is simply the set of all elements that are in *both* $A$ and
$B$. The intersection symbol is $\cap$ and you read $A \cap B$ as "$A$
intersection $B$." In set-builder notation we write

$$A \cap B = \{x \mid x \in A \text{ and } x \in B\}.$$

This is read "$A \cap B$ equals the set of all $x$ such that $x$ is an
element of $A$ and $x$ is an element of $B$." Figure 1.2 shows a set
diagram for $A \cap B$. Notice that the shaded region contains exactly
those objects common to both $A$ and $B$.

Here are some examples of unions and intersections. Suppose
first that $A = \{1, 2, 3, 4\}$ and $B = \{3, 4, 5\}$. If we list the elements
of $A$ next to the elements of $B$, the result is $A \cup B = \{1, 2, 3,$

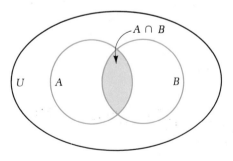

Fig. 1.2

4, 3, 4, 5}. But a symbol for a set merely tells us what elements are present in the collection it represents; so there is no point listing an element twice. Thus, $A \cup B = \{1, 2, 3, 4, 5\}$.

To find $A \cap B$, just examine the elements of $A$ one by one to decide which of them is also in $B$. Thus $1 \notin B, 2 \notin B, 3 \in B, 4 \in B$, so that $A \cap B = \{3, 4\}$.

Next we take $A = \{1, 2, 3, 4\}$ and $B = \{2, 4\}$. Note here that $B \subset A$. Reasoning as we did in the foregoing examples, we see that $A \cup B = \{1, 2, 3, 4\}$ and $A \cap B = \{2, 4\}$. Notice how $B \subset A$ affects things!

Finally, let $G = \{1, 2, 3\}$ and $H = \{4, 5\}$. Clearly, $G \cup H = \{1, 2, 3, 4, 5\}$, but how about $G \cap H$? When you examine the elements of $G$ one by one, you quickly discover that $1 \notin H, 2 \notin H, 3 \notin H$. Therefore, $G \cap H = \emptyset$. When two sets have no element in common, they are called *disjoint* sets.

*disjoint*

## Exercise set 1.2

Let $A = \{1, 2, 3, 4\}$, $B = \{1, 3, 5, 7\}$, $C = \{1, 3, 6, 8\}$, and $D = \{5, 6, 7\}$. Perform the indicated operation and list the elements of the resulting set within braces.

$(\circ)$    **1**   $A \cap B$      **2**   $A \cap C$      **3**   $A \cap D$      **4**   $B \cap C$

$(\circ)$    **5**   $B \cap D$      **6**   $C \cap D$      **7**   $A \cup B$      **8**   $A \cup C$

       **9**   $A \cup D$     **10**   $B \cup C$     **11**   $B \cup D$     **12**   $C \cup D$

If these problems don't seem very exciting to you, remember that you need to practice a new idea to make it stick. Problems 13 to 20 are a little different; so be sure to think before you write!

$(\circ)$    **13**   $A \cap A$     **14**   $A \cup A$     **15**   $D \cup D$     **16**   $D \cap D$

$(\circ)$    **17**   $A \cap \emptyset$     **18**   $B \cap \emptyset$     **19**   $A \cup \emptyset$     **20**   $B \cup \emptyset$

Now let $G, H$ be sets. One of the following four statements applies to each of problems 21 to 34. Decide which is the correct one.

**a** If this is true then $G \subset H$.

**b** If this is true then $H \subset G$.

**c** This is always true.

**d** This is true only when $G = H$.

$(\circ)$    **21**   $G \cap H = G$             **22**   $G \cap H = H$

| 23 $G \cup H = G$ | 24 $G \cup H = H$ |
|---|---|
| 25 $G \cup \emptyset = G$ | 26 $G \cap \emptyset = \emptyset$ |
| (∘)   27 $G \cap H \subset G$ | 28 $G \cap H \subset H$ |
| 29 $G \subset G \cup H$ | 30 $H \subset G \cup H$ |
| 31 $G \cup (G \cap H) = G$ | 32 $G \cap (G \cup H) = G$ |
| 33 $G \cap H = G \cup H$ | 34 $G \cup H \subset G$ |

(∘)

*Sample solutions*
*for exercise set 1.2*

1 Let's examine the elements of $A$ one by one to determine which of them are also in $B$. We have $1 \in B$, $2 \notin B$, $3 \in B$, $4 \notin B$; so $A \cap B = \{1, 3\}$.

7 Now we must list all elements that are in $A$ or $B$. We must have all elements of $A$, so that we list 1, 2, 3, 4. Because there is no point listing an element twice, we just consider those elements that $B$ adds to this list. Thus 5 and 7 are added; so $A \cup B = \{1, 2, 3, 4, 5, 7\}$. You ought to be able to do a problem like this by inspection.

13 Every element of $A$ is both in $A$ and in $A$, hence in $A \cap A$. Evidently every element of $A \cap A$ is also an element of $A$; so $A \cap A = A = \{1, 2, 3, 4\}$.

17 If $x \in A \cap \emptyset$, then $x \in A$ and $x \in \emptyset$. But because the empty set has no elements, $A \cap \emptyset = \emptyset$.

21 Let $x \in G$. Then $x \in G \cap H$; so $x \in G$ and $x \in H$. Thus, every element of $G$ is also an element of $H$. But this says that $G \subset H$; so the correct answer is choice **a**.

27 If $x \in G \cap H$, then $x \in G$ and $x \in H$; so in particular, $x \in G$. Thus, $G \cap H$ is always a subset of $G$, and the correct answer is choice **c**.

## (1.3) (Number lines)

To solve a problem, you must first understand what is involved. You might accomplish that by just thinking about the problem, or by writing it out carefully, or by viewing the problem symbolically, or even sometimes by drawing pictures. In mathematics, a picture-drawing approach is called *graphing*. For sets of real numbers, here is how it is done.

    The set $R$ of real numbers can be made to correspond with the set of points on a horizontal line $L$, called a *number line*, by fixing a point $O$, called the *origin*, on $L$ and then choosing a convenient unit of measurement. Associated with a *positive* number $r$, there

*number line*
*origin*
*positive, right*

is a point $P$ located $r$ units to the *right* of the origin. In cases when $r$ is *negative*, $-r$ is positive and the associated point $P$ is located $-r$ units to the *left* of the origin. In either case, $P$ is called the *graph* of $r$ and $r$ the *coordinate* of $P$. Of course the graph of $O$ is the origin itself. Corresponding to every real number there is one and only one point on the line $L$, and for every point on $L$ there is one and only one real number. Figure 1.3 shows the graph of $3/2$ on a number line.

Number lines can help you to visualize order properties of the real numbers. Recall from your study of elementary algebra that the real number $a$ is *less than* the real number $b$ (or, alternately, that $b$ is *greater than* $a$) if $a + c = b$ for some *positive* number $c$. Viewed geometrically, $a$ is less than $b$ if and only if the graph of $a$ lies to the left of the graph of $b$ on a number line. To symbolize this relationship we write $a < b$, which is read "$a$ is less than $b$" or alternately, $b > a$, read "$b$ is greater than $a$." The following symbols are also useful: $a \leq b$ means "$a$ is less than or equal to $b$," and $b \geq a$ means "$b$ is greater than or equal to $a$."

Once this notation has been established, we can write $b > 0$ (or $0 < b$) to denote the fact that $b$ is positive and write $b < 0$ (or $0 > b$) for $b$ negative. It is also useful to call a number $b$ *non-negative* in case it is positive *or* zero. The symbol for this is $b \geq 0$ or $0 \leq b$. An expression like $a \leq c \leq b$ means that *both* $a \leq c$ and $c \leq b$ are true. Thus to say that $1 \leq x < 3$ is equivalent to saying that $1 \leq x$ *and* $x < 3$.

The slant bar, $/$, can also be used to negate each of these symbols. For example, $a \nless b$ means "$a$ is *not* less than $b$." In this usage, you generally read $/$ as "not."

So far, we have just been graphing single real numbers. The number line, however, can be used to graph both finite and infinite sets of numbers. Look at Figure 1.4. Part **(a)** is the graph of $\{x \mid 1 \leq x \leq 3, x \in I\}$; **(b)** represents $\{x \mid 1 \leq x \leq 3, x \in R\}$; in **(c)** we have $\{x \mid -2 < x \leq 1, x \in I\}$; and in **(d)**, $\{x \mid -2 < x \leq 1, x \in R\}$. The dots in Figure 1.4 **(b)** mean that the graphs of the points 1 and 3 are to be considered part of our graph; the circle at $-2$ in **(d)** means that $-2$ is to be omitted.

It is often useful to talk about distance from the origin without

Fig. 1.3

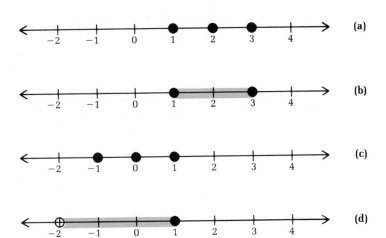

Fig. 1.4

*absolute value*

worrying about direction. We use the symbol $|a|$ to denote the *absolute value* of a—that is, the distance (without regard to direction) between the graph of *a* and the origin. For example, the graph of $-3$ is three units to the left of the origin, whereas that of $+3$ is three units to the right. Hence $|-3| = 3$ and $|+3| = 3$. It is worth noting that $|a| \geq 0$ for *every* real number *a*. We may write a more formal definition as follows.

$(\circ)$  $$|a| = \begin{cases} a \text{ if } a \geq 0 \\ -a \text{ if } a < 0 \end{cases}$$

Thus if $a = -5$, then $|a| = -a = 5 \geq 0$; if $a = 6$, $|a| = a = 6 \geq 0$. Notice how $|a|$ gives you the magnitude of the number *a*. If $a \geq 0$ then $|a|$ is *a*; and for a negative, $|a|$ is *a* multiplied by $-1$.

## Exercise set 1.3

**1**  Construct a number line with inches as the unit of measurement and plot the graphs of each of the following numbers: $1, -2, \dfrac{3}{2}, -\dfrac{5}{4}, -1\dfrac{13}{16}$.

**2**  Construct a number line with centimeters as the unit of measurement, and plot the graphs of each of the following numbers: $1, -2, -\dfrac{3}{2}, \dfrac{15}{4}, -\dfrac{7}{10}.$

*In problems 3 to 8, determine the coordinate of the point halfway between the given points using the following number line.*

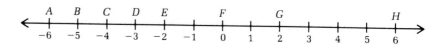

( ∘ ) 3  *A* and *B*            4  *A* and *C*

       5  *G* and *H*            6  *F* and *H*

( ∘ ) 7  *E* and *G*            8  *D* and *H*

       9  *B* and *H*           10  *A* and *H*

*Continue using the same number line in problems 11 to 18. Here we want the coordinate of the point one-fourth of the way from the first point to the second point.*

( ∘ ) 11  *A* and *B*            12  *A* and *C*

       13  *G* and *H*           14  *F* and *H*

( ∘ ) 15  *E* and *G*            16  *D* and *H*

       17  *B* and *H*           18  *A* and *H*

*Now for some practice in reading inequalities. Remember, an inequality sign always opens to the larger of the two quantities and points to the smaller. Express each of the following facts in words.*

( ∘ ) 19  $1 < 2$            20  $5 > 3$

       21  $-3 \geq -4$         22  $5 \leq 7$

       23  $5 \leq 5$            24  $5 \geq 5$

( ∘ ) 25  $1 \not> 2$            26  $4 \not< 0$

*Turning things around a bit, express the following statements in terms of symbols.*

( ∘ ) 27  −1 is less than +1

       28  +3 is greater than −4

       29  1 is less than or equal to 3

       30  4 is not less than 3

Let $a$, $b$ denote real numbers. Every statement involving a symbol of the form $<$, $>$, $\leq$, or $\geq$ can be written as one using $\not<$, $\not>$, $\not\leq$, or $\not\geq$, and vice versa. Rewrite problems 31 to 34 in a form that does not use the slant bar.

$\left(\circ\right)$  **31** $a \not< b$      **32** $a \not> b$      **33** $a \not\leq b$      **34** $a \not\geq b$

Now express each of the following in a form that uses the slant bar.

$\left(\circ\right)$  **35** $a < b$      **36** $a > b$      **37** $a \leq b$      **38** $a \geq b$

It is very important to be able to translate the symbols that indicate a set of numbers to a picture form. Use a separate number line to plot the graph of each of the following sets of numbers.

$\left(\circ\right)$  **39** $\{1, 2, 3\}$                  **40** $\{-1, +1\}$
**41** $\{x \mid 1 \leq x \leq 3, x \in R\}$          **42** $\{x \mid 1 \leq x < 3, x \in R\}$
**43** $\{x \mid -1 < x \leq 2, x \in R\}$          **44** $\{x \mid -1 < x < 2, x \in R\}$
**45** $\{x \mid 2 < x \leq 4, x \in I\}$          **46** $\{x \mid 2 < x < 4, x \in I\}$
**47** $\{x \mid x \leq 6, x \in N\}$              **48** $\{x \mid x < 6, x \in N\}$

To graph sets such as $\{x \mid x \geq 1, x \in R\}$, we use an arrow to show that the graph continues indefinitely in a particular direction. Thus the graph of the set in question looks like this.

Draw the associated graph for these next problems.

$\left(\circ\right)$  **49** $\{x \mid x < 1, x \in R\}$          **50** $\{x \mid x \leq 1, x \in R\}$
**51** $\{x \mid x \geq -3, x \in R\}$          **52** $\{x \mid x > 9, x \in R\}$

Rewrite each of the following expressions in a form that does not use absolute-value notation.

$\left(\circ\right)$  **53** $|6|$                  **54** $|-8|$                  **55** $-|14|$
$\left(\circ\right)$  **56** $-|-2|$                **57** $|x + 1|$              **58** $|2w + 3|$
**59** $|-x|$                 **60** $-|x|$

$\left(\circ\right)$

*Sample solutions*
*for exercise set 1.3*

**3**  $A$ is $-6$ units to the left of the origin and $B$ is $-5$ units to the left. The point halfway between $A$ and $B$, therefore, has coordinate equal to $-5\frac{1}{2}$.

**7**   The point halfway between $E$ and $G$ is the origin.

**11**   The distance between $A$ and $B$ is 1 unit. Hence, we wish to go one-fourth of this distance to the right of $A$. The desired point has coordinate $-5\frac{3}{4}$.

**15**   The distance between $E$ and $G$ is 4 units. The point we are seeking must lie one-fourth of this distance to the right of $E$; in other words, 1 unit to the right of $E$. Its coordinate is, therefore, $-1$.

**19**   1 is less than 2

**25**   1 is not greater than 2

**27**   $-1 < +1$

**31**   Since $a \not< b$, the graph of $a$ cannot lie to the left of the graph of $b$. Hence, it must lie on or to the right of the graph of $b$. But this says that $a \geq b$. An equally correct answer, of course, is to say that $b \leq a$.

**35**   To say that $a < b$ is to say that the graph of $a$ lies to the left of the graph of $b$; hence, it cannot lie to the right or on the graph of $b$; so $a \not\geq b$. Another correct answer would be to say that $b \not\leq a$.

**39**   We must graph 1, 2, 3 on the same number line.

**49**   You simply graph all real numbers less than 1 and use an arrow pointing off to the left to indicate that the graph continues indefinitely in that direction. The graph in question looks like this, with the circle denoting the fact that 1 is not part of the graph.

**55**   $|14| = 14$ so that $-|14| = -14$

**57**   By definition of absolute value, $|x + 1| = x + 1$ if $x + 1 \geq 0$, and $|x + 1| = -(x + 1)$ if $x + 1 < 0$.

# (1.4) (Addition and subtraction)

*sum*

*addition*

Associated with any two real numbers $a$ and $b$, there is a unique real number, $a + b$, called their *sum*. The process of forming the sum, $a + b$, is called *addition*. For your convenience, the basic properties of the addition operation are listed following. Let $a$, $b$, $c$ $\in R$.

| | | |
|---|---|---|
| $(\circ)$ | A.1 $(a + b) + c = a + (b + c)$ | *(Associative law for addition)* |
| $(\circ)$ | A.2 $a + b = b + a$ | *(Commutative law for addition)* |
| $(\circ)$ | A.3 $a + 0 = a = 0 + a$ | *(Additive identity law)* |
| $(\circ)$ | A.4 *For each real number a, there corresponds a unique real number $-a$, called negative a or the additive inverse of a, such that* $a + (-a) = 0 = (-a) + a.$      *(Additive inverse law)* | |

*associative law*

The parentheses used in the statement of the *associative law* dictate what is to be done first. Thus, to evaluate $(a + b) + c$, one first forms $a + b$ and then adds $c$ to the result. Specifically, $(1 + 2) + 3 = 3 + 3 = 6$, and $1 + (2 + 3) = 1 + 5 = 6$. Because of the associative law, an expression of the form $a + b + c$ can be interpreted either as $(a + b) + c$ or $a + (b + c)$ since the result is the same no matter where the parentheses are placed. The *commutative law*

*commutative law*

states that the order in which two numbers are added does not affect their sum. The *additive inverse law* asserts that $-a$ is the

*additive inverse law*

unique real number that produces 0 when it is added to $a$.

It is assumed that you know how to add positive numbers. To add a pair of negative numbers, you just add their absolute values, then place a minus sign in front of the result. Thus $(-2) + (-5) = -7$, since $|-2| = 2$, $|-5| = 5$ and $2 + 5 = 7$.

*difference*

If $a$, $b$ are positive numbers with $a > b$, their *difference, $a - b$,* is defined to be the unique number $d$ such that $b + d = a$. Thus $7 - 4 = 3$ since $3 + 4 = 7$; and $25 - 14 = 11$ since $11 + 14 = 25$.

You will find the notion of difference useful in connection with sums of positive and negative numbers. To form such a sum, you just take the difference between the larger and smaller of their absolute values, then prefix this number with the sign of the number having the larger absolute value. For example, $6 + (-4) = 2$ since $|6| - |-4| = 6 - 4 = 2$, and $11 + (-18) = -7$ since $|-18| - |11| = 18 - 11 = 7$. You should be able to do such computations mentally.

The definition we just gave for difference can be extended to define the *difference, $a - b$,* of *arbitrary* real numbers $a$ and $b$.

| | |
|---|---|
| $(\circ)$ | *For all real numbers a and b, their difference $a - b$ is the unique real number d such that $b + d = a$.* |

The operation of forming $a - b$ is called *subtraction,* and it is useful to note that for all real numbers $a$ and $b$,

$(\circ)$   $a - b = a + (-b)$

This follows from the fact that if $d = a + (-b)$, then

$d + b = [a + (-b)] + b = a + [(-b) + b] = a + 0 = a.$

Each of these formulations of subtraction is useful, and you should keep them both in mind.

The symbol $-a$ that appears in the *additive inverse law* is often a source of confusion. Don't be tempted to think of $-a$ as a negative number, for it need not be. If $a = 5$, then $-a = -5$, which is negative. On the other hand, if $a = -3$, then $-a$ is the *positive* number $+3$ since $(-3) + (+3) = 0$. More generally, the statement that $a + (-a) = 0$ tells us that $a$ is the negative of $-a$; in other words, that *double negative law* $a = -(-a)$. This is known as the *double negative law,* which can be stated formally as follows: For all real numbers $a$,

$(\circ)$   $a = -(-a)$

Now let $a, b$ denote real numbers. Using laws A.1 to A.4 we may write

$(a + b) + [(-a) + (-b)] = [a + (-a)] + [b + (-b)] = 0 + 0 = 0.$

But this tells us that $(-a) + (-b)$ is the negative of $a + b$. In other words, we have established that for all real numbers $a$ and $b$,

$(\circ)$   $-(a + b) = (-a) + (-b)$

These results are useful in connection with subtraction problems, especially when parentheses are involved. Here are some examples designed to illuminate the procedure:

$8 - 2 = 6$ since $2 + 6 = 8$,

$6 - 10 = 6 + (-10) = -4$,

$4 - (-2) = 4 + [-(-2)] = 4 + 2 = 6$,

$-5 - (-7) = -5 + [-(-7)] = -5 + 7 = 2$,

$-1 - 8 = -1 + (-8) = -(1 + 8) = -9$,

$6 - [(-4) + 3] = 6 - (-1) = 6 + [-(-1)] = 6 + 1 = 7.$

So far, we have been writing down statements like $a = b$ without going too deeply into the meaning of the symbol "=." If $a$ and $b$ denote real numbers, it is time to decide what we mean by the assertion $a = b$. Surely it does not say that $a$ and $b$ are the same letters of the alphabet! What it does say is that $a$ and $b$ are both symbols for the same real number. If a number is added to this number, the result is clearly independent of the symbols used. This shows that if $a = b$, then for any number $c$, $a + c = b + c$. Similar reasoning tells us that if $a + c = b + c$, then $a = b$. Mathematicians like to use the phrase "if and only if" to combine pairs of such observations into a single statement. Thus we can say that for all real numbers $a$, $b$, $c$,

$$( \circ ) \quad a = b \text{ if and only if } a + c = b + c$$

Similar reasoning shows that

$$( \circ ) \quad a = b \text{ if and only if } a - c = b - c$$

Did you notice all the mileage we got from the symbols "+" and "−"? They can be used to specify positive and negative numbers ($+3$, $-3$, and so on). The symbol "−" can also be used as a prefix to denote the additive inverse $-a$ of the real number $a$. Finally, they are used to indicate the operations of addition and subtraction. Their meaning, therefore, must always be judged from the context in which they are used.

## Exercise set 1.4

*To be certain that you have the notions of addition and subtraction firmly in hand, here are a few problems for you. Let's start with some addition problems in which you are to express each sum as an integer.*

| | | | |
|---|---|---|---|
| | **1** $3 + 11$ | **2** $6 + 4$ | **3** $14 + 8$ |
| ( $\circ$ ) | **4** $3 + 5$ | **5** $(-6) + (-8)$ | **6** $(-2) + (-3)$ |
| | **7** $(-14) + (-3)$ | **8** $(-1) + (-6)$ | **9** $(-2) + 3$ |
| | **10** $(-6) + 9$ | **11** $7 + (-4)$ | **12** $11 + (-10)$ |
| ( $\circ$ ) | **13** $(-5) + 3$ | **14** $(-6) + 4$ | **15** $3 + (-10)$ |
| | **16** $6 + (-27)$ | | |

*Express each difference as an integer in the following subtraction problems.*

| | | | | | |
|---|---|---|---|---|---|
| **17** | $13 - 4$ | **18** | $6 - 2$ | **19** | $15 - 4$ |
| **20** | $9 - 5$ | **21** | $-3 - 2$ | **22** | $-1 - 6$ |
| **23** | $-14 - 3$ | **24** | $-6 - 4$ | **25** | $3 - (-2)$ |
| **26** | $5 - (-4)$ | **27** | $6 - (-9)$ | **28** | $3 - (-1)$ |
| **29** | $14 - 15$ | **30** | $3 - 8$ | **31** | $6 - 20$ |
| **32** | $1 - 93$ | **33** | $-3 - (-1)$ | **34** | $-2 - (-8)$ |
| **35** | $-9 - (-7)$ | **36** | $-1 - (-8)$ | | |

($\circ$) applies to problem 23.

*Now, let's try some more complicated problems. Remember the definition of subtraction and the use of parentheses as you express each of the following as an integer.*

| | | | | |
|---|---|---|---|---|
| ($\circ$) **37** | $3 - 5 + 4$ | | **38** | $2 - 8 + 3$ |
| **39** | $14 - 10 + 3$ | | **40** | $2 - 8 + 6$ |
| **41** | $3 - 5 - (-1)$ | | **42** | $2 - 4 - (-6)$ |
| **43** | $-2 + 3 - 1$ | | **44** | $-5 - 3 + 1$ |
| ($\circ$) **45** | $2 - (3 - 5)$ | | **46** | $1 - (9 - 8)$ |
| **47** | $(2 - 4) - (3 - 5)$ | | **48** | $(6 - 3) - (4 - 8)$ |
| ($\circ$) **49** | $(2 - 3 + 5) - (4 - 2 + 6)$ | | **50** | $(1 - 8 + 7) - (3 - 5 + 8 - 2)$ |

**($\circ$)**

*Sample solutions for exercise set 1.4*

**5** $(-6) + (-8) = -14$

**13** $(-5) + 3 = -2$

**25** $3 - (-2) = 3 + [-(-2)] = 3 + 2 = 5$

**37** $3 - 5 + 4 = 3 + (-5) + 4 = (-2) + 4 = 2$

**45** $2 - (3 - 5) = 2 - (-2) = 2 + [-(-2)] = 2 + 2 = 4$

**49** In connection with this problem, note how the expressions inside the parentheses are evaluated first.

$$(2 - 3 + 5) - (4 - 2 + 6) = [2 + (-3) + 5] - [4 + (-2) + 6]$$
$$= 4 - 8 = -4$$

# (1.5) (Multiplication and division)

*product*

Multiplication is the operation of taking two real numbers *a* and *b*, and assigning to them a real number called their *product.* We shall use the symbols $a \cdot b$ or $ab$ to denote this product. Letting *a, b, c* represent real numbers, let's have a quick look at the properties of multiplication.

$(\circ)$  M.1  $(ab)c = a(bc)$           *(Associative law for multiplication)*

$(\circ)$  M.2  $ab = ba$             *(Commutative law for multiplication)*

$(\circ)$  M.3  $a \cdot 1 = a = 1 \cdot a$        *(Multiplicative identity law)*

$(\circ)$  M.4  *For each nonzero real number a, there corresponds a unique real number* $\dfrac{1}{a}$ *called the multiplicative inverse, or the reciprocal of a, such that* $a \cdot \left(\dfrac{1}{a}\right) = 1 = \left(\dfrac{1}{a}\right) \cdot a.$

*(Multiplicative inverse law)*

Each of the foregoing laws is related to the corresponding law of addition. It should be noted, however, that *every* real number has an additive inverse, but only nonzero real numbers have a multiplicative inverse. The associative law allows us to interpret the product $a \cdot b \cdot c$ of three numbers either as $(ab)c$ or $a(bc)$. For example, $(2 \cdot 3) \cdot 4 = 6 \cdot 4 = 24$, and $2 \cdot (3 \cdot 4) = 2 \cdot 12 = 24$; so in either interpretation, $2 \cdot 3 \cdot 4 = 24$. Laws M.2 and M.3 need no further explanation. The *multiplicative inverse law* states that for $a \neq 0$, $1/a$ is the unique number that produces 1 when it is multiplied by *a.*

You must have instant recall of the rules governing the multiplication of signed numbers to be successful in learning intermediate algebra. The results of multiplying the four combinations of positive and negative numbers are set out below.

$(\circ)$

| *a* | *b* | *ab* |
|---|---|---|
| *positive* | *positive* | *positive* |
| *negative* | *negative* | *positive* |
| *positive* | *negative* | *negative* |
| *negative* | *positive* | *negative* |

Notice that the product $ab$ is positive only when the signs of $a$ and $b$ are the same. The product $ab$ is negative if and only if $a$ and $b$ have opposite signs. For example, $(2)(4) = 8$, $(-2)(-4) = 8$, $(2)(-4) = -8$, and $(-2)(4) = -8$.

If you recall that $a = b$ means that $a$ and $b$ are different names for the same number, the following statement will be clear.

$(\circ)$    *For any real number c, a = b implies ac = bc.*

Going in the other direction, if $ac = bc$ and $c \neq 0$, then multiplication by $1/c$ produces $a = b$. This is known as the *cancellation law,* which, stated formally, reads:

*cancellation law*

$(\circ)$    *If ac = bc and c ≠ 0, then a = b.*

*factor*

    The number $a$ is called a *factor* of the product $ab$. In the set $N$ of natural numbers, 4 is a factor of 36 since $4 \cdot 9 = 36$; similarly, 12 is a factor of 36 since $12 \cdot 3 = 36$. Every natural number $n$ has $n$ and 1 as factors. If $n > 1$ and $n$ and 1 are its only natural-number factors, then $n$ is called a *prime number.* The first few prime numbers are 2, 3, 5, 7, 11, and 13. It is a fact (known as the *fundamental theorem of arithmetic*) that apart from the order of the factors, every natural number greater than 1 can be expressed in only one way as a product of prime numbers. A negative integer may be thought of as $(-1)n$ where $n$ is a natural number. Thus it is either $-1$, or it may be written as $-1$ times the prime factors of $n$. We shall have more to say about this in Section 3.2.

*prime number*

*fundamental theorem of arithmetic*

    We turn now to the division operation. Given real numbers $a$ and $b$ with $b \neq 0$, the quotient $a/b$ is defined by the rule

$(\circ)$    $$\frac{a}{b} = a \cdot \left( \frac{1}{b} \right)$$

*division*

*dividend, divisor*

The formation of $a/b$ is called *division* of $a$ by $b$; $a$ is called the *dividend* and $b$ the *divisor.* It should be noted that the division operation is restricted to nonzero divisors. Notice also that $q = a/b$ is the unique real number such that $qb = a$. (This statement often serves as the definition of the quotient.) To see this, note that

$$\left( \frac{a}{b} \right) b = \left[ a \left( \frac{1}{b} \right) \right] b = a \left[ \left( \frac{1}{b} \right) b \right] = a \cdot 1 = a,$$

and if $qb = a$, then

$$\frac{a}{b} = a\left(\frac{1}{b}\right) = (qb) \cdot \left(\frac{1}{b}\right) = q\left[b\left(\frac{1}{b}\right)\right] = q \cdot 1 = q.$$

You will find it useful to think of the quotient $a/b$ in either of the foregoing ways, depending on the occasion.

An easy way to keep track of whether the quotient $q = a/b$ is positive or negative is to remember that $qb$ must equal $a$. Thus

$$\frac{6}{2} = 3 \text{ since } (2)(3) = 6,$$

$$\frac{6}{-2} = -3 \text{ since } (-2)(-3) = 6,$$

$$\frac{-6}{2} = -3 \text{ since } (2)(-3) = -6,$$

$$\frac{-6}{-2} = 3 \text{ since } (-2)(3) = -6.$$

We shall have a closer look at all this in the next section.

## Exercise set 1.5

*Here are some multiplication problems so that you can practice the rules of multiplying signed numbers.*

$(\circ)$  **1**  $(2)(-4)$      **2**  $(3)(-5)$      **3**  $(-6)(2)$

     **4**  $(-7)(5)$      **5**  $(-6)(-2)$      **6**  $(-7)(-3)$

$(\circ)$  **7**  $(3)(-4)(2)$      **8**  $(5)(-3)(2)$      **9**  $(-6)(-2)(4)$

     **10**  $(-7)(-3)(6)$      **11**  $(3)(-5)(-10)(-1)$

     **12**  $(-1)(-2)(-3)(-4)(-5)$

*Now for some division problems.*

$(\circ)$  **13**  $\dfrac{18}{2}$      **14**  $\dfrac{27}{9}$      **15**  $\dfrac{36}{-9}$

     **16**  $\dfrac{-4}{2}$      **17**  $\dfrac{-20}{-2}$      **18**  $\dfrac{-49}{-7}$

$$(\circ) \quad 19 \quad -\frac{-25}{-5} \qquad\qquad 20 \quad -\frac{16}{-4} \qquad 21 \quad \frac{0}{5}$$

$$22 \quad \frac{0}{-6} \qquad\qquad 23 \quad \frac{4}{0} \qquad 24 \quad \frac{0}{0}$$

*In each of the following problems, a quotient is in the form $\frac{a}{b} = q$.*

*Rewrite each equation in the form $qb = a$.*

$$(\circ) \quad 25 \quad \frac{18}{2} = 9 \qquad\qquad 26 \quad \frac{16}{4} = 4 \qquad 27 \quad \frac{0}{6} = 0$$

$$(\circ) \quad 28 \quad \frac{0}{-4} = 0 \qquad\qquad 29 \quad \frac{32}{-4} = -8 \qquad 30 \quad \frac{-32}{4} = -8$$

$$31 \quad \frac{-88}{-11} = 8 \qquad\qquad 32 \quad \frac{-90}{-3} = 30$$

**(∘)**

*Sample solutions*
*for exercise set 1.5*

1 $\quad (2)(-4) = -8$

7 $\quad (3)(-4)(2) = (-12)(2) = -24$

15 $\quad \dfrac{36}{-9} = -4$

21 $\quad \dfrac{0}{5} = 0$ since $(5)(0) = 0$

25 $\quad (9)(2) = 18$

29 $\quad (-8)(-4) = 32$

# (1.6) (The distributive law and the order of operations)

So far there has been no mention of a relationship between the addition and multiplication operations. Of course there is one, and *distributive law* here is the law (known as the *distributive law*) that expresses it.

---

(∘)    If *a, b, c* are real numbers, then $a(b + c) = ab + ac$.

---

This extremely important law is the basis of some well-known facts about arithmetic. For example, everyone knows that zero times any other number is zero, but did you know that this pops right out from the distributive law? To see this, note first that $a \cdot 0 = a \cdot (0 + 0) = (a \cdot 0) + (a \cdot 0)$; so $0 + (a \cdot 0) = (a \cdot 0) + (a \cdot 0)$. Subtracting $a \cdot 0$ from both sides of the second equation, we have $0 = a \cdot 0$.

As a second example, we ask you to recall that the product of two numbers with like signs is positive and with unlike signs is negative. But the distributive law leads to a much more general observation. Let $a$, $b$ denote real numbers. Using the fact that $b + (-b) = 0$, we have

$$a[b + (-b)] = a \cdot 0 = 0,$$

and by the distributive law

$$ab + a(-b) = a[b + (-b)] = 0.$$

But this is telling us something. What property characterizes the number $-(ab)$? It is the fact that $ab + [-(ab)] = 0$. Thus our equation tells us that

---

$( \circ )$  $a(-b) = -(ab)$

---

Similar reasoning shows that $(-a)b = -(ab)$. Therefore, it is not ambiguous to write $-ab$ for $-(ab)$, and this we shall often do.

Of course this can all be formulated in the language of quotients. For $b \neq 0$, the quotient $q = a/b$ is the unique real number such that $qb = a$. It follows that

$$-q = \frac{-a}{b}, \quad -q = \frac{a}{-b}, \text{ and } q = \frac{-a}{-b}.$$

Let us now consider the value of the expression $2 \cdot 3 + 4$. This seems easy. After all, $3 + 4 = 7$, and $2 \cdot 7 = 14$; so $2 \cdot 3 + 4 = 2 \cdot 7 = 14$. On the other hand, $2 \cdot 3 = 6$ and $6 + 4 = 10$; so $2 \cdot 3 + 4 = 6 + 4 = 10$. Which is the correct answer, 14 or 10? What went wrong? If you think about this for a moment, you will see that the trouble comes from the fact that you did not know which operation to perform first, and it *does* make a difference. What we need is an agreement about the order in which operations are to be performed, and here it is.

23 ($\circ$) **The distributive law and the order of operations**

**Step 1:** If an expression involves grouping devices (such as parentheses, brackets, or braces), start with the portion enclosed within the innermost pair of such devices.

**Step 2:** Perform multiplications and divisions in order from left to right.

**Step 3:** Perform additions and subtractions in order from left to right.

**Step 4:** Repeat these steps as many times as necessary to evaluate the given expression.

Thus $2 \cdot 3 + 4 = 6 + 4 = 10$. If we wanted the other order of operations, we would use parentheses and write $2 \cdot (3 + 4) = 2 \cdot 7 = 14$. As a further example, note that

$$3[(-4) + 6(5 - 3)] = 3[(-4) + 6 \cdot 2]$$
$$= 3[(-4) + 12] = 3 \cdot 8 = 24.$$

You will get more practice in the exercises.

## Exercise set 1.6

Do the indicated operations, keeping in mind the rules for order of operations.

$(\circ)$    **1**   $7 \cdot 4 + 8$        **2**   $6 \cdot 5 + 1$

       **3**   $3 - 4 \cdot 5$        **4**   $8 - 2 \cdot 3$

$(\circ)$    **5**   $7(4 + 8)$        **6**   $6(5 + 1)$

       **7**   $(3 - 4) \cdot 5$        **8**   $(8 - 2) \cdot 3$

$(\circ)$    **9**   $2 \cdot 3 + 4 \cdot 2$        **10**   $2(-3) + (-8) \cdot (-2)$

    **11**   $(6)(-4) + 3 \cdot 6$        **12**   $(-8)(-6) - 2 \cdot 10$

    **13**   $3(4 - 2 + 6) - 1$        **14**   $(-6)(5 - 1 + 4) - 3$

The rules governing the order of operations tell you to do all products and quotients first, then do any sums and differences. When you are evaluating a fraction, you must remember that the line in the fraction is itself a kind of grouping device. In other words, the numerator $a$ and the denominator $b$ of the fraction $\dfrac{a}{b}$ must first be evaluated separately before the quotient $\dfrac{a}{b}$ is computed. As a concrete illustration of this we have $\dfrac{8 - 2}{2 + 1} + 4 = \dfrac{6}{3} + 4 = 2 + 4 = 6$.

*The remaining problems are designed to give you some practice at this sort of thing.*

**15** $\dfrac{6-4}{2}+1$

**16** $\dfrac{16+4}{4}-5$

(∘) **17** $4\left(\dfrac{10-4}{2+1}+3\right)$

**18** $(-6)\left(\dfrac{8+4}{8-4}-2\right)$

**19** $\dfrac{6-\dfrac{4+2}{3-5}}{-2+\dfrac{8+2}{4-2}}$

**20** $\dfrac{\dfrac{5+7}{5-7}+2}{\dfrac{18-4}{(7)(2)}+1}$

**21** $\dfrac{2\cdot3+4\cdot2}{2\cdot2+(-1)}$

**22** $\dfrac{6\cdot5-(4-2)}{3\cdot2+\dfrac{6-4}{5-3}}$

(∘) **23** $1-3\left[\dfrac{6+\dfrac{4+8}{6-2}}{\dfrac{4}{2}+1}\right]$

**24** $3\cdot2-4+6\left[\dfrac{\dfrac{3+6}{5-2}+\dfrac{6-12}{3-4}}{1+4\cdot2}\right]$

(∘)

*Sample solutions for exercise set 1.6*

**1** $7\cdot4+8=28+8=36$

**5** $7(4+8)=7\cdot12=84$

**9** $2\cdot3+4\cdot2=6+8=14$

**17** $4\left(\dfrac{10-4}{2+1}+3\right)=4\left(\dfrac{6}{3}+3\right)=4(2+3)=4\cdot5=20$

**23** $1-3\left[\dfrac{6+\dfrac{4+8}{6-2}}{\dfrac{4}{2}+1}\right]=1-3\left[\dfrac{6+\dfrac{12}{4}}{2+1}\right]=1-3\left[\dfrac{6+3}{3}\right]$

$\qquad\qquad\qquad =1-3\cdot\dfrac{9}{3}=1-(3)(9)\left(\dfrac{1}{3}\right)=1-9=-8$

**25 (∘) The distributive law and the order of operations**

# (1.7) (Order properties of the real numbers)

In this section we are going to have a close look at order properties of the reals. When we constructed number lines back in Section 1.3, we divided the real numbers into three subsets: the positive real numbers, the negative real numbers, and a set having 0 as its only member. We shall find it useful to let

$R^+$ denote the set of positive real numbers,

$R^-$ the set of negative real numbers,

$Q^+$ the positive rational numbers, and

$Q^-$ the set of negative rational numbers.

We have already used concepts involving order properties of the reals. In discussing the addition and multiplication operations, we informally assumed that you know the sum or product of a pair of positive numbers to be positive. Facts like these are important enough to warrant a formal statement, however, and here it is.

(∘)    O.1 The sum of two positive numbers is positive.

(∘)    O.2 The product of two positive numbers is positive.

Let $a$ and $b$ denote real numbers. Do you remember the meaning of the symbol $a < b$? On a number line, it means that the graph of $a$ lies to the left of the graph of $b$. It also means that $a + c = b$ for some positive number $c$. By definition of the subtraction operation, such a number $c$ must equal $b - a$; so we may say that

(∘)    $a < b$ if and only if $b - a \in R^+$

This formulation leads to a pair of important observations about the real numbers. Given any two real numbers $a$ and $b$, the number $b - a$ must either be positive, zero, or negative, which tells us that $a < b$, $a = b$, or $a > b$. Also, if $a < b$ and $b < c$, then $b - a$, $c - b \in R^+$. Since the sum of two positive numbers is positive, this yields

$$c - a = (c - b) + (b - a) \in R^+$$

so $a < c$. Stated formally, we have

**0.3** *Given a, b ∈ R, exactly one of the following is true: a < b, a = b, or a > b.* *(Trichotomy law)*

**0.4** *Given a, b, c ∈ R, if a < b and b < c, then a < c.* *(Transitive law)*

A useful notion is provided by intervals of real numbers. Let *a < b*. The *closed interval* [a,b] is defined by

*closed interval*

$[a,b] = \{x \mid a \le x \le b, x \in R\}$

The *open interval* (a,b) is defined by the rule

*open interval*

$(a,b) = \{x \mid a < x < b, x \in R\}$

There will also be occasions when we shall need to consider intervals of the form

$[a,b) = \{x \mid a \le x < b\}$

$(a,b] = \{x \mid a < x \le b\}$

but we will not give these intervals special names. This is all really very easy to keep track of. The numbers *a* and *b* are called the *endpoints* of the interval. Square brackets are used to indicate that an endpoint is to be considered part of the interval, and a parenthesis means that it is to be left out.

*endpoints*

You will frequently have to work with unions and intersections

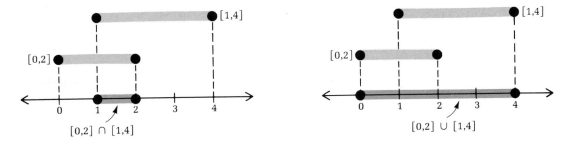

Fig. 1.5

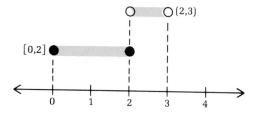

Fig. 1.6

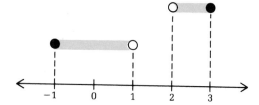

Fig. 1.7

of intervals. Some examples designed to show you how to do this follow; so let's roll up our sleeves and get familiar with the details. Suppose you are asked to express $[0,2] \cap [1,4]$ and $[0,2] \cup [1,4]$ as intervals. The first step is to visualize the problem on a number line (see Figure 1.5). Now, the points lying under the graphs of both intervals are those points in the intersection, while those lying under at least one of them form the union. Hence $[0,2] \cap [1,4]$ $= [1,2]$, and $[0,2] \cup [1,4] = [0,4]$.

Now consider $[0,2] \cap (2,3)$ and $[0,2] \cup (2,3)$. Once again, a picture (Figure 1.6) helps us to see what is happening. From the graph we see that $[0,2] \cap (2,3) = \emptyset$, and $[0,2] \cup (2,3) = [0,3)$.

Finally, let us consider $[-1,1) \cap (2,3]$ and $[-1,1) \cup (2,3]$ as pictured in Figure 1.7. Here $[-1,1) \cap (2,3] = \emptyset$, while $[-1,1] \cup (2,3]$ is not an interval. In fact, here is its graph.

## Exercise set 1.7

*Use the fact that a < b if and only if b − a is positive to fill in each blank with < or >.*

$\left( \circ \right)$

| 1 | 2____3 | 2 | 6____8 | 3 | 10____14 |
|---|--------|---|--------|---|----------|
| 4 | 5____−1 | 5 | 2____−6 | 6 | 4____−5 |

$\left( \circ \right)$

| 7 | −2____−4 | 8 | −1____−5 | 9 | 3____0 |
|---|----------|---|----------|---|--------|
| 10 | −9____0 | 11 | −3____−4 | 12 | −3____−7 |
| 13 | −2____0 | 14 | −7____0 | | |

The statement that $x$ is greater than 3 but does not exceed 4 can be written symbolically as $3 < x \le 4$. Use this type of notation to write problems 15 to 20 in symbolic form.

( ○ )  **15**  $w$ is greater than 1 and less than 5

**16**  $t$ is less than 6 and greater than $-1$

**17**  $x$ is greater than 2 but does not exceed 3

**18**  $x$ is less than 5 but not less than 1

**19**  $t$ is "properly" between 2 and 3

**20**  $t$ is greater than 5 and not greater than 6

In problems 21 to 26, translate the symbolic statement to a statement in words.

( ○ )  **21**  $-1 < x \le 4$    **22**  $-5 < x < -3$    **23**  $1 \le t < 6$

**24**  $-7 < y < 6$    **25**  $15 < t \le 16$    **26**  $-1 < x \le 0$

Now for some practice in finding unions and intersections of intervals. Express, if possible, each of the following as a single interval. If this is not possible, simply state that fact. Sketch the graph of each set on a separate number line.

( ○ )  **27**  $[1,3] \cap [2,4]$        **28**  $[-2,2] \cap (0,1)$

**29**  $[-6,2] \cap (3,5)$        **30**  $[1,2) \cap (2,3]$

( ○ )  **31**  $[1,3] \cup [2,4]$        **32**  $[-2,2] \cup (0,1)$

**33**  $[-6,2] \cup (3,5)$        **34**  $[1,2) \cup (2,3]$

**35**  $[1,2] \cup (2,3]$        **36**  $(-2,2) \cup (3,5)$

Plot the graph of each of the following sets of numbers on a separate number line. Remember, to plot the graph of a set like $\{x \mid x \ge 1, x \in R\}$, you use an arrow to show that the graph extends indefinitely in the indicated direction.

( ○ )  **37**  $[1,4] \cap N$

**38**  $(0,6) \cap N$

( ○ )  **39**  $\{x \mid x \le 4, x \in R\} \cap \{x \mid x > 1, x \in R\}$

**40**  $\{x \mid x \le 1, x \in R\} \cap \{x \mid x \ge 0, x \in R\}$

**41**  $\{x \mid x < 3, x \in R\} \cap \{x \mid x \le 2, x \in R\}$

29 ( ○ ) Order properties of the real numbers

**42** $\{x \mid x > 3, x \in R\} \cap \{x \mid x > -4, x \in R\}$

**43** $\{x \mid x > 1, x \in R\} \cup \{x \mid x < 0, x \in R\}$

**44** $\{x \mid x \geq 3, x \in R\} \cup \{x \mid x < -4, x \in R\}$

$(\circ)$     **45** $\{x \mid x \geq 1, x \in R\} \cup \{x \mid x > -5, x \in I\}$

**46** $\{x \mid x < 6, x \in R\} \cup \{x \mid x < 8, x \in N\}$

$(\circ)$

*Sample solutions for exercise set 1.7*

**1**    $2 < 3$ since $3 - 2 = 1 > 0$

**7**    $-2 > -4$ since $(-2) - (-4) = (-2) + [-(-4)]$

$$= (-2) + 4 = 2 > 0$$

**15**   To say that $w$ is greater than 1 is to say $w > 1$; to indicate that $w$ is less than 5, one writes $w < 5$. A correct answer, therefore, is $1 < w < 5$.

**21**   There are several ways to phrase this condition. Here are two solutions: $x$ is greater than $-1$ and less than or equal to 4; and $x$ is greater than $-1$ but does not exceed 4.

**27**   Let us plot the graphs of both intervals on the same number line.

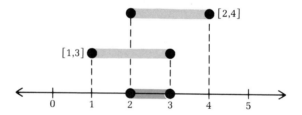

Thus $[1,3] \cap [2,4] = [2,3]$.

**31**   Again we draw a graph.

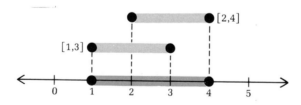

Inspection of the graph shows that we have the closed interval $[1,4]$.

**37**   We want the graph of those natural numbers that lie in the closed interval from 1 to 4. The hard part is now done, and here is the graph.

**39** Note that the set in question is the interval (1,4].

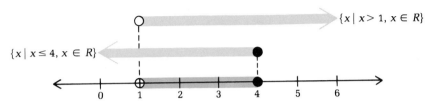

**45** Our set consists of all $x \geq 1$ as well as all integers greater than $-5$. This second set merely adds the elements $-4$, $-3$, $-2$, $-1$, and $0$ to the first set since everything else in the second set is already accounted for. Thus the graph is this.

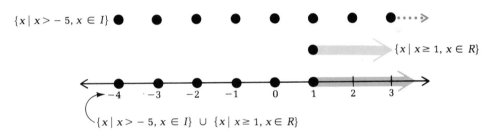

# (2)

## *Polynomials*

### (2.1) (Positive integral exponents)

The product of two numbers $a$ and $b$ is denoted by $a \cdot b$ or just $ab$. The product of two factors of $b$ is denoted by $bb$, three factors of $b$ by $bbb$, four factors of $b$ by $bbbb$, and so on. This notation soon becomes inconvenient; so people use symbols like $b^4$ to denote $bbbb$. Thus

$$2\cdot2\cdot2\cdot2\cdot2 = 2^5 = 32,$$

$$3\cdot3\cdot3\cdot3 = 3^4 = 81,$$

$$5\cdot5 = 5^2 = 25,$$

$$8^1 = 8.$$

In general, we agree to let

$(\circ)$ $\quad b^n = $ *(the product of)* n *factors of* b

*exponent*

*base, power*

for any real number $b$ and any natural number $n$. In an expression such as $b^n$, $n$ is called the *exponent*, $b$ the *base*, and $b^n$ a *power* of $b$. In particular, $b^n$ is the $n$th *power of* $b$, or $b$ *to the* $n$th *power*, or simply $b$ *to the* $n$th. We now have

$b^1 = b$       (read "$b$ to the *first*"),

$b^2 = bb$      (read "$b$ *squared*" or "$b$ to the *second*"),

$b^3 = bbb$     (read "$b$ *cubed*" or "$b$ to the *third*"),

$b^4 = bbbb$    (read "$b$ to the *fourth*"), and so on.

A number of useful rules govern operations with exponents. For example, we may simplify $2^2 \cdot 2^3$ by observing that $(2^2)(2^3) = (2 \cdot 2)(2 \cdot 2 \cdot 2) = 2 \cdot 2 \cdot 2 \cdot 2 \cdot 2 = 2^5$. More generally, for $m$ and $n \in N$,

$$2^m \cdot 2^n = (m \text{ factors of } 2)(n \text{ factors of } 2)$$

$$= (m + n \text{ factors of } 2)$$

$$= 2^{m+n}.$$

There is no magic in the number 2, and the same reasoning shows that for any real number $b$ and for any natural numbers $m$ and $n$,

( ∘ )   E.1  $b^m \cdot b^n = b^{m+n}$

*You form the product of two powers of the same base by adding the exponents and using the sum as an exponent on the same base.* Applying that law, we have $3^4 \cdot 3^5 = 3^9$, and $8^6 \cdot 8^4 = 8^{10}$.

Next we observe that for $m$ and $n \in N$, $(2^m)^n = (n \text{ factors of } 2^m) = (mn \text{ factors of } 2) = 2^{mn}$. The same procedure can be used for any real number $b$ to show that for arbitrary $m$ and $n \in N$,

( ∘ )   E.2  $(b^m)^n = b^{mn}$

*You raise the power of a number $b$ to a power by multiplying the exponents and using the product as an exponent on the original base $b$.* Thus $(3^4)^5 = 3^{20}$, $(8^2)^6 = 8^{12}$, and $(w^9)^{11} = w^{99}$.

We also have that for any natural number $n$ and for any real numbers $a$ and $b$,

( ∘ )   E.3  $(ab)^n = a^n b^n$

*A power of the product of two real numbers is the product of their corresponding powers.* This is easily seen from the fact that

$$(ab)^n = (n \text{ factors of } ab)$$

$$= (n \text{ factors of } a)(n \text{ factors of } b)$$

$$= a^n b^n.$$

Again we consider the application of this law. It says, for example, that $(2x)^3 = 2^3 \cdot x^3 = 8x^3$.

Finally, we turn to the division operation. Notice that $\dfrac{2^5}{2^3} = 2^5 \cdot \dfrac{1}{2^3}$

$= 2^2 \cdot \left( 2^3 \cdot \dfrac{1}{2^3} \right) = 2^2$. More generally, if $m$ and $n \in N$ with $m > n$,

we have $\dfrac{2^m}{2^n} = 2^m \cdot \dfrac{1}{2^n} = 2^{m-n} \cdot \left( 2^n \cdot \dfrac{1}{2^n} \right) = 2^{m-n}$. The same reason-

ing shows that for any nonzero real number $b$ and any natural numbers $m$ and $n$ with $m > n$,

---

(∘)    E.4    $\dfrac{b^m}{b^n} = b^{m-n}$

---

*To divide a power of a nonzero number by a smaller power of that same number, you subtract the smaller exponent from the larger and use the difference as an exponent on the same base.* To illustrate, you can see that $16^8/16^6 = 16^2 = 256$, and $x^{12}/x^3 = x^9$.

The preceding laws must, of course, be applied with some degree of caution. First of all, we agree that for every real number $b$, $-b^n$ denotes the additive inverse of $b^n$, and $(-b)^n$ represents the number $-b$ raised to the $n$th power. Thus $-2^4 = -16$, whereas $(-2)^4 = (-2)(-2)(-2)(-2) = +16$. Secondly, an exponent is always applied to the number immediately to its left, whether that number is represented by a numeral, a letter, or an expression enclosed within parentheses. Thus

$42^2 = (42)(42) = 1,762$, whereas $4 \cdot 2^2 = 4 \cdot 4 = 16$,

$(2x)^3 = 8 \cdot x^3$, whereas $2x^3 = 2 \cdot x^3$,

$(3 \cdot 2)^2 = 6^2 = 36$, whereas $3 \cdot 2^2 = 3 \cdot 4 = 12$.

There is nothing wrong with using parentheses to make your intent crystal clear. For example, $(42)^2$ clearly means $(42)(42)$, whereas $4(2^2)$ clearly means $4 \cdot 2^2$, and there can be no confusion.

This brings us to a third issue. The process of forming $b^n$ is a multiplication process and must be treated as such. In an ungrouped expression, you should first form any powers, then do multiplications and divisions, and finally any additions and subtractions. For example,

$$2 + 3^4 = 2 + 81 = 83,$$

$$6 - 9 \cdot 4^2 = 6 - 9(16) = 6 - 144 = -138,$$

$$\frac{5^2 - 4^2}{3} = \frac{25 - 16}{3} = \frac{9}{3} = 3.$$

## Exercise set 2.1

Express each of the following in a form that does not use exponents.

$(\circ)$  **1** $x^5$      **2** $a^3$      **3** $(2b)^3$

**4** $(3x)^2$      **5** $-x^3$      **6** $(-x)^3$

In the next batch, put each expression in a form that involves exponents.

$(\circ)$  **7** $xxx$      **8** $yyyy$      **9** $w$

$(\circ)$  **10** $(-a)(-a)$      **11** $ssttt$      **12** $-(-2a)(-2a)(-2a)$

Write each of the following as a product in which each variable occurs only once and in which there are no exponents applied to a grouped expression. Thus $(3b^2)^2$ would be put in the form $9b^4$.

$(\circ)$  **13** $x \cdot x^2$      **14** $y^2 \cdot y^5$      **15** $\dfrac{w^6}{w^4}$

**16** $\dfrac{(2t)^3}{(2t)^2}$      **17** $\dfrac{(ab)^3}{ab}$      **18** $(2x^2)^3$

**19** $(3x^2 y)^2$      **20** $\dfrac{(x^5)^2}{x}$

Now for some problems involving order of operations. Simplify each of the following, and keep in mind which operations come first.

$(\circ)$  **21** $-4^2$      **22** $(-4)^2$

**23** $(3 + 2)^2$      **24** $3 + 2^2$

$(\circ)$  **25** $5 + 4 \cdot 3^2$      **26** $3^2 + 2^3$

**27** $(-3)2^2 + 4^2 - 1$      **28** $-6^2 \cdot 3 + 5^2 \cdot 2$

**29** $4 + (3 + 1)^2$      **30** $5 - (3 - 5)^3$

$(\circ)$  **31** $\dfrac{(-4)^2 + 2}{(-4)^2 - 7} + 1$      **32** $\dfrac{(-3)^2 + (-2)^4}{-2^2 - 1}$

35 $(\circ)$ Positive integral exponents

1  xxxxx     7  $x^3$     11  $s^2 t^3$

13  $x^3$     15  $w^2$     21  $-16$

Notice how, in problems 25 and 31, the squaring operations are performed first, and the expressions are then evaluated by using the usual conventions regarding order of operations.

25  $5 + 4 \cdot 3^2 = 5 + 4 \cdot 9 = 5 + 36 = 41$

31  $\dfrac{(-4)^2 + 2}{(-4)^2 - 7} + 1 = \dfrac{16 + 2}{16 - 7} + 1 = \dfrac{18}{9} + 1 = 2 + 1 = 3$

# (2.2) (Some basic definitions)

*polynomials*

*algebraic expression*

In this section we introduce some terminology about *polynomials*. Let us begin with the notion of an *algebraic expression*, which is nothing more than a meaningful collection of numerals, constants, variables, and signs of operation. Each of the expressions $2x + 5$ and $x^2 + 3xy + 2y^2 - 4$ is an example of a polynomial. The poly-

*terms*

*coefficient*

*constant term*

nomial $2x + 5$ is the sum of the *terms* $2x$ and 5. The number 2 in the term $2x$ is called its *coefficient*, and the term 5, which contains no variable, is called the *constant term*. The polynomial $x^2 + 3xy + 2y^2 - 4$ is the sum of the four terms $x^2$, $3xy$, $2y^2$, and $-4$, whose coefficients are, respectively, 1, 3, 2, and $-4$.

A *term*, then, may be defined as either a number or a number times positive integral powers of one or more variables. Furthermore, we consider $x^n$ the same as $1x^n$ and 0 the same as $0x^n$. Any factor or group of factors of a term may be thought of as the *coefficient* of the remaining factors. Thus, in the term $12x^3yz^2$, $12x^3$ is the coefficient of $yz^2$, $y$ is the coefficient of $12x^3z^2$, and 12 is the coefficient of $x^3yz^2$. Unless otherwise specified, however, when we speak of the *coefficient* of a term, we shall be referring to its numerical factor. The coefficient of $12x^3yz^2$ is 12, that of $-3x$ is $-3$, and 1 is the coefficient of $x^2yz$ since $x^2yz = 1x^2yz$.

*like, unlike terms*

Terms such as $2x$, $3x$, and $-5x$ or $-7xy^2$ and $8xy^2$ are called *like* terms because they differ only in their coefficients. Terms that are not like terms are called *unlike* terms.

We can now define a *polynomial* to be either a term or the sum of a finite number of unlike terms. A polynomial that consists of

*monomial*

*binomial*

a single, nonzero term is called a *monomial;* if it consists of two nonzero terms, it is called a *binomial;* and three nonzero terms a *trinomial*. Polynomials are also often classified by their *degree*. If a term consists of a number, it is called a *constant term* and assigned degree 0 (except that for special reasons the constant term 0 is not assigned a degree). A term involving a single variable is assigned a degree equal to the exponent of that variable. The degree of a term involving more than one variable is simply the sum of the exponents of the variables. Thus $2x^3$ has degree 3, 4 has degree 0, and $12x^3yz^2$ has degree 6 $(= 3 + 1 + 2)$. *The degree of a polynomial is defined to be the degree of its highest degree term.*

It is useful to allow differences as well as sums of terms to appear in a polynomial. To do this, we simply agree that an expression like $3 - 2x$ shall have the same meaning as $3 + (-2)x$. More generally, we shall regard the expressions $-ax^n$ and $+(-a)x^n$ as equivalent.

A polynomial in a single variable is often represented by a symbol like $P(x)$. This is read "$P$ of $x$" and represents a polynomial in the variable $x$. Thus $Q(x)$ might be a second polynomial in $x$, and $R(y)$ would represent a polynomial in the variable $y$. Hold on, though. We get a little added bonus from this agreement. If, for example, $P(x) = 2x^2 - 3$, then $x$ represents some unspecified real number. We can let

$P(1) = 2 \cdot 1^2 - 3 = 2 - 3 = -1,$

$P(2) = 2 \cdot 2^2 - 3 = 8 - 3 = 5,$

$P(3) = 2 \cdot 3^2 - 3 = 18 - 3 = 15,$ and so on.

---

(∘) *In general, if P(x) is a polynomial in x, and if c is any number, then P(c) is the number obtained by replacing x with c in the polynomial P(x).*

---

These have all been examples of *real* polynomials—real in the sense that all coefficients and all variables have been restricted to be real numbers. Unless otherwise specified, a seemingly homeless polynomial should always be considered real in the aforementioned sense.

Have you asked yourself for an example of an expression that is *not* a polynomial? The thought should at least have crossed your mind. Here are two such examples: The expression $3 + 4/x$ is not a polynomial because the term $4/x$ involves a variable in its denominator; the expression $(6 + 5x)/x^2$ is not a polynomial either, and for much the same reason. The problem is that $x^2$ appears in the denominator. Just remember the definition of polynomial and you should have little difficulty recognizing one when you see one. A polynomial is either a term or a sum of terms, and a term is either a number or a number times positive integral powers of one or more variables. Thus a polynomial cannot involve division by a variable.

## Exercise set 2.2

Find the degree and coefficient of each of the following terms.

$(\circ)$  **1** $3x^2$      **2** $-6x$      **3** $-8$

     **4** $6$      **5** $-7y^3$      **6** $2z^6$

$(\circ)$  **7** $5xy$      **8** $10xyz$      **9** $4xz^2$

     **10** $-7x^2yz^3$      **11** $2s^3t^2u^4v$      **12** $10w^3x^2y$

Determine whether each of the following is a monomial, binomial, or trinomial, and in each case give its degree.

$(\circ)$  **13** $62x^4$      **14** $7x^2$

     **15** $x^3 - 3x + 8$      **16** $-x^6 + 14$

$(\circ)$  **17** $w^4 + 6w$      **18** $v^3 + 6v$

     **19** $23t + t^2 + t^{54}$      **20** $16t^3 + t^5 - t^2$

$(\circ)$  **21** $8x^2 - 6xy + y^2$      **22** $5xz - xy$

     **23** $z^3 + 3wz$      **24** $x^2y + 2xy^2$

     **25** $5x^2y - 3xyz^3 + 2x^4$      **26** $6x + 3y - 4z$

In each of the following problems, find $P(-1)$, $P(0)$, and $P(2)$.

$(\circ)$  **27** $P(x) = 5x + 6$      **28** $P(x) = -4x^3 + 16$

     **29** $P(w) = 5w^2 + 1$      **30** $P(s) = -s^2 - \dfrac{3}{2}s + 5$

     **31** $P(y) = 2$      **32** $P(x) = 3$

Here are a few more problems on which to test your skills.

$(\circ)$  **33** Find $P(2)$ and $Q(2)$ if $P(x) = -5x^2 + 6$ and $Q(x) = 3x + 5$.

     **34** Find $P(-1)$ and $Q(-1)$ if $P(x) = 4 - 7x + x^2$ and $Q(x) = x + x^5 + x^6$.

     **35** Find $P(4)$ and $Q(-1)$ if $P(w) = 3$ and $Q(w) = 6 - 4w^2$.

**36** Find $P(-2)$ and $Q(2)$ if $P(s) = 5 - 3s$ and $Q(s) = s^2 - 6$.

Suppose now that $P(x) = 4x + 3$ and $Q(y) = 2y^4 + y$. To evaluate $P[Q(-1)]$, you first find $Q(-1)$ and then replace $x$ with $Q(-1)$ in $P(x)$. Thus $Q(-1) = 2(-1)^4 + (-1) = 2 - 1 = 1$, and $P[Q(-1)] = P(1) = 4 \cdot 1 + 3 = 4 + 3 = 7$.

*Use the foregoing example as a model when you do the next few problems.*

$(\circ)$   **37**   If $P(x) = x + 3$ and $Q(y) = 6y - 18$, find $P[Q(0)]$.

**38**   If $P(x) = x^2 - 4$ and $Q(y) = 3y$, find $P[Q(1)]$.

**39**   If $P(x) = 2$ and $Q(t) = 1 + 6t + t^{10}$, evaluate $P[Q(2)]$.

**40**   If $P(s) = 6$ and $Q(z) = 400z + 60z^5$, evaluate $P[Q(2)]$.

$(\circ)$   **41**   If $P(z) = 3z + 4$ and $Q(w) = w^2$, evaluate $P[Q(1)]$ and $Q[P(1)]$.

**42**   If $P(u) = 6 - 5u + u^3$ and $Q(v) = 4v + 8$, evaluate $P[Q(-2)]$ and $Q[P(-2)]$.

*In the remaining problems you are given a polynomial P in a single variable. Find all real numbers c such that $P(c) = 0$. Take $P(x) = x + 6$, for example: the only real number c such that $P(c) = 0$ is $c = -6$.*

$(\circ)$   **43**   $P(x) = x - 5$                **44**   $P(x) = x + 3$

**45**   $P(y) = y + 9$                **46**   $P(s) = s - 23$

$(\circ)$   **47**   $P(w) = w^2 - 1$                **48**   $P(x) = x^2$

**49**   $P(x) = x^2 + 1$                **50**   $P(x) = x^2 + 2$

$(\circ)$

*Sample solutions*
*for exercise set 2.2*

**1**   This term has degree 2 and coefficient 3.

**9**   Here the coefficient is 4. To obtain the degree you observe that the exponent of $x$ is 1, the exponent of $z$ is 2, and $1 + 2 = 3$. Hence the degree of this term is 3.

**13**   This is a monomial of degree 4.

**17**   Here we have a binomial of degree 4.

**21**   This one is a trinomial. To obtain its degree, notice that each term is of second degree; so the polynomial has degree 2.

**27**   Since $P(x) = 5x + 6$, we see that $P(-1) = 5(-1) + 6 = -5 + 6 = 1$, $P(0) = 5(0) + 6 = 0 + 6 = 6$, and $P(2) = 5(2) + 6 = 10 + 6 = 16$.

**33**   $P(2) = -5(2^2) + 6 = -20 + 6 = -14$ and $Q(2) = 3 \cdot 2 + 5 = 6 + 5 = 11$

**39** $(\circ)$ **Some basic definitions**

**37**  To obtain $P[Q(0)]$, we observe that $Q(0) = 6 \cdot 0 - 18 = -18$, and $P[Q(0)] = P(-18) = -18 + 3 = -15$.

**41**  Notice that $Q(1) = 1^2 = 1$, and $P[Q(1)] = P(1) = 3 \cdot 1 + 4 = 7$. On the other hand, $P(1) = 7$ (as we have just seen), so that $Q[P(1)] = Q(7) = 7^2 = 49$. You should observe that $P[Q(1)]$ need not equal $Q[P(1)]$.

**43**  The only real number $c$ such that $c - 5 = 0$ is $c = +5$.

**47**  To say that $c^2 - 1 = 0$ is equivalent to saying that $c^2 = 1$. The only way to ensure this is to have $c = 1$ or $c = -1$.

# (2.3) (Addition and subtraction)

Any variables that occur in a polynomial represent unspecified real numbers. Since you are familiar with addition and subtraction of real numbers, you should not be shocked to learn that there are corresponding notions of addition and subtraction for polynomials. Given the polynomials $P$ and $Q$, we are going to define a polynomial, $P + Q$, called their *sum*. How are we going to do this? The easiest way of seeing what is happening is to think of the variables as real numbers and then to use the distributive law for the reals. Thus

$$2x + 3x = (2 + 3)x = 5x,$$

$$6y + (-4)y = [6 + (-4)]y = 2y,$$

$$3xw^2 + 5xw^2 + (-2)xw^2 = [3 + 5 + (-2)]xw^2 = 6xw^2.$$

*combining like terms*

In other words, the sum of two or more like terms is a like term whose coefficient is the sum of the coefficients of the original terms. This process is often called *combining like terms*. It is worth mentioning that if $x$ is replaced by any real number, then the expression $5x$ represents the same number as $2x + 3x$. Expressions that have this property are called *equivalent expressions*. A similar observation can be made about any sum of like terms.

*equivalent expressions*

As we pointed out in the last section, differences of terms are treated by agreeing that

$$-ax^n = + (-a)x^n.$$

Thus, $2x^3 - 5x^3 = (2 - 5)x^3 = (-3)x^3 = -3x^3$, and $6xz - 8xz + 4xz = (6 - 8 + 4)xz = 2xz$. Incidentally, this device is useful to avoid multiple minus signs and parentheses. Thus, negative coefficients are usually indicated by writing $-ax^n$ with $a$ positive, rather than $(-a)x^n$. To illustrate this concretely, the polynomial $(-1)x^4 + 6x^3 - (-5)x^2 + (-3)x + (-1)$ would usually be written in the equivalent form $-x^4 + 6x^3 + 5x^2 - 3x - 1$.

We are now ready to define the sum of the polynomials $P$ and $Q$. *This is the polynomial $P + Q$ formed by adding all terms from $P$ and $Q$, and then combining any like terms.* Let's nail the idea down with an example. If $P(x) = x^4 + 5x + 3$ and $Q(x) = x^3 - 3x + 2$, then

$$P(x) + Q(x) = (x^4 + 5x + 3) + (x^3 - 3x + 2)$$
$$= x^4 + x^3 + (5x - 3x) + (3 + 2)$$
$$= x^4 + x^3 + 2x + 5.$$

Let's have another example:

$$(7xw^2 - 5x^2 w + 3w) + (xw^2 + 5x^2 w + 6w)$$
$$= (7xw^2 + xw^2) + (-5x^2 w + 5x^2 w) + (3w + 6w)$$
$$= 8xw^2 + 9w.$$

Once you get some practice you should be able to do most of these computations mentally.

Now for the subtraction process. Recall that for real numbers $a$ and $b$, $a - b = a + (-b)$. We are going to apply the same process to polynomials. For a polynomial $P$, we define $-P$ (read *negative P*) to be the polynomial obtained by replacing each coefficient of $P$ by its additive inverse. If the convention mentioned earlier regarding negative coefficients is observed—in other words, if $P$ is written as the sum or difference of terms with positive coefficients—the formation of $-P$ is especially easy. You just change the sign in front of each term. Minus signs get changed to plus signs, and plus signs get changed to minus signs. Thus $-(2x^3 - 6x + 1) = -2x^3 + 6x - 1$, and $-(8x^5 y - 4x^3 z + 6y - 3z) = -8x^5 y + 4x^3 z - 6y + 3z$.

Given the polynomials $P$ and $Q$, we now define

---

$(\circ)$   $P - Q = P + (-Q)$

---

To illustrate this definition we observe that

$$(x^3 + 5x + 1) - (x^2 - 2x + 1) = (x^3 + 5x + 1) + (-x^2 + 2x - 1)$$
$$= x^3 - x^2 + 7x.$$

And, for a look at another example, we see that

$$(y^4 - 3y^3z + 7yz^3) - (-y^4 - 3y^3z - 7yz^3 + 1)$$

$$= (y^4 - 3y^3z + 7yz^3) + (y^4 + 3y^3z + 7yz^3 - 1)$$

$$= 2y^4 + 14yz^3 - 1.$$

Parentheses, brackets, and braces are used to indicate the order of operations in computations with real numbers. They are used in much the same way in connection with polynomials. Remember one thing, though: when faced with an expression involving multiple grouping devices, start with the innermost such device, perform the indicated operation or operations on the expression it encloses, and then work your way outward. The next examples illustrate this sequence of operations:

$$(6x + 3) + [2 - (4x - 1)] = (6x + 3) + (2 - 4x + 1)$$

$$= (6x + 3) + (-4x + 3)$$

$$= 2x + 6;$$

$$(6x + 3) - [2 - (4x - 1)] = (6x + 3) - (2 - 4x + 1)$$

$$= (6x + 3) - (-4x + 3)$$

$$= (6x + 3) + (4x - 3) = 10x.$$

## Exercise set 2.3

Find the sum of each of the following pairs of polynomials.

( ∘ )   **1**  $x; 2x$                                         **2**  $5w; 7w$

**3**  $6st^2; (-8)st^2$                             **4**  $-6x^3y; 7x^3y$

( ∘ )   **5**  $x^2 + 2x + 1; x^3 + 3x^2 + 2$

**6**  $x^6 + x^3; x^6 + x^2 + 1$

**7**  $t^4 - (-2)t^3 + \dfrac{1}{2}t + 1; t^4 - 2t^3 + 3 - \dfrac{1}{2}t$

**8**  $2x^2 + x + 1 + 3x^3; (-3)x^3 - (-2)x - 4$

( ∘ )   **9**  $6x^2y + 3xz - 5xyz; -6xy^2 - 3xy + 5xyz$

**10**  $t^5u - 2t + 3tu^3; -2t - t^5u + 2tu^3$

**11**  $0; x^2 + 2xy + y^2$

**12**  $0; x^2 - 3xy - 4y^2$

*For each polynomial P, compute $-P$ and observe the convention concerning negative coefficients by expressing $-P$ as the sum or difference of terms with positive coefficients.*

**13** $x - x^2$

**14** $9x - x^3 + 1$

( ∘ ) **15** $w^2 + (-3)w + 1$

**16** $2t^3 + (-6)t + (-7)$

**17** $-\dfrac{3}{2} + 6yz - (-2)y^2z^2$

**18** $4s - 6s^2t - (-2)s^3t^2$

**19** $-3xyz + 2x^2z - 5y$

**20** $8 - 6a^2b - ab^3$

*Now for some subtraction problems. Remember the definition $P - Q = P + (-Q)$. For problems 21 to 32, use the polynomials in problems 1 to 12. Subtract the second polynomial from the first one.*

*In problems 33 to 42, perform the indicated operations, expressing each result as a polynomial in the appropriate variables.*

( ∘ ) **33** $(3x - 1) - (x^2 - 3x - 1) + (x^2 + x)$

**34** $(2 - 5y) - (5y - 4) + (y^2 + 10y - 6)$

**35** $(a^2 - 3ab) - (a + ab) - (b^2 + ab)$

**36** $(a - b - c) - (a + b + c) - (b + c - a)$

**37** $(a^3 + 3a^2b) - (a^2b + ab^2) - (3ab^2 + b^3)$

**38** $(s^3wz + 3swz) + (z^3 - 3swz) - (swz + s^3wz)$

**39** $(x^2yz + z^3) - (xyz - z^3)$

**40** $(xy + yz + zx) + (xy - yz - 2zx)$

**41** $(a^2 - ab + c) + (2ab - c + c^2)$

**42** $(x^4 + 6x^2yz^3 + z^4) + (x^3 - x^4 - 5x^2yz^3)$

*In problems 43 to 46, treat a, b, c, and d as part of the coefficient and any other letters as variables.*

( ∘ ) **43** $(ax + by) + (cx + dy)$

**44** $a^2xy^2 + (xy^3 - 3axy^2) + (bxy^3 - axy^2) + 2axy^2$

**45** $3awx + (6abwx - b^2wx^2) + 5bwx^2$

**46** $(a^2x^2 + 2abx + b^2) + (b^2y^2 - 2abx + b^2) + (ax - by^2)$

*In problems 47 to 50, form the sum $R(x) = P(x) + Q(x)$, and for the indicated value of c, verify that $R(c) = P(c) + Q(c)$.*

( ∘ ) **47** $P(x) = x + 3;\ Q(x) = x^2 - x;\ c = 4$

**48** $P(x) = x^3 - 3x + 1;\ Q(x) = x^3 - 4;\ c = 2$

**43 ( ∘ ) Addition and subtraction**

**49**  $P(x) = x^5 - x^4 + x^3 - x; Q(x) = x^4 + x + 1; c = -1$

**50**  $P(x) = 2x^3 - 3x + 1; Q(x) = x + x^2 + (-3)x^3; c = -2$

*For the remaining problems, as with all problems involving multiple grouping devices, remember to start with the innermost devices and then work your way outward. Express each of the following as a polynomial in the indicated variables.*

( ∘ )  **51**  $2 - [3 + (x^2 - 4)]$

**52**  $(6x + 3) - [x - (3x + 5)]$

**53**  $(2x^2 + 1) - [x^2 - (3x + x^2 + 1)]$

**54**  $(2x^2 + x^3 - 3) \div [(-3x + 4) - (2x + 5)]$

( ∘ )  **55**  $-\{(x + 3) - [(x + 2) - (x - 1)]\}$

**56**  $-\{x - [(x^2 + 2x + 1) - (x^2 - 2x + 1)]\}$

**57**  $2x - [(3y - z) - (z - 2y)]$

**58**  $(5x^2 y - xy^2) - [(3xy^2 - x^2 y + xy) + (xy - x^2 y)]$

**59**  $a - \{a - [(a - b) - (2a - b) - b]\}$

**60**  $sw - \{(sw - w) - \{sw - [(w^2 - w - s) - (sw + w)] - w\}\}$

( ∘ )

*Sample solutions for exercise set 2.3*

**1**  $x + 2x = (1 + 2)x = 3x$

**5**  $(x^2 + 2x + 1) + (x^3 + 3x^2 + 2) = x^3 + (x^2 + 3x^2) + 2x + (1 + 2)$

$$= x^3 + 4x^2 + 2x + 3$$

**9**  $(6x^2 y + 3xz - 5xyz) + (-6xy^2 - 3xy + 5xyz)$

$= 6x^2 y - 6xy^2 + 3xz - 3xy + (-5xyz + 5xyz)$

$= 6x^2 y - 6xy^2 + 3xz - 3xy$

**15**  If $P = w^2 + (-3)w + 1 = w^2 - 3w + 1$, then $-P = -w^2 + 3w - 1$.

**21**  $x - 2x = x + (-2)x = -x$

**25**  $(x^2 + 2x + 1) - (x^3 + 3x^2 + 2) = (x^2 + 2x + 1) + (-x^3 - 3x^2 - 2)$

$$= -x^3 + (x^2 - 3x^2) + 2x + (1 - 2)$$

$$= -x^3 - 2x^2 + 2x - 1$$

**29**  $(6x^2 y + 3xz - 5xyz) - (-6xy^2 - 3xy + 5xyz)$

$= (6x^2 y + 3xz - 5xyz) + (6xy^2 + 3xy - 5xyz)$

$= 6x^2 y + 6xy^2 + 3xz + 3xy - 10xyz$

**33**   Using the fact that $P - Q = P + (-Q)$, we have

$$(3x - 1) - (x^2 - 3x - 1) + (x^2 + x) = (3x - 1) + (-x^2 + 3x + 1) + (x^2 + x)$$
$$= (-x^2 + x^2) + (3x + 3x + x)$$
$$+ (-1 + 1)$$
$$= 7x.$$

**43**   Remembering to treat $a$, $b$, $c$, $d$ as part of the coefficient, we write

$$(ax + by) + (cx + dy) = (ax + cx) + (by + dy)$$
$$= (a + c)x + (b + d)y.$$

**47**   We first note that

$$R(x) = P(x) + Q(x) = (x + 3) + (x^2 - x)$$
$$= x^2 + (x - x) + 3 = x^2 + 3.$$

Hence $R(4) = 4^2 + 3 = 16 + 3 = 19$. Note finally that $P(4) = 4 + 3 = 7$, $Q(4) = 4^2 - 4 = 16 - 4 = 12$, and $P(4) + Q(4) = 7 + 12 = 19 = R(4)$.

**51**   
$$2 - [3 + (x^2 - 4)] = 2 - [x^2 + (3 - 4)]$$
$$= 2 - [x^2 - 1]$$
$$= 2 + [-x^2 + 1]$$
$$= -x^2 + (2 + 1)$$
$$= -x^2 + 3$$

**55**   Notice how we start with the innermost grouping devices, and systematically work our way outward.

$$-\{(x + 3) - [(x + 2) - (x - 1)]\} = -\{(x + 3) - [(x + 2) + (-x + 1)]\}$$
$$= -\{(x + 3) - [(x - x) + (2 + 1)]\}$$
$$= -[(x + 3) - 3]$$
$$= -[x + (3 - 3)]$$
$$= -x$$

# (2.4) (Multiplication)

Operations with polynomials are not limited to addition and subtraction. Multiplication of real numbers can also be carried over to a multiplication operation for polynomials. Given polynomials $P$ and $Q$, pretend all variables are real numbers, and then use the various properties of the reals to compute a polynomial $R = P \cdot Q$, called the *product* of $P$ and $Q$. Well, that's the theory of it, anyway. Now let's develop some working procedures.

We begin with the product of the monomials $2x$ and $3x^2$. Applying the associative and commutative laws for multiplication, we find that $(2x)(3x^2) = 2 \cdot x \cdot 3 \cdot x^2 = 2 \cdot 3 \cdot x \cdot x^2 = 6x^3$. Similarly,

$$8(9w^3) = 8 \cdot 9 \cdot w^3 = 72w^3,$$

$$(5xy^3)(-4x^2 y) = 5 \cdot x \cdot y^3 \cdot (-4) \cdot x^2 \cdot y$$
$$= 5 \cdot (-4) \cdot x \cdot x^2 \cdot y^3 \cdot y = -20x^3 y^4.$$

The basic idea for the multiplication of two monomials is this:

$$a(bx^n) = abx^n$$

$$(ax^n)(bx^m) = abx^{n+m}$$

Here $a$ and $b$ are real numbers, $x$ is a variable, and $n$ and $m$ are natural numbers.

Multiplication of a polynomial with more than one term by a monomial requires the use of the distributive law. For example,

$$3(5x - 4) = 3(5x) + 3(-4) = 15x - 12,$$

$$(5x)(2x^3 - 6x + 1) = (5x)(2x^3) + (5x)(-6x) + (5x)(1)$$
$$= 10x^4 - 30x^2 + 5x,$$

$$(6xy^2)(5x^2 y - 3xy^3) = (6xy^2)(5x^2 y) + (6xy^2)(-3xy^3)$$
$$= 30x^3 y^3 - 18x^2 y^5.$$

With some practice you should be able to skip the middle step and obtain $(-4xz^3)(x^2 - 2yz) = -4x^3 z^3 + 8xyz^4$ directly.

How about the product of two polynomials that both contain more than one term? Again, the distributive law is used, but now it is used more than one time. But watch yourself. The distributive law gets quite a workout, and so will you! See if you can follow the steps in the following example:

$$(2x + 5)(3x - 7) = 2x(3x - 7) + 5(3x - 7)$$
$$= 6x^2 - 14x + 15x - 35$$
$$= 6x^2 + x - 35.$$

Notice what we did. With $Q(x) = 3x - 7$, the distributive law was used to write $(2x + 5)Q(x) = 2xQ(x) + 5Q(x)$. Then the rule for multiplying a polynomial by a monomial was applied twice. Finally, like terms were combined, and the answer was written as a polynomial. Here's another example.

$$(2x - 7y)(x^2 - 3xy + 2y^2) = 2x(x^2 - 3xy + 2y^2) - 7y(x^2 - 3xy + 2y^2)$$
$$= 2x^3 - 6x^2y + 4xy^2 - 7x^2y + 21xy^2 - 14y^3$$
$$= 2x^3 - 6x^2y - 7x^2y + 4xy^2 + 21xy^2 - 14y^3$$
$$= 2x^3 - 13x^2y + 25xy^2 - 14y^3.$$

A few types of binomial products occur so frequently that you will find it useful to be able to recognize them on sight. They are represented by the following formulas:

$(\circ)$   **BP.1**   $(x + a)(x - a) = x^2 - a^2$

$(\circ)$   **BP.2**   $(x + a)^2 = x^2 + 2ax + a^2$

$(\circ)$   **BP.3**   $(x + a)(x + b) = x^2 + (a + b)x + ab$

$(\circ)$   **BP.4**   $(ax + b)(cx + d) = acx^2 + (ad + bc)x + bd$

## Exercise set 2.4

*Express each answer as a polynomial. Any letters that appear in exponents are to be regarded as representing positive integers. We begin with some products of monomials.*

$(\circ)$   **1**   $x \cdot x^2$        **2**   $x^3 \cdot x^4$

     **3**   $(2w)(-3w^2)$        **4**   $(5v)(-4v^2)$

     **5**   $(2xy)(3xz)$        **6**   $(5wx)(-6w^2)$

     **7**   $(5ab)(6a^2b)$        **8**   $(7abcd)(-2ad)$

     **9**   $(6)(2x^2yz)$        **10**   $(-3x^3yw)^2$

$(\circ)$   **11**   $(2y^n)(8y^{n+2})$        **12**   $(2x^{2n})(-4x^{n+1})$

*Here are a few involving a monomial times a polynomial.*

$(\circ)$   **13**   $(2x)(x^2 + 5)$        **14**   $(6x^3)(2x - 3)$

     **15**   $(-5w)(-w^2 + 1)$        **16**   $(-3v)(12v^9 - 8v^7)$

     **17**   $(x^2y)(xy + 2xy^2)$        **18**   $(3xz)(4x - 2z)$

$(\circ)$   **19**   $(3ax)(ax + 2a^2y)$        **20**   $(5at)(4ax - 5t)$

     **21**   $(2x)(x^2 - 3xy + 2y^2)$        **22**   $(-3y)(2x^2 + 5xy - y^2)$

**23** $(2x^n)(x^{2n} - 3x^n + 2)$    **24** $(-3y^n)(2y^{2n} + 5y^n - 1)$

*The next problems involve the product of two binomials.*

( ∘ )    **25** $(x + 2)^2$    **26** $(w - 6)^2$

**27** $(x - 7y)^2$    **28** $\left(x - \dfrac{1}{2}y\right)^2$

**29** $(x - 3)(x + 3)$    **30** $(x + 7)(x - 7)$

**31** $(2x + 5y)(2x - 5y)$    **32** $(4x + 3y)(4x - 3y)$

**33** $(x + 3)(x - 2)$    **34** $(y + 5)(y - 4)$

**35** $(2x + 3)(7x - 5)$    **36** $(6x + 3)(x - 1)$

( ∘ )    **37** $(x - 3xy)(2x + 5xy)$    **38** $(2w + 3w^2 x)(3w + 2w^2 x)$

**39** $(x + 2)(2x^2 - 5)$    **40** $(y - 3)(3y^2 + 4)$

**41** $(2x + 3xy)(2x^2 - 4y)$    **42** $(3x^2 y - 5y)(2x + 5xy)$

*We turn next to some products that are a bit more complicated. The basic ideas still apply, though.*

( ∘ )    **43** $(x + 3)(x^2 - 2x + 1)$    **44** $(x - 2)(3x^2 + x + 2)$

**45** $(2y - 3)(y^2 + y + 1)$    **46** $(5w + 6)(3w^2 - 2w + 1)$

**47** $(x + 2y)(2x^2 - xy + y^2)$    **48** $(5x - 2y)(x^2 - 2xy + 2y^2)$

**49** $(2t^{2n} - 5)(t^{3n} + 3t^n + 1)$    **50** $(3z^{2n} + z^n + 1)(z^n - 4)$

The product of three or more polynomials may be obtained by using the fact that multiplication of polynomials is associative. Thus, to obtain the product $2x(x - 3y)(x + 3y)$, you might write

$$2x[(x - 3y)(x + 3y)] = (2x)(x^2 - 9y^2)$$

$$= 2x^3 - 18xy^2.$$

Of course, it is equally correct to write $[(2x)(x - 3y)](x + 3y)$ and carry out the indicated operations. The choice is yours. More generally, any product $P \cdot Q \cdot R$ of three polynomials may be computed either as $P \cdot (Q \cdot R)$ or $(P \cdot Q) \cdot R$.

*The foregoing observations should help you with problems 51 to 56.*

**51** $(2x)(3x + 5)(3x - 5)$

**52** $(3w)(2w + 1)(2w - 1)$

( ∘ )    **53** $(x + 1)(x + 2)(x + 3)$

**54** $(x - 1)(x - 2)(x - 3)$

**55** $(2x + 3y)(x - 2y)(3x - y)$

**56**  $(s + 4t)(2t - s)(3s - t)$

*If you are still uncertain about the order of operations, take time out to review Section 1.6. The following problems involve both multiplication of polynomials and order of operations.*

( ◦ )  **57**  $(x + 3)[2 - x(3 + x)]$

**58**  $(y - 4)[3y - (y + 2)(y + 1)]$

**59**  $[4 - (x + 2)^2][2 + x(x + 1)]$

**60**  $[x(x + 1) + (x - 2)(x - 1)][1 - x(x - 3)]$

**61**  $[y - (y + 3)(y - 2)]^2$

**62**  $[9 + x + (x - 4)(x + 3)]^2$

*In problems 63 to 66, treat a, b as part of the coefficient of any polynomials in which they appear.*

( ◦ )  **63**  Establish a rule to write the product $(ax + by)(ax - by)$ directly as a polynomial.

**64**  Establish a rule to write the product $(ax + by)^2$ directly as a polynomial.

**65**  Establish a rule to write the product $(x + ay)(x + by)$ directly as a polynomial.

**66**  Establish a rule to write the product $(ax + y)(bx + y)$ directly as a polynomial.

( ◦ )

*Sample solutions for exercise set 2.4*

**1**  $x \cdot x^2 = x^3$

**11**  We do not know the value of $n$, but we do know that it represents some unspecified positive integer. Thus $(2y^n)(8y^{n+2}) = 16y^{n+n+2} = 16y^{2n+2}$.

**13**  $(2x)(x^2 + 5) = (2x)(x^2) + (2x)(5) = 2x^3 + 10x$

**19**  $(3ax)(ax + 2a^2 y) = (3ax)(ax) + (3ax)(2a^2 y)$
$$= 3a^2 x^2 + 6a^3 xy$$

**25**  Of course you can use the procedure for multiplication of polynomials to obtain $(x + 2)^2$. But it is a lot easier to apply equation BP.2 (page 47) with $a = 2$ and write directly $(x + 2)^2 = x^2 + 4x + 4$.

**37**  $(x - 3xy)(2x + 5xy) = x(2x + 5xy) + (-3xy)(2x + 5xy)$
$$= 2x^2 + 5x^2 y - 6x^2 y - 15x^2 y^2$$
$$= 2x^2 - x^2 y - 15x^2 y^2$$

**43**  $(x + 3)(x^2 - 2x + 1) = x(x^2 - 2x + 1) + 3(x^2 - 2x + 1)$

$$= x^3 - 2x^2 + x + 3x^2 - 6x + 3$$

$$= x^3 + x^2 - 5x + 3$$

**53**  We begin by grouping the product $(x + 1)(x + 2)(x + 3)$ as $(x + 1)[(x + 2)(x + 3)]$. Working inside the brackets we now have

$(x + 1)[(x + 2)(x + 3)] = (x + 1)(x^2 + 5x + 6)$

$$= x(x^2 + 5x + 6) + 1(x^2 + 5x + 6)$$

$$= x^3 + 5x^2 + 6x + x^2 + 5x + 6$$

$$= x^3 + 6x^2 + 11x + 6.$$

**57**  We first simplify the expression enclosed by the brackets by observing that $2 - x(3 + x) = 2 - 3x - x^2$, so that

$(x + 3)[2 - x(3 + x)] = (x + 3)(2 - 3x - x^2)$

$$= x(2 - 3x - x^2) + 3(2 - 3x - x^2)$$

$$= 2x - 3x^2 - x^3 + 6 - 9x - 3x^2$$

$$= -x^3 - 6x^2 - 7x + 6.$$

**63**  $(ax + by)(ax - by) = (ax)(ax - by) + (by)(ax - by)$

$$= a^2x^2 - abxy + abxy - b^2y^2 = a^2x^2 - b^2y^2$$

Hence the desired rule is $(ax + by)(ax - by) = a^2x^2 - b^2y^2$.

## (2.5) (Monomial factors)

We know from the last section that $(2x)(x^2 + 3x + 1) = 2x^3 + 6x^2 + 2x$. Suppose instead that we start with the polynomial $2x^3 + 3x^2 + 2x$ and write this polynomial as the product of $2x$ and $x^2 + 3x + 1$. This process is called *factoring*, and it a useful tool in dealing with polynomials.

*factoring*

Let's begin by studying the factorization of monomials. A typical problem might be to find the appropriate term in the expression $8x^4 = (2x)(?)$. The question now is, $2x$ times *what* equals $8x^4$. Well, 2 times 4 equals 8, and $x$ times $x^3$ equals $x^4$, so that $8x^4 = (2x)(4x^3)$.

Let's have another example: $6x^2y^7 = (2xy^3)(?)$. The key observations here are that $2 \cdot 3 = 6$, $x \cdot x = x^2$, and $y^3 \cdot y^4 = y^7$, so that $6x^2y^7 = (2xy^3)(3xy^4)$.

You can use this same technique on an arbitrary polynomial. Suppose you were asked to find the missing term in the expression $8x^6 - 6x^4 + 2x^3 = (2x^2)(?)$. Looking at the problem term by term, we have

$$(2x^2)(4x^4) = 8x^6$$

$$(2x^2)(-3x^2) = -6x^4$$

$$(2x^2)(x) = 2x^3.$$

By the distributive law, $8x^6 - 6x^4 + 2x^3 = (2x^2)(4x^4 - 3x^2 + x)$. With a little practice you can factor some expressions just by inspection. Thus the (?) in $25w^8z^2 - 15w^6z^3 + 50w^2z^4 = (5w^2z^2)(?)$ is easily seen to be $(5w^6 - 3w^4z + 10z^2)$.

## Exercise set 2.5

You may assume that any letter appearing in an exponent represents a positive integer. In problems 1 to 10, you are to replace (?) with the missing monomial factor.

(∘)   1   $12x^4 = 3x(?)$       2   $16x^6 = (8x^4)(?)$

3   $6y^3 = 3(?)$            4   $18w^2 = (-2)(?)$

5   $15x^4y^7 = (5x^2y^3)(?)$      6   $-9x^3y^5 = (-3x^3y^4)(?)$

(∘)   7   $64a^3b^2cd^4 = (?)(8a^2bcd^2)$    8   $-12s^2t^2u^2 = (?)(-4st^2u)$

9   $8x^{2n} = (4x^n)(?)$        10   $20w^{n^2+n} = (5w^n)(?)$

In the next few problems, (?) is to be replaced by a suitable polynomial factor.

(∘)   11   $6x^4 + 10x^3 = (2x)(?)$

12   $18w^5 + 9w^2 = (3w^2)(?)$

13   $9t^3 + 18t^2 + 3t = (3t)(?)$

14   $50x^8 + 25x^6 + 5x^3 = (5x^2)(?)$

15   $x^{2n} + 3x^n = (x^n)(?)$

16   $4x^{3n} + 5x^{2n} - x^n = (x^n)(?)$

(∘)   17   $6x^{3k} - 2x^k = (x^{k-2})(?)$     $(k > 2)$

18   $12y^{2k+1} + 6y^k + 3y^{k-3} = (3y^{k-3})(?)$     $(k > 3)$

19   $15x^4y^2 + 20x^5y^2 = (5x^2y^2)(?)$

20   $12y^3z^4 - 8y^2z^2 = (2yz^2)(?)$

(∘)   21   $10x^6y^4z^2 + 25x^3yz = (2x^3y^3z + 5)(?)$

22   $8a^4b^2c - 4a^2b^4c = (-4a^3b^2 + 2ab^4)(?)$

23   $6x^2yz^3 - 8x^3y^4z^2 = (xyz^2)(?)$

24   $36s^3t^2w^4 + 18s^2t^3w^2 = (9s^2t^2w^2)(?)$

**25**   $-14axy^2 + 28a^2xy^3 + 7ax = (7ax)(?)$

**26**   $16bwx^3y^2 + 4bw^2x^2y^2 - 8b^3w^3xy^2 = (-2bwxy^2)(?)$

Notice how the distributive law is used to write each of the following examples as the product of two polynomials:

$x(x - 2) + 3(x - 2) = (x + 3)(x - 2),$

$x^2(x + 1) - 3x(x + 1) + 4(x + 1) = (x^2 - 3x + 4)(x + 1),$

$x^3(x - 2xy) - 2y^3(x - 2xy) = (x^3 - 2y^3)(x - 2xy).$

*This should give you an idea of how to handle the remaining group of problems. In each case fill in the (?) with an appropriate polynomial factor.*

(∘)   **27**   $2x(x + 1) + 3(x + 1) = (?)(x + 1)$

**28**   $5y^2(y - 6) + 3y(y - 6) = (?)(y - 6)$

**29**   $5w^2(2w + 3) + w(2w + 3) + 2w + 3 = (?)(2w + 3)$

**30**   $(7t^5)(t - 2) + (3t^3)(t - 2) + t(t - 2) = (?)(t - 2)$

**31**   $x^2(x + y) + y(x + y) = (?)(x + y)$

**32**   $6x^2w(x^2 - xy) + 3xyw^2(x^2 - xy) = (?)(x^2 - xy)$

**33**   $(a + b)^2 + (a - b)(a + b) + a(a + b) = (?)(a + b)$

**34**   $(x + y + z)^2 + (x + y - z)(x + y + z) + (x - y)(x + y + z) = (?)(x + y + z)$

(∘)

*Sample solutions for exercise set 2.5*

**1**   Use the facts that $3 \cdot 4 = 12$ and $x \cdot x^3 = x^4$ to write $12x^4 = (3x)(4x^3)$.

**7**   $64a^3b^2cd^4 = (8abd^2)(8a^2bcd^2)$

**11**   We have that $6x^4 = (2x)(3x^3)$, and $10x^3 = (2x)(5x^2)$. Thus we have $6x^4 + 10x^3 = (2x)(3x^3 + 5x^2)$.

**17**   Here we observe that $x^{k-2} \cdot x^{2k+2} = x^{3k}$, and $x^{k-2} \cdot x^2 = x^k$; so $6x^{3k} - 2x^k = x^{k-2}(6x^{2k+2} - 2x^2)$.

**21**   $10x^6y^4z^2 + 25x^3yz = (2x^3y^3z + 5)(5x^3yz)$

**27**   Using the distributive law, we see that $2x(x + 1) + 3(x + 1) = (2x + 3)(x + 1)$.

# (2.6) (Factoring quadratics)

*quadratic*

A second-degree polynomial is often called a *quadratic* polynomial. Quite frequently, you will have to express a quadratic polynomial with integer coefficients as the product of two first-degree polynomials, each of which has integer coefficients. This kind of factoring is important for solving second-degree equations. It is easy to develop formulas for this by rewriting equations BP.1 to BP.4.

( ○ ) **BP.1a** $x^2 - a^2 = (x + a)(x - a)$

( ○ ) **BP.2a** $x^2 + 2ax + a^2 = (x + a)^2$

( ○ ) **BP.3a** $x^2 + (a + b)x + ab = (x + a)(x + b)$

( ○ ) **BP.4a** $acx^2 + (ad + bc)x + bd = (ax + b)(cx + d)$

Let's illustrate with a few examples. Some quadratics can be factored by inspection alone. We can see right away that $2x^2 + 4x = 2x(x + 2)$. If we are called upon to factor $x^2 + 6x + 9$, we use equation BP.2a with $a = 3$ to write $x^2 + 6x + 9 = (x + 3)^2$. Similarly, a difference of squares is easy to factor. By BP.1a, $x^2 - 16 = (x + 4)(x - 4)$, $9y^2 - 25x^2 = (3y + 5x)(3y - 5x)$, and so on.

But what about a quadratic like $x^2 + 5x + 6$? In view of BP.3a, any factorization must look like $(x + a)(x + b)$ where $ab = 6$ and $a + b = 5$. The positive factors of 6 are 1, 2, 3, and 6. Since $1 + 6 = 7$ and $2 + 3 = 5$, we see from BP.3a that $x^2 + 5x + 6 = (x + 2)(x + 3)$.

Let's now consider $x^2 - 6xy - 16y^2$. Any factorization must be of the form $(x + ay)(x + by)$ where $ab = -16$ and $a + b = -6$. Thus we want to examine the factors of 16 for a pair whose difference is $-6$. We observe that

$16 = 16 \cdot 1$ with $16 - 1 = 15$

$16 = 8 \cdot 2$ with $8 - 2 = 6$

$16 = 4 \cdot 4$ with $4 - 4 = 0$.

This shows that our factorization looks like $(x \quad 8y)(x \quad 2y)$ with the correct signs yet to be filled in. Observing that $-8 + 2 = -6$ and $(-8)(2) = -16$, we are finally able to write $x^2 - 6xy - 16y^2 = (x - 8y)(x + 2y)$.

In order to illustrate factorizations based on BP.4a, consider the polynomial $4x^2 + 16x + 15$. We may write 4 as $4 \cdot 1$ or as $2 \cdot 2$; so any possible factorization must look like $(4x + ?)(x + ?)$ or $(2x + ?)(2x + ?)$. On the other hand, the positive factors of 15 are 1, 3, 5, and 15; so any possible factorization must look like $(? + 1)(? + 15)$ or $(? + 3)(? + 5)$. The next step is to examine the possible combinations of factors systematically until you find the correct one. You usually do such calculations in your head, but the table following will allow you to visualize the procedure. We begin with factorizations of the type $(4x + ?)(x + ?)$ and, in each case, compute the coefficient of $x$ in the product.

| Factors $(4x + b)(x + d)$ | Coefficient of $x$ $4d + b$ |
|---|---|
| $(4x + 1)(x + 15)$ | $4 \cdot 15 + 1 = 61$ |
| $(4x + 3)(x + 5)$ | $4 \cdot 5 + 3 = 23$ |
| $(4x + 5)(x + 3)$ | $4 \cdot 3 + 5 = 17$ |
| $(4x + 15)(x + 1)$ | $4 \cdot 1 + 15 = 19$ |

None of these combinations will work; so let's move on to factorizations of the type $(2x + ?)(2x + ?)$. The coefficient of $x$ in $(2x + 1)(2x + 15)$ is $2 \cdot 15 + 1 \cdot 2 = 32$; so this possibility must be discarded. On the other hand, the coefficient of $x$ in $(2x + 3)(2x + 5)$ is $3 \cdot 2 + 2 \cdot 5 = 16$. We conclude that $4x^2 + 16x + 15 = (2x + 3)(2x + 5)$.

If the coefficients of a quadratic have a common integral factor, you should factor this number from the quadratic before you do anything else. Thus $4x^2 - 24x + 36 = 4(x^2 - 6x + 9) = 4(x - 3)^2$.

Of course, not every quadratic can be factored as the product of first-degree polynomials with integer coefficients. For example, neither $x^2 + 1$ nor $x^2 - 2$ can be written in this way. We shall have more to say about this later.

## Exercise set 2.6

In problems 1 to 32, express each quadratic as a product of first-degree factors with integer coefficients. The first group of problems can be done either by inspection or by using equation BP.1a or BP.2a on page 53.

$(\circ)$   1   $x^2 - 3x$        2   $2w^2 + 6w$

      3   $x^2 - 9$        4   $y^2 - 25$

$(\circ)$   5   $x^2 + 10x + 25$        6   $y^2 - 4y + 4$

      7   $9x^2 - 16y^2$        8   $25u^2 - 49v^2$

**9** $x^2 - 4xy + 4y^2$

**10** $s^2 + 10st + 25t^2$

Now for a group of problems in which you must use the fact that $x^2 + (a + b)x + ab = (x + a)(x + b)$.

( ∘ ) **11** $x^2 + 7x + 10$

**12** $y^2 - 9y + 14$

**13** $w^2 - 6w - 16$

**14** $t^2 + 11t + 18$

**15** $2x^2 - 26x + 84$

**16** $4x^2 + 12x + 8$

**17** $x^2 + 9xy + 14y^2$

**18** $a^2 + 9ab + 18b^2$

**19** $x^2 - xy - 6y^2$

**20** $x^2 + 3xy - 28y^2$

Problems 21 to 30 should provide you with a little variety.

( ∘ ) **21** $4x^2 + 4x - 3$

**22** $10x^2 - 8x - 2$

**23** $7w^2 + 23w + 6$

**24** $8y^2 + 6y - 9$

**25** $2t^2 - 18$

**26** $3x^2 - 18x$

**27** $9z^2 + 54z$

**28** $2x^2 + 20x + 18$

( ∘ ) **29** $4x^2 + 12xy + 5y^2$

**30** $6s^2 + 11st + 3t^2$

**31** $16a^2 + 28ab - 30b^2$

**32** $24g^2 + 14gh - 20h^2$

Sometimes a polynomial of higher degree can be treated as a quadratic. For example, $6x^4 + 12x^3 + 6x^2$ has a monomial factor of $6x^2$. Once $6x^2$ is factored out, what is left is a quadratic. Thus

$$6x^4 + 12x^3 + 6x^2 = (6x^2)(x^2 + 2x + 1)$$

$$= (6x^2)(x + 1)^2.$$

In general, if a polynomial with more than one term has a monomial factor, that monomial should be factored from the polynomial before anything else is done. A second example of this type is $y^4 + 6y^2 + 5$. If $y^2$ is viewed as the variable, this expression becomes a quadratic. Once this is recognized you have $y^4 + 6y^2 + 5 = (y^2 + 5)(y^2 + 1)$.

Use this idea to factor each of the following.

( ∘ ) **33** $x^6 - 9$

**34** $w^6 + 2w^3 + 1$

**35** $y^8 - y^7$

**36** $x^6 - 36x^4$

**37** $x^5 + 6x^3 + 9x$

**38** $2y^4 - 5y^2 - 12$

**39** $x^{2n} - 7x^n + 12$    $(n \in N)$

**40** $z^{4n} + 5z^{2n} - 14$    $(n \in N)$

( ∘ ) **41** $t^4 + 6t^2u^4 + 9u^8$

**42** $y^6 + 6y^3z^4 + 9z^8$

**43** $x^6 + 10x^3y^2 + 21y^4$

**44** $s^8 - 9s^4t^2 + 18t^4$

You can even think of an expression as being a quadratic polynomial with respect to some other expressions. For example, if $2x - 3$ is

**55** ( ∘ ) **Factoring quadratics**

replaced by $w$ in the expression $(2x - 3)^2 - 2(2x - 3) - 3$, the result is the quadratic $w^2 - 2w - 3$. The fact that $w^2 - 2w - 3 = (w - 3)(w + 1)$ now shows that the original expression factors as $[(2x - 3) - 3][(2x - 3) + 1]$, which in turn equals $(2x - 6)(2x - 2)$. Consequently,

$$(2x - 3)^2 - 2(2x - 3) - 3 = (2x - 6)(2x - 2)$$

$$= 4(x - 3)(x - 1).$$

*Here are a few factorization problems based on this general idea.*

$(\circ)$    **45**   $(x + 1)^2 + 6(x + 1) + 9$     *Hint:* Set $w = x + 1$.

**46**   $(x^2 + 1)^2 - 4$     *Hint:* Set $w = x^2 + 1$.

**47**   $(7x - 5)^2 + 6(7x - 5) - 16$

**48**   $(3x^4 + 1)^2 - 7(3x^4 + 1) + 6$

$(\circ)$    **49**   $(2x + 5y)^2 + 3(2x + 5y) - 40$

**50**   $(3x^2 - 2y)^2 + 2(3x^2 - 2y) - 15$

**51**   $(6x - 5y)^2 + 3(6x - 5y)(x + 3y) + 2(x + 3y)^2$

**52**   $(2x + 3y)^2 - 4(2x + 3y)(x - 2y) - 12(x - 2y)^2$

The product $ab$ of the real numbers $a$ and $b$ is 0 if and only if $a = 0$ or $b = 0$. This fact may be used to determine that $P(x) = (x - a)(x - b)$ has the value 0 only when $x$ is replaced by $a$ or $b$.

$(\circ)$    **53**   For what values of $x$ is it true that $(x + 3)(x - 1) = 0$?

**54**   For what values of $x$ is it true that $(x - 2)(x + 5) = 0$?

**55**   Let $P(x) = x^2 + 6x + 9$. What real numbers $a$ have the property that $P(a) = 0$?

**56**   Let $Q(x) = x^2 + 9x + 20$. For what real numbers $a$ is it true that $Q(a) = 0$?

$(\circ)$

*Sample solutions for exercise set 2.6*

**1**   By inspection, $x^2 - 3x = x(x - 3)$.

**5**   Using equation BP.2a with $a = 5$, we see that $x^2 + 10x + 25 = (x + 5)^2$.

**11**   To factor $x^2 + 7x + 10$ in the form $(x + a)(x + b)$ we must find two positive integers whose product is 10 and whose sum is 7. The possibilities are 10 and 1 or 5 and 2. Since $10 + 1 = 11$ and $5 + 2 = 7$, we see that the desired factorization must be $(x + 5)(x + 2)$.

**21**   On the one hand, any factorization of $4x^2 + 4x - 3$ must look like $(4x + ?)(x + ?)$ or $(2x + ?)(2x + ?)$. On the other hand, it must be of the form $(?\ 3)(?\ 1)$ with an appropriate choice of signs. The next step is to examine the possibilities one by one and, in each case, compute the coefficient of $x$ in the product.

| Factors | Coefficient of x |
|---------|------------------|
| $(4x - 1)(x + 3)$ | $-1 + 12 = 11$ |
| $(4x + 1)(x - 3)$ | $1 - 12 = -11$ |
| $(4x - 3)(x + 1)$ | $-3 + 4 = 1$ |
| $(4x + 3)(x - 1)$ | $3 - 4 = -1$ |
| $(2x + 3)(2x - 1)$ | $6 - 2 = 4$    (Stop here!) |
| $(2x - 3)(2x + 1)$ | |

This shows that $4x^2 + 4x - 3 = (2x + 3)(2x - 1)$.

29  Any factorization must be of the form $(4x + ?)(x + ?)$ or $(2x + ?)(2x + ?)$. It must also look like $(?\ 5y)(?\ y)$. Examining the possibilities shows that $4x^2 + 12xy + 5y^2 = (2x + 5y)(2x + y)$.

33  Treating $x^6 - 9$ as a polynomial in $x^3$ leads to the factorization $(x^3 + 3)(x^3 - 3)$.

41  If we consider $t^2$ and $u^4$ to be the variables, we have

$$t^4 + 6t^2 u^4 + 9u^8 = (t^2)^2 + 6t^2 u^4 + 9(u^4)^2$$

$$= (t^2 + 3u^4)^2.$$

45  With $w = x + 1$, the expression becomes $w^2 + 6w + 9$. The factorization $w^2 + 6w + 9 = (w + 3)^2$ now shows that $(x + 1)^2 + 6(x + 1) + 9 = [(x + 1) + 3]^2 = (x + 4)^2$.

49  Here we set $w = 2x + 5y$ to obtain the quadratic $w^2 + 3w - 40$. This factors as $(w + 8)(w - 5)$, thus showing that the original expression factors as $(2x + 5y + 8)(2x + 5y - 5)$.

53  It is true that $x + 3 = 0$ only when $x = -3$, and $x - 1 = 0$ only when $x = 1$. Hence $(x + 3)(x - 1) = 0$ only when $x$ is replaced by $-3$ or $1$.


# (2.7) (More factoring)

A few other factoring problems come up often enough to warrant their study. We begin by considering the factorizations

$$x^3 + a^3 = (x + a)(x^2 - ax + a^2),$$

$$x^3 - a^3 = (x - a)(x^2 + ax + a^2).$$

Both formulas can be checked by direct computation. The following examples show how to use them.

$$x^3 - 8 = x^3 - 2^3$$

$$= (x - 2)(x^2 + 2x + 2^2)$$

$$= (x - 2)(x^2 + 2x + 4),$$

$$27y^3 + 1 = (3y)^3 + 1$$

$$= (3y + 1)[(3y)^2 - 3y + 1]$$

$$= (3y + 1)(9y^2 - 3y + 1).$$

Of course, these ideas can apply to sums and differences of cubes of more general expressions. For example,

$$(2x + 3y)^3 - 27y^3 = [(2x + 3y) - 3y][(2x + 3y)^2 + 3y(2x + 3y) + (3y)^2]$$

$$= (2x)(4x^2 + 12xy + 9y^2 + 6xy + 9y^2 + 9y^2)$$

$$= (2x)(4x^2 + 18xy + 27y^2).$$

Sometimes an expression can be grouped in such a way that the grouped expressions all have a common factor. For example, we can factor $ax + by + bx + ay$ by writing it as

$$(ax + bx) + (ay + by) = (a + b)x + (a + b)y$$

$$= (a + b)(x + y).$$

Here's another example.

$$x^3 - 2x^2 + 4x - 8 = x^2(x - 2) + 4(x - 2)$$

$$= (x^2 + 4)(x - 2).$$

Suppose we wish to factor $2x^2 - 15xy^2 - 6x^2y + 5xy$. If we tried to write this as $(2x^2 - 15xy^2) + (-6x^2y + 5xy) = x(2x - 15y^2) + xy(-6x + 5)$, we would get nowhere. On the other hand, the grouping $(2x^2 - 6x^2y) + (5xy - 15xy^2)$ leads to the factorization

$$2x^2(1 - 3y) + 5xy(1 - 3y) = (2x^2 + 5xy)(1 - 3y)$$

$$= x(2x + 5y)(1 - 3y).$$

In problems where you have to hit upon the exact grouping from a lot of possibilities, only practice, and a lot of it, can give you the experience to make your choices wisely and swiftly.

## Exercise set 2.7

Express each of the following as the product of a first-degree polynomial and a quadratic.

$(\circ)$　1　$x^3 + 1$　　　　　　　　　　　2　$w^3 + 8$

　　　　3　$8t^3 - 27$　　　　　　　　　　4　$125y^3 - 64$

　　　　5　$27a^3 - b^3$　　　　　　　　　6　$64c^3 + 27d^3$

　　　　7　$(x - 1)^3 + 1$　　　　　　　　8　$(x + 1)^3 - 1$

　　　　9　$(x - 1)^3 - 8$　　　　　　　10　$(y - 2)^3 + 8$

$(\circ)$　11　$(x - y)^3 - 8y^3$　　　　　　12　$(y - 2z)^3 + 8z^3$

　　　13　$(s + 2t)^3 + 64(s - t)^3$　　　14　$(u - 2v)^3 - 8(u + v)^3$

Factor each of the following expressions.

$(\circ)$　15　$x^3 + 2x^2 + x$　　　　　　　16　$y^3 + 2y^2 - 3y$

　　　17　$x^9 - 8$　　　　　　　　　　18　$t^{12} + 1$

　　　19　$8x^6 - 216(y + 3)^3$　　　　　20　$343(w - 1)^3 + w^{12}$

Express each of the following as a polynomial in x.

$(\circ)$　21　$(x - 1)^3 - (x + 1)^3$

　　　22　$(x + 2)^3 - (x + 1)^3$

　　　23　$(x + a)^3 - (x - a)^3$　　　$(a \in R)$

　　　24　$(ax + b)^3 + (ax - b)^3$　　　$(a, b \in R)$

Grouping is required to factor the remaining problems.

$(\circ)$　25　$x^3 + 5x^2 - 3x - 15$　　　　26　$y^3 + 4y^2 + 8y + 32$

　　　27　$t^6 - 2t^4 - 8t^2 + 16$　　　　28　$z^6 - 4z^5 + 2z - 8$

　　　29　$a^2 - bc + ac - ab$　　　　　30　$a^2b - 2c^2 - 2abc + ac$

　　　31　$sx - 3xy + x^2 - 3sy$　　　　32　$s^3 - 2t^2 + 2ts^2 - st$

　　　33　$3x^2y^2 - 4xy + 12y^3 - x^3$

　　　34　$8x^2y^2 - 15xy - 20y^3 + 6x^3$

The method of grouping sometimes involves a more complicated factorization. For example, the expression $2x^2 + 6x + 5xz - 3z^2 - 3z$

may be grouped as $(2x^2 + 5xz - 3z^2) + (6x - 3z)$. This leads to the factorization $(x + 3z)(2x - z) + 3(2x - z) = (x + 3z + 3)(2x - z)$.

*Use this procedure on problems 35 to 40.*

(∘) **35** $x^2 - 3xy + 2y^2 + x - 2y$

**36** $x^2 - 4xy + 4y^2 + 2x - 4y$

**37** $s^2 - 9st + 14t^2 + 3st^2 - 6t^3$

**38** $a^3 + a^2 + 3a^2b + 8ab + 15b^2$

**39** $a^3 + a - 3b - 27b^3$

**40** $8c^3 + 10c + d^3 + 5d$

**(∘)**

*Sample solutions for exercise set 2.7*

**1** $x^3 + 1 = (x + 1)(x^2 - x + 1)$

**11** $(x - y)^3 - 8y^3 = (x - y)^3 - (2y)^3$

$$= [(x - y) - 2y][(x - y)^2 + 2y(x - y) + (2y)^2]$$
$$= (x - 3y)(x^2 - 2xy + y^2 + 2xy - 2y^2 + 4y^2)$$
$$= (x - 3y)(x^2 + 3y^2)$$

**15** The polynomial $x^3 + 2x^2 + x$ has the monomial factor $x$. Once this is factored out, the remaining quadratic can easily be factored. Here are the details: $x^3 + 2x^2 + x = x(x^2 + 2x + 1) = x(x + 1)^2$.

**21** The idea behind this one is to replace $x$ with $x - 1$ and $a$ with $x + 1$ in the formula $x^3 - a^3 = (x - a)(x^2 + ax + a^2)$. Upon doing this, you find that

$$(x - 1)^3 - (x + 1)^3 = [(x - 1) - (x + 1)][(x - 1)^2 + (x + 1)(x - 1) + (x + 1)^2]$$
$$= (x - 1 - x - 1)(x^2 - 2x + 1 + x^2 - 1 + x^2 + 2x + 1)$$
$$= (-2)(3x^2 + 1)$$
$$= -6x^2 - 2.$$

**25** The fact that the coefficient of $x^2$ is five times the coefficient of $x^3$ and the constant coefficient five times the coefficient of $x$ suggests the grouping $(x^3 + 5x^2) + (-3x - 15)$, which leads to the factorization $x^2(x + 5) - 3(x + 5) = (x^2 - 3)(x + 5)$.

**35** $(x^2 - 3xy + 2y^2) + (x - 2y) = (x - 2y)(x - y) + (x - 2y)(1)$
$$= (x - 2y)(x - y + 1)$$

# (3)

# *Fractions*

## (3.1) (Basic definitions)

So far, in working with polynomials we have formed sums, differences, and products of certain algebraic expressions. To perform
these operations we just pretended that every letter in sight was
really a number and then used our knowledge of the arithmetic
of the real number system. Now we will apply this same technique
in order to develop properties of quotients of algebraic expressions.

*fractions*    Such quotients are called *fractions*. Expressions such as

$$\frac{a}{b}, \frac{2x+y}{3}, \frac{6}{x}, \text{ and } \frac{x^2-6}{3x+4y}$$

are all examples of fractions. An expression like $x^2 + 3x - 2$ may
be represented as a fraction by writing it in the equivalent form
$(x^2 + 3x - 2)/1$. Of course, whenever we write down a fraction,
$A/B$, whose numerator and denominator represent algebraic expressions, we must exclude any value of a variable that would make
the denominator, $B$, equal zero, whether we specifically say this
or not. For example, when we consider the fraction $x/(x + 1)$,

we have agreed to exclude the number $-1$ as a possible replacement for $x$.

If the fraction $a/b$ equals the fraction $c/d$, then multiplication of both fractions by $bd$ will produce $ad = bc$. On the other hand, if $ad = bc$ and $b$, $d \neq 0$, then multiplication through by $1/bd$ will show that $a/b = c/d$. We conclude that

( ∘ )   F.1   $\dfrac{a}{b} = \dfrac{c}{d}$ if and only if $ad = bc$

Thus $2/3 = 6/9$ since $(2)(9) = (3)(6)$, and $4/2 = 64/32$ since $(4)(32) = (2)(64)$. The same reasoning applies to quotients of algebraic expressions: so we see that

$$\frac{x}{x+1} = \frac{x^2 - x}{x^2 - 1} \qquad (x \neq 1, -1)$$

since $x(x^2 - 1) = (x + 1)(x^2 - x)$.

This brings us to what is called the *fundamental principle of fractions*. It says that if $a$, $b$, and $c$ are real numbers and if $b$, $c \neq 0$, then

( ∘ )   F.2   $\dfrac{a}{b} = \dfrac{ac}{bc}$

This principle is simple to state and simple to prove. In fact, its proof consists of observing that $a(bc) = b(ac)$. Naturally, it also applies to quotients of algebraic expressions. Indeed, if $A$, $B$, $C$ are expressions and if all variables are restricted so that neither $B$ nor $C$ can equal zero, then the fundamental principle asserts that

( ∘ )   F.2   $\dfrac{A}{B} = \dfrac{AC}{BC}$

To illustrate this truth we observe that if $x \neq 0$ or $-1$, then

$$\frac{x+2}{x} = \frac{(x+2)(x+1)}{x(x+1)}.$$

If you wonder why this principle is important, be patient. You will find out in the next few sections.

Finally, we ask you to consider the possible signs of a fraction $A/B$. A minus sign can be placed in the numerator, in the denominator, or in front of the entire fraction. From the definition of a fraction we see that

$$-\frac{A}{B} = -\left(A \cdot \frac{1}{B}\right) = (-A) \cdot \frac{1}{B} = \frac{-A}{B}$$

and by law F.2, the fundamental principle of fractions,

$$\frac{-A}{B} = \frac{A}{-B}.$$

It follows that

$$-\frac{A}{B} = \frac{-A}{B} = \frac{A}{-B}.$$

This disposes of the placement of a single minus sign. Moreover, we can use this equation together with the fundamental principle to consider the placement of two minus signs. The situation is that

$$\frac{A}{B} = \frac{-A}{-B} = -\frac{-A}{B} = -\frac{A}{-B}.$$

As a matter of notational convenience we shall try to avoid minus signs in denominators. Thus $\dfrac{3}{-5}$ will usually be written as $-\dfrac{3}{5}$ or as $\dfrac{-3}{5}$. There will, of course, be times when this is not possible. For example, we can express no real preference between $\dfrac{a-b}{1-a}$ and $\dfrac{b-a}{a-1}$.

## Exercise set 3.1

*In problems 1 to 8, you will be given a fraction, then five lettered choices of fractions that may or may not be equivalent to the given fraction. You may assume all denominators are different from 0. Use law F.1 to determine which of the choices is equivalent to the given fraction.*

**( ○ ) 1** $\dfrac{3}{2}$    **a** $\dfrac{6}{4}$    **b** $\dfrac{-3}{-2}$    **c** $\dfrac{15}{-10}$    **d** $\dfrac{20}{30}$    **e** $\dfrac{-12}{-9}$

**2** $\dfrac{3}{4}$    **a** $\dfrac{3}{7}$    **b** $\dfrac{-3}{4}$    **c** $\dfrac{15}{20}$    **d** $\dfrac{16}{24}$    **e** $\dfrac{27}{35}$

**3** $\dfrac{12}{20}$    **a** $\dfrac{3}{7}$    **b** $\dfrac{13}{21}$    **c** $-\dfrac{12}{20}$    **d** $\dfrac{3x}{5x}$    **e** $\dfrac{9}{16}$

**4** $\dfrac{21}{36}$    **a** $\dfrac{-7}{12}$    **b** $\dfrac{14}{24}$    **c** $\dfrac{41}{72}$    **d** $\dfrac{3}{5}$    **e** $\dfrac{-7y}{-12y}$

**( ○ ) 5** $\dfrac{x+1}{x^2+1}$    **a** $\dfrac{x+2}{x^2+2}$    **b** $\dfrac{x^2+x}{x^3+1}$    **c** $\dfrac{x^2+x}{x^3+x}$

**d** $\dfrac{x^3+x^2-x-1}{x^4-1}$    **e** $\dfrac{x^2-1}{x^3-x^2+x-1}$

**6** $\dfrac{x}{x+2}$    **a** $\dfrac{x+1}{x+3}$    **b** $\dfrac{xy}{xy+2}$    **c** $\dfrac{x^2}{x^2+2x}$

**d** $\dfrac{x^2-2x}{x^2-4}$    **e** $\dfrac{x^2+x}{x^2+3x+2}$

**7** $\dfrac{y-5}{y+2}$    **a** $\dfrac{2y-10}{2y+4}$    **b** $\dfrac{5-y}{-2-y}$    **c** $\dfrac{y+5}{y-2}$

**d** $\dfrac{y^2-5}{y^2+2}$    **e** $\dfrac{y-5x}{y+2x}$

**8** $\dfrac{z+3}{z-4}$    **a** $\dfrac{z^3+3z}{z^3-4z}$    **b** $\dfrac{1+3z}{1-4z}$    **c** $\dfrac{z^2+4z+3}{z^2-3z-4}$

**d** $\dfrac{z^2+5z+6}{z^2-6z+8}$    **e** $\dfrac{yz+3y}{-4y+yz}$

In problems 9 to 16, fill in each blank with whatever it takes to make the given fractions equivalent. For example, the blank in $\dfrac{1-x}{x-y}$

$= \dfrac{\quad}{y-x}$ would be filled in by $x-1$. In other words, you are to make use of the fact that $\dfrac{a}{-b} = \dfrac{-a}{b} = -\dfrac{a}{b}$.

**( ○ ) 9** $\dfrac{4}{8} = \dfrac{\quad}{-8}$          **10** $\dfrac{36}{10} = \dfrac{\quad}{-10}$

**11** $\dfrac{a}{a-b} = \dfrac{}{b-a}$

**12** $\dfrac{-a}{a-b-c} = \dfrac{}{b+c-a}$

**13** $-\dfrac{x-1}{x^2+x-1} = \dfrac{}{1-x-x^2}$

**14** $-\dfrac{xy-z^2}{6xy-4yz-1} = \dfrac{}{1+4yz-6xy}$

**15** $\dfrac{a-1}{a+1} = \dfrac{1-a}{}$

**16** $\dfrac{ab-b^2}{-a+b} = \dfrac{b(b-a)}{}$

*In the remaining problems, specify any real values of the variables that must be excluded in order to make the assertion of the problem true.*

$( \circ )$ **17** $\dfrac{x^2}{x} = \dfrac{x}{1}$

**18** $\dfrac{x^2}{2x} = \dfrac{2x}{4}$

**19** $\dfrac{1}{x+1} = \dfrac{x-1}{x^2-1}$

**20** $\dfrac{y}{2y} = \dfrac{1}{2}$

$( \circ )$ **21** $\dfrac{1}{x^2+1} = \dfrac{y}{x^2 y + y}$

**22** $\dfrac{1}{w^2+2} = \dfrac{w^2}{w^4+2w^2}$

**23** $\dfrac{x-2}{x+1} = \dfrac{(x-2)(x-3)}{(x+1)(x-3)}$

**24** $\dfrac{x^2-4}{3x+5} = \dfrac{(x^2-4)(x+y)}{(3x+5)(x+y)}$

**25** $\dfrac{6}{x-1} = \dfrac{6(x+3y)(x-z)}{(x-1)(x+3y)(x-z)}$

**26** $\dfrac{x-y}{x+y} = \dfrac{(x-y)^2(x+y)}{(x-y)(x+y)^2}$

**27** $\dfrac{x-yz}{x-yz} = \dfrac{(x-yz)^2}{(x-yz)^2}$

**28** $\dfrac{a}{b} = \dfrac{a(a-c)}{b(a-c)}$

$( \circ )$

*Sample solutions for exercise set 3.1*

**1** Choices **(a)** and **(b)** are the only ones equivalent to the given fractions. This is seen as follows:

**a** $\dfrac{3}{2} = \dfrac{6}{4}$ since $(3)(4) = (2)(6) = 12$

**b** $\dfrac{3}{2} = \dfrac{-3}{-2}$ since $(3)(-2) = (2)(-3) = -6$

**c** $\dfrac{3}{2} \neq \dfrac{15}{-10}$ since $(3)(-10) = -30$ and $(2)(15) = +30$

**d** $\dfrac{3}{2} \neq \dfrac{20}{30}$ since $(3)(30) = 90$ and $(2)(20) = 40$

**e** $\dfrac{3}{2} \neq \dfrac{-12}{-9}$ since $(3)(-9) = -27$ and $(2)(-12) = -24$

5   Choices **(c)**, **(d)**, and **(e)** are all equivalent to the given fraction. Here are **(b)** and **(e)** worked out.

**b** $\dfrac{x+1}{x^2+1} \neq \dfrac{x^2+x}{x^3+1}$   since   $(x+1)(x^3+1) = x^4 + x^3 + x + 1$   whereas

$(x^2+1)(x^2+x) = x^4 + x^3 + x^2 + x$

**e** $\dfrac{x+1}{x^2+1} = \dfrac{x^2-1}{x^3-x^2+x-1}$ since $(x^2+1)(x^2-1) = x^4 - 1$ and

$(x+1)(x^3-x^2+x-1) = (x^4 - x^3 + x^2 - x) + (x^3 - x^2 + x - 1)$

$$= x^4 - 1$$

9   $\dfrac{4}{8} = \dfrac{-4}{-8}$

11   $\dfrac{a}{a-b} = \dfrac{-a}{b-a}$

17   We must have $x \neq 0$ since the denominator cannot be 0.

21   Since $x^2 + 1$ is never 0, the only restriction is that $y \neq 0$.

# (3.2) (Least common multiples)

Recall that an integer $a$ is a *factor* of the integer $b$, or that $a$ *divides* $b$, or that $b$ is a *multiple* of $a$, when $b = ac$ for some integer $c$. Certain integers $n$ can be factored in the form $n = ab$, with $a \neq \pm 1$ and $b \neq \pm 1$ ($\pm 1$ means $+1$ or $-1$). The result of such factoring is called a *proper factorization* of $n$. Thus $6 = 2 \cdot 3$ is a proper factorization of 6, whereas $6 = (-6)(-1)$ is not. Of course there

*proper factorization*

are integers $n > 1$ that cannot be properly factored. Such integers are called *primes* (see Section 1.5); in a very real sense they form the building blocks for the set of integers. It is a fact that every integer $n > 1$ can be factored into a product of powers of primes. The process is illustrated in the following examples:

$$24 = 2 \cdot 12 = 2 \cdot 2 \cdot 6 = 2 \cdot 2 \cdot 2 \cdot 3 = 2^3 \cdot 3,$$

$$120 = 2 \cdot 60 = 2 \cdot 2 \cdot 30 = 2 \cdot 2 \cdot 2 \cdot 15 = 2 \cdot 2 \cdot 2 \cdot 3 \cdot 5 = 2^3 \cdot 3 \cdot 5,$$

$$2{,}025 = 3 \cdot 675 = 3 \cdot 3 \cdot 225 = 3 \cdot 3 \cdot 3 \cdot 75 = 3 \cdot 3 \cdot 3 \cdot 3 \cdot 25 = 3^4 \cdot 5^2.$$

In general, we shall say that a nonzero integer $n$ has been *completely factored* in case $n = \pm 1$ or $n$ has been written as plus or minus a product of powers of primes. We shall not speak of completely factoring any other kind of real number.

*common multiple*

*least common multiple*

*LCM*

Any integer $c$ that is a multiple of both integers $a$ and $b$ is called a *common multiple* of $a$ and $b$. The smallest positive integer that is a common multiple of $a$ and $b$ is called their *least common multiple* (abbreviated LCM). How do you find the LCM of $a$ and $b$? The process is easy: Write $a$ and $b$ in completely factored form, take each prime to the highest power it appears, then multiply these numbers together. The resulting product is the desired LCM. To find the LCM of 24 and 36, for example, we write $24 = 2^3 \cdot 3$ and $36 = 2^2 \cdot 3^2$; so LCM $(24, 36) = 2^3 \cdot 3^2 = 72$.

Naturally, the process can be extended to find LCMs of three or more integers. Suppose we must establish the LCM of $-26$, 20, and 39. Here we have $26 = 2 \cdot 13$, $20 = 2^2 \cdot 5$ and $39 = 3 \cdot 13$; so LCM $(-26, 20, 39) = 2^2 \cdot 3 \cdot 5 \cdot 13 = 780$.

Let us now consider least common multiples of polynomials with integer coefficients. If $P$ and $Q$ are polynomials of this type, we say that $P$ is a *factor* of $Q$, or that $P$ *divides* $Q$, or that $Q$ is a *multiple* of $P$ if $PR = Q$ for some polynomial $R$ with integer coefficients. We say that $P = AB$ is a *proper factorization* of the polynomial $P$ in case $A$ and $B$ are polynomials with integer coefficients other than $\pm 1$. A polynomial $P$ is said to be *prime* if $P \neq \pm 1$ and $P$ has no proper factorization. A *complete factorization* of a polynomial with integer coefficients is what you might expect—it expresses the polynomial as $\pm 1$ or as a product of powers of primes. The process of finding an LCM for a set of polynomials is much the same as it is for integers: write each polynomial in completely factored form, then multiply the highest powers of each prime to appear in any of these factorizations. The product of the resulting polynomials is the desired LCM.

*complete factorization*

To illustrate, let us find an LCM for $x^2 - 4$ and $(x^2 + x - 6)(x - 2)^2$. Factoring, we have

$$x^2 - 4 = (x - 2)(x + 2),$$

$$(x^2 + x - 6)(x - 2)^2 = (x + 3)(x - 2)(x - 2)^2$$

$$= (x + 3)(x - 2)^3.$$

It follows that an LCM of the original polynomials is $(x - 2)^3(x + 2)(x + 3)$.

As a second example, let $P = (6x - 12y)(x^2 + y^2)^2$ and $Q = (4x^2 - 16y^2)(x^2 + y^2)$. Writing

$$P = 2 \cdot 3 \cdot (x - 2y)(x^2 + y^2)^2,$$

$$Q = 2 \cdot 2 \cdot (x^2 - 4y^2)(x^2 + y^2)$$

$$= 2^2(x - 2y)(x + 2y)(x^2 + y^2),$$

we see that their LCM is $2^2 \cdot 3(x - 2y)(x + 2y)(x^2 + y^2)^2$. Notice that in both cases the LCM was left in factored form. It will often be convenient to do just that.

## Exercise set 3.2

Express each of the following in completely factored form.

( ◦ )   **1**   8          **2**   12          **3**   15

     **4**   25          **5**   28          **6**   36

( ◦ )   **7**   35          **8**   99          **9**   72

( ◦ )   **10**   75        **11**   $2x - 4$        **12**   $6y + 9$

     **13**   $2x^2 + 2$        **14**   $2x^2 - 4x + 2$        **15**   $42x^2 - 84x$

( ◦ )   **16**   $42x^4 - 84x^2$        **17**   $(a^2 - 4b^2)(a + 2b)$

     **18**   $(a^2 - b^2)(12a + 12b)$

In each problem, find the LCM of the indicated numbers.

( ◦ )   **19**   4; 6        **20**   5; 9        **21**   $-60$; 18

     **22**   $-64$; 20        **23**   18; 12; $-10$        **24**   27; $-63$; $-15$

( ◦ )   **25**   40,817; 3,179        **26**   1,573; 6,591

     **27**   1,400; 286; 10,780     **28**   9,163; 2,093; 99,127

In problems 29 to 36, find an LCM for the indicated polynomials. You may leave your answer in factored form.

(○)

$(\circ)$   **29**   $6x^2y;\ 8xy^3$      **30**   $-4ab^3;\ 6a^2b^4$

**31**   $51x^2y^4z;\ 34xy^3z^3;\ 17x^5$      **32**   $63s^2t^3;\ 9st^4;\ 14s^3$

$(\circ)$   **33**   $(a-b)^2;\ a^2-b^2$      **34**   $(x-2y)^2;\ x^2-4y^2$

**35**   $a^2+3ab+2b^2;\ a^3+3a^2b+3ab^2+b^3$

**36**   $a^2-5a+6;\ a^2-4a+4$

$(\circ)$

*Sample solutions*
*for exercise set 3.2*

**1**   $8 = 2\cdot4 = 2\cdot2\cdot2 = 2^3$

**9**   $72 = 2\cdot36 = 2\cdot2\cdot18 = 2\cdot2\cdot2\cdot9 = 2^3\cdot3^2$

**11**   $2x-4 = (2)(x-2)$

**17**   $(a^2-4b^2)(a+2b) = (a-2b)(a+2b)(a+2b)$
$$= (a-2b)(a+2b)^2$$

**19**   $4 = 2^2$ and $6 = 2\cdot3$ so LCM $(4, 6) = 2^2\cdot3 = 12$

**25**   We find that $40{,}817 = 7^4\cdot17$ and $3{,}179 = 11\cdot17^2$. Their LCM is, therefore, $7^4\cdot11\cdot17^2 = 7{,}632{,}779$.

**29**   The LCM of 6 and 8 is 24 since $6 = 2\cdot3$ and $8 = 2^3$. It follows that the LCM of $6x^2y$ and $8xy^3$ is $24x^2y^3$.

**33**   The polynomial $(a-b)^2$ is already in factored form, and $a^2-b^2 = (a+b)(a-b)$. The answer is $(a+b)(a-b)^2$.

## (3.3) (Using the fundamental principle)

If $a$ and $b$ are integers, the fraction $a/b$ is said to be in *lowest terms* if $a$ and $b$ have no common prime factors. The process of replacing a fraction of this kind with an equivalent fraction in lowest terms is called *reducing* the fraction to lowest terms. It is accomplished by completely factoring the numerator and denominator and then using the fundamental principle of fractions (Section 3.1) to remove any common prime factors. Thus

$$\frac{20}{45} = \frac{2^2\cdot5}{3^2\cdot5} = \frac{2^2}{3^2} = \frac{4}{9}$$

whereas

$$\frac{24}{90} = \frac{2^3\cdot3}{2\cdot3^2\cdot5} = \frac{2^2\cdot2\cdot3}{3\cdot5\cdot2\cdot3} = \frac{2^2}{3\cdot5} = \frac{4}{15}.$$

Similarly, for $A$ and $B$, polynomials with integer coefficients, we say that $A/B$ is in *lowest terms* if $A$ and $B$ have no common prime polynomial factors with integer coefficients. The process of reducing such a fraction to lowest terms is carried out much as it is for integers: completely factor both numerator and denominator, then use the fundamental principle of fractions to remove any common prime factors. Thus

$$\frac{x^2}{2x} = \frac{x}{2} \qquad (x \neq 0).$$

It is essential that you include the fact that $x \neq 0$ in the foregoing equivalence. If $x = 0$, $x^2/2x$ is not even defined; so it cannot be equivalent to $x/2$. Similarly,

$$\frac{x - 2}{2 - x} = \frac{x - 2}{(-1)(x - 2)} = -1 \qquad (x \neq 2).$$

Further examples are provided by:

$$\frac{x^2 - 3x + 2}{x^2 - 2x + 1} = \frac{(x - 2)(x - 1)}{(x - 1)(x - 1)} = \frac{x - 2}{x - 1} \qquad (x \neq 1);$$

$$\frac{y(x - y) + 2(x - y)}{x - y} = \frac{(y + 2)(x - y)}{x - y} = y + 2 \qquad (x \neq y);$$

$$\frac{6x^3 - 18x^2 + 12x}{24x} = \frac{(6x)(x^2 - 3x + 2)}{(6x)(4)} = \frac{x^2 - 3x + 2}{4} \qquad (x \neq 0).$$

Notice how it was not really necessary to factor the numerator completely in that last example; we factored just far enough so that the common factors could be removed.

A word of caution: You might be tempted to write $(x^2 - 4)/x = x - 4$ $(x \neq 0)$, but to see that this statement is incorrect you need only replace $x$ with 2 and notice that $(2^2 - 4)/2 \neq 2 - 4$. Actually, the fraction $(x^2 - 4)/x$ is already in lowest terms, which is shown by the factorization

$$\frac{x^2 - 4}{x} = \frac{(x + 2)(x - 2)}{x}.$$

There will be occasions when you will want to replace two or more fractions with equivalent fractions that have the same denominator. If the given fractions are $a/b$ and $c/d$, then $bd$ will certainly work as a common denominator. Thus we have

$$\frac{a}{b} = \frac{ad}{bd} \quad \text{and} \quad \frac{c}{d} = \frac{bc}{bd}.$$

If we can find a common multiple of $b$ and $d$ that is smaller in some sense than $bd$, there will usually be fewer computations involved. In particular, if $a$, $b$, $c$, and $d$ are integers or polynomials with integral coefficients, we may use the LCM of $b$ and $d$ as the common denominator. This LCM is called the *least common denominator* (abbreviated LCD) of $a/b$ and $c/d$. Of course, the same sort of thing applies to three or more fractions.

*least common denominator*

Let's illustrate this process with some examples. The LCD of $3/8$ and $5/6$ is 24. Noting that $24/8 = 3$ and $24/6 = 4$, we have for equivalent fractions

$$\frac{3}{8} = \frac{3 \cdot 3}{8 \cdot 3} = \frac{9}{24} \quad \text{and} \quad \frac{5}{6} = \frac{5 \cdot 4}{6 \cdot 4} = \frac{20}{24}.$$

As a second example, consider

$$\frac{5x^2 y}{12(x + 2y)^2} \quad \text{and} \quad \frac{7xy}{8x^2 + 16xy}.$$

In factored form, the denominators are $2^2 \cdot 3 \cdot (x + 2y)^2$ and $2^3 \cdot x \cdot (x + 2y)$; so the LCD is $24x(x + 2y)^2$. The LCD should be left in this factored form, which is usually easier to work with. Since $[12(x + 2y)^2](2x) = (8x^2 + 16xy)[3(x + 2y)] = 24x(x + 2y)^2$, for $x \neq 0$ or $-2y$ we have

$$\frac{5x^2 y}{12(x + 2y)^2} = \frac{(5x^2 y)(2x)}{12(x + 2y)^2 (2x)} = \frac{10x^3 y}{24x(x + 2y)^2}$$

and

$$\frac{7xy}{8x^2 + 16xy} = \frac{(7xy)[3(x + 2y)]}{(8x^2 + 16xy)[3(x + 2y)]} = \frac{21xy(x + 2y)}{24x(x + 2y)^2}.$$

Thus we have replaced the given fractions with equivalent fractions that have the same denominator. Of course, we could have used the product of the denominators as a common denominator but to do that would have led to more complicated computations.

## Exercise set 3.3

Reduce the following to lowest terms.

( ∘ ) 1   $\dfrac{4}{18}$      2   $\dfrac{5}{75}$      3   $\dfrac{18}{27}$

( ∘ ) 4   $\dfrac{27}{108}$      5   $\dfrac{27}{-60}$      6   $\dfrac{-18}{52}$

7   $\dfrac{40}{32}$      8   $\dfrac{-60}{44}$      9   $\dfrac{48}{32}$

( ∘ ) 10   $\dfrac{40}{21}$      11   $\dfrac{6x^2\,y}{8x^3\,y^4}$      12   $\dfrac{24\,w^2\,xz}{8wz}$

13   $\dfrac{36x^5\,y^3}{24x^2\,y^2}$      14   $\dfrac{70a^4\,b^2\,c}{42ab^2}$

( ∘ ) 15   $\dfrac{4x^2\,y}{8xy^2 - 10xy}$      16   $\dfrac{16xy^4}{12x^2\,y - 8xy^3}$

17   $\dfrac{18s^2\,tu^3}{9s^2\,u^2 + 12su^3 - 18su}$      18   $\dfrac{4ab^2\,c^3}{2abc^3 - 8ab^2\,c + 10abc^2}$

19   $\dfrac{6x^3\,y^2 - 8xy^2}{4x^5\,y + 12x^4\,y^2}$      20   $\dfrac{28w^2\,z^2 - 10wz^3}{28wz^3 + 10w^2\,z^2}$

( ∘ ) 21   $\dfrac{x^2 - 4}{2 - x}$      22   $\dfrac{2x + 4y}{x^2 + 5xy + 6y^2}$

23   $\dfrac{(x - 2y)^2 - 9z^2}{x - 2y + 3z}$      24   $\dfrac{3x - 6z}{x^3 - 8z^3}$

25   $\dfrac{x^2 - 4x + 4}{x^2 + x - 6}$      26   $\dfrac{x^2 - 9x + 14}{x^3 - 7x^2}$

( ∘ ) 27   $\dfrac{a^2 + 7ab + 10b^2}{a^2 + 5ab + 6b^2}$      28   $\dfrac{g^4 + 6g^2\,h - 16h^2}{g^4 - 4h^2}$

29   $\dfrac{2x^2 - 11xy + 12y^2}{10x^2 - 13xy - 3y^2}$      30   $\dfrac{4t^2 + 4t - 15}{-12t^2 + 36t - 27}$

31   $\dfrac{x - 2y}{x^2 - 2xy - 2y + x}$      32   $\dfrac{x^2 + 3}{x^5 + 3x^3 - 2x^2 - 6}$

In problems 33 to 40, replace (?) with expressions that will make the fractions equivalent.

**( ∘ )** **33** $\dfrac{4}{6} = \dfrac{(?)}{30}$    **34** $\dfrac{3}{7} = \dfrac{(?)}{42}$

**35** $\dfrac{10}{15} = \dfrac{(?)}{9}$    **36** $\dfrac{45}{36} = \dfrac{(?)}{32}$

**( ∘ )** **37** $\dfrac{6x}{5y^2} = \dfrac{(?)}{20x^2 y^3}$    **38** $\dfrac{5t}{4u} = \dfrac{(?)}{8tu^2}$

**39** $\dfrac{x+y}{x-y} = \dfrac{(?)}{x^2 - y^2}$    **40** $\dfrac{x}{x+2y} = \dfrac{(?)}{x^2 + 3xy + 2y^2}$

In each of the remaining problems, find the LCD of the given fractions, then replace each fraction with an equivalent fraction that has the LCD as its denominator.

**( ∘ )** **41** $\dfrac{2}{3} ; \dfrac{1}{2}$    **42** $\dfrac{4}{9} ; \dfrac{5}{12}$    **43** $\dfrac{3}{14} ; \dfrac{2}{21}$

**44** $\dfrac{5}{8} ; \dfrac{1}{6}$    **45** $\dfrac{5}{2x} ; \dfrac{4}{3x^2 y}$    **46** $\dfrac{x+2y}{4x} ; \dfrac{x}{6x^2 y}$

**47** $\dfrac{1}{x+y} ; \dfrac{1}{x-y}$    **48** $\dfrac{1}{x+3} ; \dfrac{x}{x^2 - 9}$

**49** $\dfrac{1}{x+y} ; \dfrac{1}{x^2 - xy + y^2} ; \dfrac{x^2 + y^2}{x^3 + y^3}$

**50** $\dfrac{1}{x+1} ; \dfrac{1}{y+2} ; \dfrac{x+y}{xy + y + 2x + 2}$

**( ∘ )** **51** $\dfrac{6}{x^2 + x} ; \dfrac{2x}{x^2 + 6x + 5}$

**52** $\dfrac{5}{w+3} ; \dfrac{w+2}{w^3 + 2w^2 - 3w}$

**53** $\dfrac{6}{5(x-1)(x-2)^2} ; \dfrac{x}{3(x-1)^2 (x-2)}$

**54** $\dfrac{1}{4x(x+1)(x+2)^3} ; \dfrac{-1}{3(x+1)^3}$

**55** $\dfrac{x+y}{(x^2 - xy)(x-y)^2} ; \dfrac{1}{(x^2 - y^2)^2}$

**56** $\dfrac{1}{s^4 - 1} ; \dfrac{s^2}{(s-1)^3 (s+1)}$

73 ( ∘ ) Using the fundamental principle

1   $\dfrac{4}{18} = \dfrac{2^2}{2\cdot 3^2} = \dfrac{2\cdot 2}{3^2\cdot 2} = \dfrac{2}{3^2} = \dfrac{2}{9}$

5   $\dfrac{27}{-60} = \dfrac{3^3}{-2^2\cdot 3\cdot 5} = -\dfrac{3^2\cdot 3}{2^2\cdot 5\cdot 3} = -\dfrac{3^2}{2^2\cdot 5} = -\dfrac{9}{20}$

11   $\dfrac{6x^2 y}{8x^3 y^4} = \dfrac{3\cdot 2x^2 y}{4xy^3\cdot 2x^2 y} = \dfrac{3}{4xy^3}$     $(x, y \neq 0)$

15   $\dfrac{4x^2 y}{8xy^2 - 10xy} = \dfrac{2xy\cdot 2x}{2xy(4y - 5)} = \dfrac{2x}{4y - 5}$     $(x, y \neq 0; 4y - 5 \neq 0)$

21   $\dfrac{x^2 - 4}{2 - x} = \dfrac{(x - 2)(x + 2)}{-(x - 2)} = \dfrac{x + 2}{-1} = -x - 2$     $(x \neq 2)$

27   $\dfrac{a^2 + 7ab + 10b^2}{a^2 + 5ab + 6b^2} = \dfrac{(a + 5b)(a + 2b)}{(a + 3b)(a + 2b)} = \dfrac{a + 5b}{a + 3b}$     $(a \neq -3b, -2b)$

33   $6\cdot 5 = 30$ so $\dfrac{4}{6} = \dfrac{4\cdot 5}{6\cdot 5} = \dfrac{20}{30}$

37   $(5y^2)(4x^2 y) = 20x^2 y^3$ so $\dfrac{6x}{5y^2} = \dfrac{6x\cdot 4x^2 y}{5y^2\cdot 4x^2 y} = \dfrac{24x^3 y}{20x^2 y^3}$     $(x, y \neq 0)$

41   The LCD of $\dfrac{2}{3}$ and $\dfrac{1}{2}$ is 6; so we may write $\dfrac{2}{3} = \dfrac{4}{6}$ and $\dfrac{1}{2} = \dfrac{3}{6}$.

51   In factored form, the denominators are $x(x + 1)$ and $(x + 5)(x + 1)$. The LCD is therefore $x(x + 1)(x + 5)$. Note now that for $x \neq 0, -1$, or $-5$,

$$\dfrac{6}{x(x + 1)} = \dfrac{6(x + 5)}{x(x + 1)(x + 5)}$$

and

$$\dfrac{2x}{(x + 5)(x + 1)} = \dfrac{2x^2}{x(x + 1)(x + 5)}.$$

## (3.4) (Addition and subtraction)

Let's turn our attention to the addition of fractions. We first consider the sum of two fractions, $a/b$ and $c/b$, that have the same denominator. If real numbers are substituted for any variables, then both fractions are real numbers; so we may use the distributive law to write

$$\frac{a}{b} + \frac{c}{b} = a\left(\frac{1}{b}\right) + c\left(\frac{1}{b}\right) = (a + c)\left(\frac{1}{b}\right) = \frac{a + c}{b} \qquad (b \neq 0).$$

*To add two fractions with the same denominator, we place the sum of the numerators over the given denominator.* In other words, we *define*

( ∘ )  F.3  $\dfrac{a}{b} + \dfrac{c}{b} = \dfrac{a + c}{b} \qquad (b \neq 0)$

For example,

$$\frac{4}{3} + \frac{7}{3} = \frac{11}{3},$$

$$\frac{6}{x - 3} + \frac{2}{x - 3} = \frac{8}{x - 3} \qquad (x \neq 3),$$

$$\frac{x^2 - y}{x - y} + \frac{y - y^2}{x - y} = \frac{x^2 - y^2}{x - y} = \frac{(x + y)(x - y)}{x - y} = x + y \qquad (x \neq y).$$

In case the fractions $a/b$ and $c/d$ do not have the same denominator, we simply use the methods of Section 3.3 to replace them with equivalent fractions that *do* have the same denominator, then proceed as above. If we decide to use $bd$ as a common denominator, we have $\dfrac{a}{b} = \dfrac{ad}{bd}$ and $\dfrac{c}{d} = \dfrac{bc}{bd}$; so

$$\frac{a}{b} + \frac{c}{d} = \frac{ad}{bd} + \frac{cd}{bd} = \frac{ad + bc}{bd} \qquad (b, d \neq 0).$$

This leads us to define

( ∘ )  F.4  $\dfrac{a}{b} + \dfrac{c}{d} = \dfrac{ad + bc}{bd} \qquad (b, d \neq 0)$

When adding fractions whose denominators are integers or polynomials with integer coefficients, it will generally profit you to use the LCD of the fractions as your common denominator. It is time to look at some examples:

$$\frac{5}{6} + \frac{3}{8} = \frac{20}{24} + \frac{9}{24} = \frac{29}{24}$$

and

$$\frac{3}{x} + \frac{2}{x+1} = \frac{3(x+1)}{x(x+1)} + \frac{2x}{x(x+1)} = \frac{3x+3+2x}{x(x+1)}$$

$$= \frac{5x+3}{x(x+1)} \qquad (x \neq 0, -1).$$

Now consider this one:

$$\frac{1}{t-t^2} + \frac{t+1}{(t-1)^2} + \frac{(-1)}{t^3 - 2t^2 + t}.$$

The LCD of these fractions is easily seen to be $t(t-1)^2 = t^3 - 2t^2 + t$; so we have

$$\frac{1-t}{t(t-1)^2} + \frac{t(t+1)}{t(t-1)^2} + \frac{(-1)}{t(t-1)^2} = \frac{1-t+t^2+t-1}{t(t-1)^2}$$

$$= \frac{t^2}{t(t-1)^2} = \frac{t}{(t-1)^2} \qquad (t \neq 0, 1).$$

In summary, here is the procedure for finding the sum of two or more fractions:

**Step 1:** Find a common denominator for the fractions. Any common denominator will work, but it is usually best to find their LCD (where this is defined).

**Step 2:** Replace each fraction with an equivalent fraction over the common denominator.

**Step 3:** Add the numerators of the fractions obtained in Step 2.

**Step 4:** Place the sum obtained in Step 3 over the common denominator.

**Step 5:** If necessary, reduce to lowest terms.

Subtraction is handled by defining

$\dfrac{a}{b} - \dfrac{c}{d} = \dfrac{a}{b} + \dfrac{-c}{d}$

Thus

$$\frac{a}{b} - \frac{c}{d} = \frac{ad - bc}{bd}.$$

For example,

$$\frac{3}{x - 4} - \frac{2x + 7}{x^2 - 3x - 4} = \frac{3(x + 1)}{(x - 4)(x + 1)} - \frac{2x + 7}{(x - 4)(x + 1)}$$

$$= \frac{3x + 3 - 2x - 7}{(x - 4)(x + 1)}$$

$$= \frac{x - 4}{(x - 4)(x + 2)}$$

$$= \frac{1}{x + 2} \qquad (x \neq 4, -2).$$

Suppose you want to find $\dfrac{1}{x - 2} - \dfrac{2x}{4 - x^2}$. You may use $x^2 - 4$ as a common denominator, but an extra difficulty arises because the denominator of the second fraction is $4 - x^2$. Using the fact that $\dfrac{-2x}{4 - x^2} = \dfrac{+2x}{x^2 - 4}$, however, you may proceed as follows:

$$\frac{1}{x - 2} - \frac{2x}{4 - x^2} = \frac{x + 2}{x^2 - 4} - \frac{2x}{4 - x^2} = \frac{x + 2}{x^2 - 4} + \frac{-2x}{4 - x^2}$$

$$= \frac{x + 2}{x^2 - 4} + \frac{2x}{x^2 - 4} = \frac{3x + 2}{x^2 - 4} \qquad (x \neq 2, -2).$$

## Exercise set 3.4

*Perform the indicated operations, and express your answer as a fraction in lowest terms. Be sure to indicate any restrictions on the variables.*

$(\circ)$ **1** $\dfrac{2}{7} + \dfrac{3}{14}$  **2** $\dfrac{3}{5} + \dfrac{5}{3}$  **3** $\dfrac{5}{8} - \dfrac{7}{12}$

**4** $\dfrac{3}{2} - \dfrac{7}{6}$

**5** $\dfrac{2}{x} + \dfrac{3}{x}$

**6** $\dfrac{1}{x+1} + \dfrac{x}{x+1}$

(∘) **7** $\dfrac{3}{2x} - \dfrac{5}{x}$

**8** $\dfrac{8}{3t} - \dfrac{5}{2t}$

**9** $\dfrac{2x}{x+y} - \dfrac{3}{x+y}$

**10** $\dfrac{2t}{t-x} + \dfrac{x}{t-x}$

**11** $\dfrac{8}{ax} - \dfrac{a}{x}$

**12** $\dfrac{3}{bx} - \dfrac{2}{x}$

(∘) **13** $\dfrac{2}{y+x} + \dfrac{1}{y-x}$

**14** $\dfrac{a}{x} + \dfrac{a^2}{3}$

**15** $\dfrac{2}{x-1} + \dfrac{1}{x+1}$

**16** $\dfrac{1}{a} - \dfrac{2a}{a+1}$

**17** $\dfrac{5}{b} + \dfrac{4}{b^2+b}$

**18** $1 + \dfrac{3}{x^2+1}$

**19** $\dfrac{1}{x-2} - \dfrac{x}{x-1}$

**20** $\dfrac{3x-1}{x+2} - \dfrac{x+2}{3x-1}$

*The remaining problems involve LCDs that are a little harder to obtain.*

(∘) **21** $\dfrac{3}{x^2-1} + \dfrac{2}{x^2-3x+2}$

**22** $\dfrac{1}{x^2-4x+4} - \dfrac{1}{x^2-x-2}$

**23** $\dfrac{a^2}{a^2-4b^2} - \dfrac{b}{a+2b}$

**24** $\dfrac{2}{a^2+5ab+6b^2} + \dfrac{3}{a^2+6ab+9b^2}$

**25** $\dfrac{1}{a-1} + \dfrac{1}{a^3-1}$

**26** $\dfrac{1}{x+2} + \dfrac{1}{x^3+8}$

(∘) **27** $\dfrac{a}{a^2-2ab+b^2} + \dfrac{b}{a^2-4ab+3b^2} + \dfrac{2b^2}{(a-b)^2(a-3b)}$

**28** $\dfrac{1}{x^2-x} + \dfrac{2x}{x^2-3x+2} + \dfrac{1}{x^3-3x^2+2x}$

**29** $\dfrac{s}{s^2+4s+3} + \dfrac{t}{4s^2+11s-3}$

**30** $\dfrac{h+k}{h^2-2hk+k^2} + \dfrac{k}{h^2-hk} - \dfrac{2hk}{h^3-2h^2k+hk^2}$

(∘)

*Sample solutions for exercise set 3.4*

**1** $\dfrac{2}{7} + \dfrac{3}{14} = \dfrac{4}{14} + \dfrac{3}{14} = \dfrac{7}{14} = \dfrac{1}{2}$

**7** $\dfrac{3}{2x} - \dfrac{5}{x} = \dfrac{3}{2x} - \dfrac{10}{2x} = \dfrac{3-10}{2x} = \dfrac{-7}{2x} \qquad (x \neq 0)$

**15** The LCD of the two fractions is $(x-1)(x+1)$; so

$$\frac{2}{x-1} + \frac{1}{x+1} = \frac{2(x+1)}{(x-1)(x+1)} + \frac{x-1}{(x-1)(x+1)}$$

$$= \frac{2x+2+x-1}{(x-1)(x+1)}$$

$$= \frac{3x+1}{(x-1)(x+1)} \qquad (x \neq -1, 1).$$

**21** When we factor the two denominators, we find that $x^2 - 1 = (x+1)(x-1)$ and $x^2 - 3x + 2 = (x-2)(x-1)$; so the LCD is $(x-1)(x+1)(x-2)$. We now have

$$\frac{3}{x^2-1} + \frac{2}{x^2-3x+2} = \frac{3(x-2)}{(x-1)(x+1)(x-2)} + \frac{2(x+1)}{(x-1)(x+1)(x-2)}$$

$$= \frac{3x-6+2x+2}{(x-1)(x+1)(x-2)}$$

$$= \frac{5x-4}{(x-1)(x+1)(x-2)} \qquad (x \neq 1, 2, -1).$$

**27** We begin by finding the LCD of the three fractions. Noting that $a^2 - 2ab + b^2 = (a-b)^2$ and $a^2 - 4ab + 3b^2 = (a-b)(a-3b)$, we see that the LCD is $(a-b)^2(a-3b)$. We therefore may compute as follows:

$$\frac{a}{a^2-2ab+b^2} + \frac{b}{a^2-4ab+3b^2} + \frac{2b^2}{(a-b)^2(a-3b)}$$

$$= \frac{a(a-3b)}{(a-b)^2(a-3b)} + \frac{b(a-b)}{(a-b)^2(a-3b)} + \frac{2b^2}{(a-b)^2(a-3b)}$$

$$= \frac{a^2-3ab+ab-b^2+2b^2}{(a-b)^2(a-3b)} = \frac{a^2-2ab+b^2}{(a-b)^2(a-3b)}$$

$$= \frac{(a-b)^2}{(a-b)^2(a-3b)} = \frac{1}{a-3b} \qquad (a \neq b, 3b).$$

# (3.5) (Multiplication and division)

Now that we have completed our discussion of addition and subtraction of fractions, let us consider the multiplication and division operations. Let $a/b$ and $c/d$ be fractions: if real numbers are substituted for any variables, then the fractions represent quotients of real numbers. Using the associative and commutative laws of multiplication, we have $(bd)\left(\dfrac{1}{b}\cdot\dfrac{1}{d}\right)=\left(b\cdot\dfrac{1}{b}\right)\left(d\cdot\dfrac{1}{d}\right)=1$; so by the multiplicative inverse law,

(1) $\quad\dfrac{1}{b}\cdot\dfrac{1}{d}=\dfrac{1}{bd}$

We know that $\dfrac{a}{b}=a\cdot\dfrac{1}{b}$ and $\dfrac{c}{d}=c\cdot\dfrac{1}{d}$; so these same laws together with (1) will show that $\dfrac{a}{b}\cdot\dfrac{c}{d}=\left(a\cdot\dfrac{1}{b}\right)\left(c\cdot\dfrac{1}{d}\right)=(ac)$

$\left(\dfrac{1}{b}\cdot\dfrac{1}{d}\right)=(ac)\left(\dfrac{1}{bd}\right)=\dfrac{ac}{bd}$. This leads us to *define*

---

(∘)   F.6   $\dfrac{a}{b}\cdot\dfrac{c}{d}=\dfrac{ac}{bd}$

---

Naturally, $ac/bd$ should then be reduced to lowest terms. Thus

$$\dfrac{2}{9}\cdot\dfrac{3}{8}=\dfrac{2\cdot3}{9\cdot8}=\dfrac{2\cdot3}{2\cdot3\cdot4\cdot3}=\dfrac{1}{4\cdot3}=\dfrac{1}{12}$$

and

$$\dfrac{x^2-xy}{x+y}\cdot\dfrac{x^2-y^2}{x}=\dfrac{(x^2-xy)(x^2-y^2)}{(x+y)(x)}$$

$$=\dfrac{x(x-y)(x-y)(x+y)}{(x+y)(x)}$$

$$=(x-y)^2\qquad(x\neq0,-y).$$

Let's summarize what we have just learned. To find the product of two or more fractions, a new fraction is formed in the following way:

Step 1: Use the product of the numerators as the new numerator.

Step 2: Use the product of the denominators as the new denominator.

Step 3: Reduce to lowest terms.

Suppose now that you want to divide a fraction $a/b$ by a fraction $c/d$ $(b, c, d\neq0)$. The notation

$$\frac{\dfrac{a}{b}}{\dfrac{c}{d}}$$

is definitely cumbersome; so let us introduce you to the division symbol $\div$. The expression $a \div b$ is read "$a$ divided by $b$" and means the same thing as $a/b$. With this notation, the fraction we have just displayed takes on the more encouraging form $a/b \div c/d$.

Establishing workable notation is all well and good, but we must still actually see how to perform the division operation. Suppose all variables have been substituted for so that we are dealing with real numbers. Setting $t = \dfrac{c}{d}$ we observe that $\dfrac{1}{t} = \dfrac{d}{c}$, since $\dfrac{c}{d} \cdot \dfrac{d}{c}$ $= \dfrac{cd}{dc} = 1$, and $\dfrac{1}{t}$ is the unique number to have the property that $\dfrac{c}{d} \cdot \dfrac{1}{t} = t \cdot \dfrac{1}{t} = 1$. Hence $\dfrac{a}{b} \div \dfrac{c}{d} = \dfrac{a}{b} \div t = \dfrac{a}{b} \cdot \dfrac{1}{t} = \dfrac{a}{b} \cdot \dfrac{d}{c}$. This leads us to *define*

---

$(\circ)$    F.7   $\dfrac{a}{b} \div \dfrac{c}{d} = \dfrac{a}{b} \cdot \dfrac{d}{c} \;(b, c, d \neq 0)$

---

*Thus to divide $a/b$ by $c/d$ you invert the fraction $c/d$ and multiply.* We have

$$\frac{5}{4} \div \frac{3}{8} = \frac{5}{4} \cdot \frac{8}{3} = \frac{5 \cdot 2 \cdot 4}{3 \cdot 4} = \frac{10}{3}$$

and

$$\frac{2a^3}{5b^2} \div \frac{6a}{25b^4} = \frac{2a^3}{5b^2} \cdot \frac{25b^4}{6a} = \frac{a^2 \cdot 5b^2 \cdot 2a \cdot 5b^2}{3 \cdot 2a \cdot 5b^2}$$

$$= \frac{5a^2 b^2}{3} \qquad (a, b \neq 0).$$

# Exercise set 3.5

*Express each product as a single fraction in lowest terms.*

(∘) 1   $\dfrac{4}{12} \cdot \dfrac{8}{6}$

2   $\dfrac{10}{45} \cdot \dfrac{6}{14}$

3   $\dfrac{21}{4} \cdot \dfrac{8}{3}$

4   $\dfrac{16}{5} \cdot \dfrac{10}{2}$

5   $\dfrac{3}{4} \cdot \dfrac{8}{15} \cdot \dfrac{5}{9}$

6   $\dfrac{7}{6} \cdot \dfrac{3}{14} \cdot \dfrac{2}{21}$

(∘) 7   $\dfrac{7x}{12y} \cdot \dfrac{3y^2}{14x^3}$

8   $\dfrac{3a^2 b}{4ab} \cdot \dfrac{8ab^3}{9a^2 b^4}$

9   $\dfrac{7x^2 y^2}{2x} \cdot \dfrac{1}{21y}$

10   $\dfrac{16gh}{3} \cdot \dfrac{1}{4gh^2}$

(∘) 11   $\dfrac{4x + 16}{3x + 15} \cdot \dfrac{2x + 10}{x + 4}$

12   $\dfrac{8x - 16}{8x} \cdot \dfrac{6x^2}{x - 2}$

13   $\dfrac{3x + 9}{2x} \cdot \dfrac{8x^2}{x^2 - 9}$

14   $\dfrac{5x + 25}{3} \cdot \dfrac{x^2}{x^2 - 25}$

(∘) 15   $\dfrac{a^2 - a - 6}{a^2 - 25} \cdot \dfrac{a^2 + 4a - 5}{a^2 - 6a + 9}$

16   $\dfrac{b^2 - 9}{4b^2 - 1} \cdot \dfrac{4b^2 - 8b + 3}{4b^2 - 9b - 9}$

17   $\dfrac{8c^2 + 6cd - 9d^2}{4c^2 - 4cd + d^2} \cdot \dfrac{2c^2 + cd - d^2}{4c^2 - 9d^2}$

18   $\dfrac{6g^2 - 5gh - 4h^2}{6g^2 + 6gh} \cdot \dfrac{g^2 - gh - 2h^2}{2g^2 - 3gh - 2h^2}$

19   $\dfrac{s^2 + 2st - 3t^2}{s^2 - 2st - 15t^2} \cdot \dfrac{2s^2 - 9st - 5t^2}{s^2 - 2st + t^2} \cdot \dfrac{s^2 - st}{2s + 2}$

20   $\dfrac{x^2 - 4}{4x^2 - 9x + 2} \cdot \dfrac{16x^2 - 1}{(x + 2)^3} \cdot \dfrac{x^2 + 5x + 6}{4x^2 + 11x - 3}$

*Now look at some division problems. Again, express your answer as a single fraction in lowest terms.*

(∘) 21   $\dfrac{3}{10} \div \dfrac{3}{2}$

22   $\dfrac{7}{16} \div \dfrac{21}{8}$

**23** $\dfrac{14x^2y}{9xz} \div \dfrac{42y^4}{3x^2z}$

**24** $\dfrac{-5ab}{3c} \div \dfrac{-20b^2}{9bc}$

$(\circ)$ **25** $\dfrac{x-2y}{x+y} \div \dfrac{5x-10y}{x^2+xy}$

**26** $\dfrac{a^2-bc}{ab+bc} \div \dfrac{a^2c-bc^2}{a+c}$

**27** $\dfrac{x^2-4x}{x+3} \div \dfrac{x}{x^2-9}$

**28** $\dfrac{w^3+2w^2}{w^3+2w^2+w} \div \dfrac{w+2}{w+1}$

**29** $\dfrac{x^2-5x+6}{x^2-2x+1} \div \dfrac{x-2}{x^2-1}$

**30** $\dfrac{4z^2-4z+1}{9z^2-8z-1} \div \dfrac{2z^2-z}{z^2-2z+1}$

$(\circ)$ **31** $\dfrac{x^2-3xy-4y^2}{-x^2+2xy+3y^2} \div \dfrac{x^2-xy-2y^2}{-x^2+3xy}$

**32** $\dfrac{a^3+b^3}{a^2-b^2} \div \dfrac{a^2-ab+b^2}{a+b}$

**33** $\dfrac{xy+x-3y-3}{xy+2x-2y-4} \div \dfrac{y+1}{x-2}$

**34** $\dfrac{x^3+x^2-3x-3}{xy^2+y^2-x-1} \div \dfrac{x+1}{y+1}$

**35** $\dfrac{x^3+3x^2+3x+1}{(x+2)^5} \div \dfrac{x^2+2x+1}{(x+2)^3}$

**36** $\dfrac{(x-1)^2(x-2)(x-3)^3}{(x^2-1)(x+2)^2} \div \dfrac{(x-1)(x-2)(x-3)}{(x+1)(x+2)}$

**37** $\dfrac{x+3}{y-2} \cdot \dfrac{x^2-6x}{xy-3x} \div \dfrac{x^2+3x}{y^2-4}$

**38** $\dfrac{x^2-1}{x+2} \cdot \dfrac{x^3-x^2}{xy+2x-y-2} \div \dfrac{x+4}{y^2-4}$

$(\circ)$

*Sample solutions for exercise set 3.5*

**1** $\dfrac{4}{12} \cdot \dfrac{8}{6} = \dfrac{32}{72} = \dfrac{2^5}{2^2 \cdot 3 \cdot 2 \cdot 3} = \dfrac{2^3 \cdot 2^2}{2^3 \cdot 3^2} = \dfrac{4}{9}$. In this particular problem, it makes more sense to reduce each fraction to lowest terms first and then to multiply. The computations are $\dfrac{4}{12} \cdot \dfrac{8}{6} = \dfrac{1}{3} \cdot \dfrac{4}{3} = \dfrac{4}{9}$.

**7** $\dfrac{7x}{12y} \cdot \dfrac{3y^2}{14x^3} = \dfrac{(7x)(3y^2)}{(12y)(14x^3)} = \dfrac{7 \cdot 3 \cdot x \cdot y^2}{2^2 \cdot 3 \cdot 2 \cdot 7 \cdot x^3 \cdot y} = \dfrac{y}{8x^2}$  $(x, y \neq 0)$

**11**

$$\frac{4x + 16}{3x + 15} \cdot \frac{2x + 10}{x + 4} = \frac{4(x + 4)(2)(x + 5)}{3(x + 5)(x + 4)} = \frac{8}{3} \qquad (x \neq -5, -4)$$

**15**

$$\frac{a^2 - a - 6}{a^2 - 25} \cdot \frac{a^2 + 4a - 5}{a^2 - 6a + 9}$$

$$= \frac{(a - 3)(a + 2)(a + 5)(a - 1)}{(a - 5)(a + 5)(a - 3)(a - 3)}$$

$$= \frac{(a + 2)(a - 1)(a - 3)(a + 5)}{(a - 5)(a - 3)(a - 3)(a + 5)}$$

$$= \frac{(a + 2)(a - 1)}{(a - 5)(a - 3)} \qquad (a \neq 3, 5, -5)$$

**21**

$$\frac{3}{10} \div \frac{3}{2} = \frac{3}{10} \cdot \frac{2}{3} = \frac{1}{5}$$

**25**

$$\frac{x - 2y}{x + y} \div \frac{5x - 10y}{x^2 + xy} = \frac{x - 2y}{x + y} \cdot \frac{x^2 + xy}{5x - 10y}$$

$$= \frac{(x - 2y)(x)(x + y)}{(x + y)(5)(x - 2y)}$$

$$= \frac{x}{5} \qquad (x \neq 0, -y, 2y)$$

**31**

$$\frac{x^2 - 3xy - 4y^2}{-x^2 + 2xy + 3y^2} \div \frac{x^2 - xy - 2y^2}{-x^2 + 3xy}$$

$$= \frac{x^2 - 3xy - 4y^2}{-x^2 + 2xy + 3y^2} \cdot \frac{-x^2 + 3xy}{x^2 - xy - 2y^2}$$

$$= \frac{(x - 4y)(x + y)}{(-x + 3y)(x + y)} \cdot \frac{x(-x + 3y)}{(x - 2y)(x + y)}$$

$$= \frac{x(x - 4y)}{(x - 2y)(x + y)} \qquad (x \neq 3y, -y, 0, 2y)$$

## (3.6) (Division by $x - a$)

Consider the fraction $157/6$. The actual division is carried out as follows:

$$
\begin{array}{r}
26 \\
6\overline{)157} \\
12 \\
\hline
37 \\
36 \\
\hline
1 \quad \text{(remainder)}
\end{array}
$$

What this says is that $157/6 = 26 + (1)/6$ or that $157 = 6(26) + 1$. It is a fact that for any $a, b \in N$, one can find non-negative integers $q$ and $r$ with $r < b$ such that

$$
\frac{a}{b} = q + \frac{r}{b}.
$$

Another way to write this same fact is $a = bq + r$.

Naturally, this statement applies to fractions of the form $\dfrac{P(x)}{x - a}$, with $P(x)$ a polynomial of degree $n \geq 1$. The process is almost identical. To illustrate, let us consider the fraction $(x^2 - 3x + 6)/(x + 2)$. Our goal is to carry out the division of $x^2 - 3x + 6$ by $x + 2$, thus finding a polynomial $Q(x)$ and a number $r$ (called the *remainder*) such that

*remainder*

$$
\frac{x^2 - 3x + 6}{x + 2} = Q(x) + \frac{r}{x + 2} \quad (x \neq -2).
$$

Step by step, here is the procedure. Start with a long division set-up:

$$
x + 2\overline{)x^2 - 3x + 6}
$$

Now divide the highest-degree term of the dividend by $x$, and note that the answer is $x$:

$$
\begin{array}{r}
x \phantom{^2 - 3x + 6} \\
x + 2\overline{)x^2 - 3x + 6}
\end{array}
$$

This result tells us that the first term of $Q(x)$ must be $x$. At this point we subtract $x(x + 2) = x^2 + 2x$ from the dividend $x^2 - 3x + 6$:

$$\begin{array}{r}
x \phantom{{}-3x+6} \\
x + 2 \overline{\smash{\big)}\, x^2 - 3x + 6} \\
\underline{x^2 + 2x} \phantom{{}+6} \\
- 5x \phantom{{}+6}
\end{array}$$

We then bring down the next term from the dividend:

$$\begin{array}{r}
x \phantom{{}-3x+6} \\
x + 2 \overline{\smash{\big)}\, x^2 - 3x + 6} \\
\underline{x^2 + 2x} \phantom{{}+6} \\
- 5x + 6
\end{array}$$

Dividing the highest-degree term of $-5x + 6$ by $x$, we see that $-5$ is the next term of $Q(x)$:

$$\begin{array}{r}
x - 5 \phantom{{}+6} \\
x + 2 \overline{\smash{\big)}\, x^2 - 3x + 6} \\
\underline{x^2 + 2x} \phantom{{}+6} \\
- 5x + 6
\end{array}$$

Now we multiply $x + 2$ by $-5$ and subtract this from $-5x + 6$, leaving a remainder of 16:

$$\begin{array}{r}
x - 5 \phantom{{}+\ 6} \\
x + 2 \overline{\smash{\big)}\, x^2 - 3x + \ 6} \\
\underline{x^2 + 2x} \phantom{{}+\ 6} \\
- 5x + \ 6 \\
\underline{- 5x - 10} \\
16 \qquad (\textit{remainder})
\end{array}$$

This says that $\dfrac{x^2 - 3x + 6}{x + 2} = x - 5 + \dfrac{16}{x + 2}$ $(x \neq -2)$, or alternately that $x^2 - 3x + 6 = (x + 2)(x - 5) + 16$. Check it!

If $x - a$ happens to be a factor of $P(x)$, then the division process merely factors $P(x)$. Thus

$$\begin{array}{r}
x + 2 \phantom{{}-\ 6} \\
x - 3 \overline{\smash{\big)}\, x^2 - \ x - 6} \\
\underline{x^2 - 3x} \phantom{{}-6} \\
2x - 6 \\
\underline{2x - 6}
\end{array}$$

shows that $x^2 - x - 6 = (x - 3)(x + 2)$.

Given a polynominal $P(x)$ of degree $n \geq 1$, you now know how to write the fraction $P(x)/(x - a)$ in the form

**(1)** $\dfrac{P(x)}{x-a} = Q(x) + \dfrac{r}{x-a}$  $(x \neq a)$

with $Q(x)$ a polynomial of degree $n-1$ and $r$ a real number called the *remainder*. Multiplication through by $x-a$ will produce

**(2)** $P(x) = (x-a)\,Q(x) + r.$

Equation (1) is not defined at $x = a$, but equation (2) is. Let's now substitute $a$ for $x$ in (2) and see what happens:

$P(a) = (a-a)\,Q(a) + r = 0 \cdot Q(a) + r = r.$

*remainder theorem*

Thus the remainder on dividing $P(x)$ by $x-a$ is necessarily $P(a)$. This very important result is called the *remainder theorem*. To illustrate it, we note that in our first example, $P(x) = x^2 - 3x + 6$, $P(-2) = (-2)^2 - 3(-2) + 6 = 4 + 6 + 6 = 16$, and 16 was the remainder on dividing $P(x)$ by $x + 2$. To find $P(a)$, the division process is often easier than direct substitution.

You can also use the remainder theorem as a tool in factoring polynomials, for it says that if $P(c) = 0$, then $x - c$ must be a factor of $P(x)$. Thus if $P(x) = x^5 - 1$, clearly $P(1) = 0$; so $x - 1$ must be a factor of $P(x)$. You then use long division to determine $x^5 - 1 = (x-1)(x^4 - x^3 + x^2 - x + 1)$.

## Exercise set 3.6

*Express each of the fractions* $\dfrac{P(x)}{x-a}$ *in the form* $Q(x) + \dfrac{r}{x-a}$ *with $Q(x)$ a polynomial and $r$ a real number.*

$(\circ)$  1  $\dfrac{x^2 + 3x + 4}{x + 1}$

2  $\dfrac{x^2 - x - 5}{x + 2}$

3  $\dfrac{y^2 - 6y + 5}{y}$

4  $\dfrac{2w^2 - 5w + 3}{w}$

5  $\dfrac{t^2 - 3t - 6}{t - 4}$

6  $\dfrac{a^2 - 8a - 4}{a - 6}$

$(\circ)$  7  $\dfrac{x^3 + x + 12}{x + 2}$

8  $\dfrac{x^3 - 9x + 8}{x - 3}$

**9** $\dfrac{x^3 + 4x^2 + 5x + 4}{x + 1}$

**10** $\dfrac{p^3 - 2p^2 - 49p + 3}{p - 8}$

( ∘ ) **11** $\dfrac{3x^2 - 11x + 12}{x - 2}$

**12** $\dfrac{2x^2 + 7x + 4}{x + 5}$

**13** $\dfrac{4z^3 - 4z^2 - z + 3}{z - 1}$

**14** $\dfrac{9v^5 + 18v^4 - 4v + 1}{v + 2}$

In problems 15 to 20, verify that for the given value of c, P(c) is the remainder when dividing $P(x)$ by $x - c$. Thus if $P(x) = x^2 + 3x + 2$ and $c = 1$, we have $P(1) = 1 + 3 + 2 = 6$ and

$$
\begin{array}{r}
x + 4 \\
x - 1\overline{)x^2 + 3x + 2} \\
\underline{x^2 - x} \\
4x + 2 \\
\underline{4x - 2} \\
6 \quad [= P(1)]
\end{array}
$$

( ∘ ) **15** $P(x) = x^2 - 4x + 2;\ c = 1$     **16** $P(w) = w^2 + 2w + 4;\ c = -2$

**17** $P(t) = 2t^2 - 5t + 1;\ c = 2$     **18** $P(a) = 3a^2 - 40;\ c = 4$

**19** $P(x) = x^3 + x^2 - 3;\ c = -2$     **20** $P(x) = 2x^3 + x^2 + x + 1:\ c = 1$

For problems 21 to 28, use the remainder theorem to decide if $x - c$ is a factor of the polynomial $P(x)$. If it is, use the division process to write $P(x) = (x - c)\,Q(x)$. If not, state that fact. Thus if $P(x) = x^4 + x + 1$, then $P(1) = 3$ shows that $x - 1$ is not a factor of $P(x)$. On the other hand, if $P(x) = x^4 - 2x + 1$, then $P(1) = 1 - 2 + 1 = 0$; so $x - 1$ is a factor. To obtain the factorization, we use division and write

$$
\begin{array}{r}
x^3 + x^2 + x - 1 \\
x - 1\overline{)x^4 \qquad\qquad - 2x + 1} \\
\underline{x^4 - x^3} \\
x^3 \\
\underline{x^3 - x^2} \\
x^2 - 2x \\
\underline{x^2 - x} \\
-x + 1 \\
-x + 1
\end{array}
$$

This shows that $x^4 - 2x + 1 = (x - 1)(x^3 + x^2 + x - 1)$. Here, then, are some problems for you to try your hand at. Remember, though, to test $x - c$ you compute $P(c)$, not $P(-c)$.

$(\circ)$    **21**   $x^3 - 3x + 1;\ x - 1$        **22**   $x^3 - 2x^2 + 3;\ x - 2$

$(\circ)$    **23**   $y^3 + 6y^2 - 16;\ y + 2$       **24**   $w^3 + w^2 + 2w + 2;\ w + 1$

       **25**   $t^6 - t^4 + 2t + 3;\ t - 1$       **26**   $h^4 - h^3 - 3h - 2;\ h - 2$

       **27**   $2s^3 + 6s^2 - s - 3;\ s + 3$      **28**   $3x^3 + 4x^2 + 15x;\ x + 3$

$(\circ)$

*Sample solutions*
*for exercise set 3.6*

**1**
$$
\begin{array}{r}
x + 2 \phantom{00000} \\
x + 1\overline{)\,x^2 + 3x + 4} \\
\underline{x^2 + \phantom{0}x} \phantom{0000} \\
2x + 4 \\
\underline{2x + 2} \\
2
\end{array}
$$

Hence, $\dfrac{x^2 + 3x + 4}{x + 1} = x + 2 + \dfrac{2}{x + 1}$    $(x \neq -1)$.

**7**
$$
\begin{array}{r}
x^2 - 2x + \phantom{0}5 \phantom{0000} \\
x + 2\overline{)\,x^3 \phantom{0000} + \phantom{0}x + 12} \\
\underline{x^3 + 2x^2} \phantom{000000000} \\
- 2x^2 + \phantom{0}x \phantom{0000} \\
\underline{- 2x^2 - 4x} \phantom{0000} \\
5x + 12 \\
\underline{5x + 10} \\
2
\end{array}
$$

So, we may write $\dfrac{x^3 + x + 12}{x + 2} = x^2 - 2x + 5 + \dfrac{2}{x + 2}$    $(x \neq -2)$.

**11**   We have

$$
\begin{array}{r}
3x - \phantom{0}5 \phantom{0000} \\
x - 2\overline{)\,3x^2 - 11x + 12} \\
\underline{3x^2 - \phantom{0}6x} \phantom{00000} \\
- \phantom{0}5x + 12 \\
\underline{- \phantom{0}5x + 10} \\
2
\end{array}
$$

when $\dfrac{3x^2 - 11x + 12}{x - 2} = 3x - 5 + \dfrac{2}{x - 2}$    $(x \neq 2)$.

**15**   We begin by observing that $P(1) = 1 - 4 + 2 = -1$. Dividing $P(x)$ by $x - 1$, we obtain

$$\begin{array}{r} x\ -\ 3 \\ x-1\overline{)x^2-4x+2} \\ \underline{x^2-\ \ x} \\ -\,3x+2 \\ -\,3x+3 \\ \hline -\,1 \end{array}$$

21  $P(1) = 1 - 3 + 1$; so $x - 1$ is not a factor.

23  $P(-2) = -8 + 24 - 16 = 0$; so $y + 2$ is a factor.

$$\begin{array}{r} y^2+4y\ -\ 8 \\ y+2\overline{)y^3+6y^2\qquad -16} \\ \underline{y^3+2y^2} \\ 4y^2 \\ \underline{4y^2+8y} \\ -\,8y-16 \\ -\,8y-16 \\ \hline \end{array}$$

Thus $y^3 + 6y^2 - 16 = (y + 2)(y^2 + 4y - 8)$.

# (3.7) (Complex fractions)

Sometimes, the numerator or denominator (or both) of a fraction may *themselves* involve one or more fractions. The original fraction is then called a *complex fraction*.

*complex fraction*

$$\dfrac{1+\dfrac{1}{2}}{5} \quad \text{and} \quad \dfrac{x+1}{2-\dfrac{1}{x-1}} \quad \text{and} \quad \dfrac{x+\dfrac{1}{y}}{x-\dfrac{1}{1+\dfrac{1}{y}}}$$

are all complex fractions.

We shall see how to deal with complex fractions by considering some examples. Let's begin with

$$\dfrac{1+\dfrac{1}{2}}{5}.$$

If we multiply both numerator and denominator of this fraction by 2, the resulting fraction is no longer complex. Thus

$$\frac{1 + \dfrac{1}{2}}{5} \cdot \frac{2}{2} = \frac{2 + 1}{10} = \frac{3}{10}.$$

Alternatively, we could write both numerator and denominator as single fractions, then take the product of the numerator by the reciprocal of the denominator. To illustrate this procedure we write

$$\frac{1 + \dfrac{1}{2}}{5} = \frac{\dfrac{3}{2}}{\dfrac{5}{1}} = \frac{3}{2} \cdot \frac{1}{5} = \frac{3}{10}.$$

Similarly, to solve $\dfrac{x + 1}{2 - \dfrac{1}{x - 1}}$, we have

$$\frac{x + 1}{2 - \dfrac{1}{x - 1}} \cdot \frac{x - 1}{x - 1} = \frac{(x + 1)(x - 1)}{2(x - 1) - 1} = \frac{(x + 1)(x - 1)}{2x - 2 - 1}$$

$$= \frac{(x + 1)(x - 1)}{2x - 3} \qquad \left( x \neq 1, \ \frac{3}{2} \right).$$

For the alternative procedure, we write

$$\frac{x + 1}{2 - \dfrac{1}{x - 1}} = \frac{x + 1}{\dfrac{2x - 2}{x - 1} - \dfrac{1}{x - 1}} = \frac{\dfrac{x + 1}{1}}{\dfrac{2x - 3}{x - 1}}$$

$$= \frac{x + 1}{1} \cdot \frac{x - 1}{2x - 3} = \frac{(x + 1)(x - 1)}{2x - 3} \qquad \left( x \neq 1, \ \frac{3}{2} \right).$$

To summarize, suppose we are given a complex fraction $a/b$ such that neither $a$ nor $b$ involves complex fractions. We must *clearing factor* find an expression $c$ (called a *clearing factor* for $a/b$) such that $ac/bc$ is not complex. Any common multiple of the denominators of the fractions appearing in $a$ or $b$ will do; in particular, the LCD of these fractions will work just fine as a clearing factor. Step by step, here is the procedure:

**Step 1:** Find a clearing factor $c$ for $\dfrac{a}{b}$.

**Step 2:** Form $\dfrac{ac}{bc}$.

**Step 3:** Simplify, and reduce to lowest terms.

Thus a clearing factor for $\dfrac{\dfrac{1}{x-y} - \dfrac{1}{x+y}}{1 + \dfrac{x}{x-y}}$ is the LCD of the

fractions $\dfrac{1}{x-y}$, $\dfrac{1}{x+y}$, $\dfrac{x}{x-y}$, which is $(x-y)(x+y)$. We therefore have

$$\frac{\dfrac{1}{x-y} - \dfrac{1}{x+y}}{1 + \dfrac{x}{x-y}} \cdot \frac{(x-y)(x+y)}{(x-y)(x+y)} = \frac{(x+y)-(x-y)}{x^2 - y^2 + x(x+y)} = \frac{x+y-x+y}{2x^2 + xy - y^2}$$

$$= \frac{2y}{(2x-y)(x+y)} \quad (x \neq -y,\ y;\ 2x \neq y).$$

The alternative procedure for simplifying the complex fraction $a/b$ involves writing both $a$ and $b$ as single fractions and then using the fact that $a/b = a \cdot (1)/b$.

For more complicated complex fractions, your best bet is to remember that a fraction bar is a grouping device—just as parentheses or brackets are grouping devices. The general procedure for dealing with grouping devices is to start with the innermost device and work your way outward. In the present context, we start with a fraction bar that groups a complex fraction whose numerator and denominator do not involve complex fractions, simplify this fraction, and then repeat the procedure as many times as needed. For example, consider

$$\frac{1 + \dfrac{1}{\dfrac{x}{y} + 3}}{1 - \dfrac{2}{\dfrac{x}{3y} + 1}}.$$

This is a complex fraction whose numerator and denominator themselves involve complex fractions. We start by simplifying $\dfrac{1}{\dfrac{x}{y} + 3}$

and $\dfrac{2}{\dfrac{x}{3y} + 1}$, which represent the innermost grouping devices. Thus

$$\frac{1}{\dfrac{x}{y} + 3} \cdot \frac{y}{y} = \frac{y}{x + 3y} \quad \text{and} \quad \frac{2}{\dfrac{x}{3y} + 1} \cdot \frac{3y}{3y} = \frac{6y}{x + 3y}.$$

We now have

$$\frac{1 + \dfrac{1}{\dfrac{x}{y} + 3}}{1 - \dfrac{2}{\dfrac{x}{3y} + 1}} = \frac{1 + \dfrac{y}{x + 3y}}{1 - \dfrac{6y}{x + 3y}}$$

and this last fraction can be simplified by using the clearing factor $x + 3y$:

$$\frac{1 + \dfrac{y}{x + 3y}}{1 - \dfrac{6y}{x + 3y}} \cdot \frac{x + 3y}{x + 3y} = \frac{x + 3y + y}{x + 3y - 6y}$$

$$= \frac{x + 4y}{x - 3y} \qquad (x \neq 3y, -3y, 0; \, y \neq 0).$$

## Exercise set 3.7

In problems 1 to 10, use both methods to simplify each complex fraction.

$(\circ)$  1  $\dfrac{\dfrac{3}{5}}{\dfrac{9}{10}}$

2  $\dfrac{\dfrac{4}{7}}{\dfrac{8}{21}}$

3  $\dfrac{\dfrac{7}{10}}{\dfrac{3}{10}}$

4  $\dfrac{\dfrac{3}{7}}{\dfrac{6}{7}}$

5  $\dfrac{\dfrac{8}{15}}{\dfrac{4}{5}}$

6  $\dfrac{\dfrac{9}{30}}{\dfrac{3}{20}}$

$(\circ)$  7  $\dfrac{\dfrac{3}{5x}}{\dfrac{9}{10x^2}}$

8  $\dfrac{\dfrac{7}{12y^2}}{\dfrac{3}{4y}}$

9  $\dfrac{\dfrac{7x}{3}}{\dfrac{14}{9x}}$

10  $\dfrac{\dfrac{6}{x+y}}{\dfrac{2}{x^2-y^2}}$

In problems 11 to 22, use either method to express each complex fraction as a single fraction in lowest terms.

$(\circ)$  11  $\dfrac{1-\dfrac{1}{2}}{1+\dfrac{1}{4}}$

12  $\dfrac{3+\dfrac{5}{3}}{2-\dfrac{7}{9}}$

13  $\dfrac{\dfrac{1}{2}+\dfrac{1}{3}}{\dfrac{5}{6}-\dfrac{1}{4}}$

14  $\dfrac{\dfrac{7}{8}-\dfrac{3}{4}}{\dfrac{5}{3}-\dfrac{3}{2}}$

**15** $\dfrac{1 - \dfrac{1}{a}}{1 + \dfrac{1}{2a}}$

**16** $\dfrac{3 + \dfrac{1}{4x}}{2 + \dfrac{1}{6x}}$

$(\circ)$ **17** $\dfrac{\dfrac{1}{a} - \dfrac{1}{2a^2}}{\dfrac{1}{4a^2} - 1}$

**18** $\dfrac{\dfrac{1}{3x} - \dfrac{1}{x^2}}{1 - \dfrac{6}{x} + \dfrac{9}{x^2}}$

**19** $\dfrac{\dfrac{1}{ab} + \dfrac{1}{bc} + \dfrac{1}{ca}}{\dfrac{a + b + c}{a}}$

**20** $\dfrac{\dfrac{1}{a} + \dfrac{1}{b} + \dfrac{1}{ab}}{1 + \dfrac{1 + b}{a}}$

**21** $\dfrac{x + \dfrac{x}{y}}{\dfrac{1}{y} - \dfrac{1}{y^2}}$

**22** $\dfrac{1 - \dfrac{x}{x + y}}{\dfrac{1}{x} - \dfrac{1}{x + y}}$

*Some of the remaining problems are a little more complicated. Remember to treat the fraction bars as grouping devices and you will be all right. Again, you are to express each complex fraction as a single fraction in lowest terms.*

$(\circ)$ **23** $\dfrac{1}{1 - \dfrac{1}{1 + \dfrac{1}{2}}}$

**24** $\dfrac{1}{1 + \dfrac{1}{1 + \dfrac{1}{3}}}$

**25** $\dfrac{\dfrac{1}{2} + \dfrac{3}{1 - \dfrac{5}{3}}}{2 + \dfrac{4 - \dfrac{2}{5}}{5 - \dfrac{4}{5}}}$

**26** $\dfrac{1 + \dfrac{2 + \dfrac{1}{4}}{1 + \dfrac{3}{8}}}{1 - \dfrac{1 + \dfrac{2}{3}}{\dfrac{1}{2} + \dfrac{1}{6}}}$

**27** $1 + \dfrac{x}{1 + \dfrac{x}{1 + x}}$

**28** $1 - \dfrac{y}{1 - \dfrac{1}{1 + y}}$

**29** $\dfrac{3}{t + \dfrac{3}{t + 2}}$

**30** $1 + \dfrac{a + \dfrac{1}{1 + a}}{a + \dfrac{1}{1 + \dfrac{1}{a}}}$

$(\circ)$  **31** $\dfrac{2 - \dfrac{3}{1 + \dfrac{a}{b}}}{1 + \dfrac{1}{1 + \dfrac{a}{b}}}$

**32** $\dfrac{1 + \dfrac{1}{\dfrac{a}{b} + 3}}{1 - \dfrac{1}{\dfrac{a}{3b} + 1}}$

**33** $\dfrac{s + 1 - \dfrac{12}{s + 2}}{s - 1 - \dfrac{4}{s + 2}}$

**34** $\dfrac{b + 5 - \dfrac{8}{b - 2}}{b - 4 + \dfrac{1}{b - 2}}$

**35** $\dfrac{\dfrac{a - b}{a + b} + \dfrac{a + b}{a - b}}{\dfrac{a^2 + b^2}{(a - b)^2}}$

**36** $\dfrac{\dfrac{a - 1}{a + 1} - \dfrac{a + 1}{a - 1}}{1 + \dfrac{1}{1 - \dfrac{1}{a^2}}}$

**37** $\dfrac{1}{1 + \dfrac{1}{1 + \dfrac{1}{1 + \dfrac{1}{2}}}}$

**38** $\dfrac{1}{3 - \dfrac{1}{3 - \dfrac{1}{3 - \dfrac{7}{3}}}}$

1  Using the alternative method introduced in the text, we have

$$\frac{\dfrac{3}{5}}{\dfrac{9}{10}} = \frac{3}{5} \cdot \frac{10}{9} = \frac{2}{3}.$$

On the other hand, 10 will work as a clearing factor; so we may write

$$\frac{\dfrac{3}{5}}{\dfrac{9}{10}} \cdot \frac{10}{10} = \frac{3 \cdot 2}{9} = \frac{2}{3}.$$

7  Using both methods, we have

$$\frac{\dfrac{3}{5x}}{\dfrac{9}{10x^2}} = \frac{3}{5x} \cdot \frac{10x^2}{9} = \frac{2x \cdot 3 \cdot 5x}{3 \cdot 3 \cdot 5x} = \frac{2x}{3} \qquad (x \neq 0)$$

and

$$\frac{\dfrac{3}{5x}}{\dfrac{9}{10x^2}} \cdot \frac{10x^2}{10x^2} = \frac{3 \cdot 2x}{9} = \frac{2x}{3} \qquad (x \neq 0).$$

11  An appropriate clearing factor is 4; so we have

$$\frac{1 - \dfrac{1}{2}}{1 + \dfrac{1}{4}} \cdot \frac{4}{4} = \frac{4 - 2}{4 + 1} = \frac{2}{5}.$$

17  For this one, note that $4a^2$ is the LCD of the fractions $\dfrac{1}{a}$, $\dfrac{1}{2a^2}$, and $\dfrac{1}{4a^2}$; so it is our choice for a clearing factor. Proceeding as usual, we have

$$\frac{\dfrac{1}{a} - \dfrac{1}{2a^2}}{\dfrac{1}{4a^2} - 1} \cdot \frac{4a^2}{4a^2} = \frac{4a - 2}{1 - 4a^2} = \frac{2(2a - 1)}{(1 - 2a)(1 + 2a)}$$

$$= \frac{-2}{1 + 2a} \qquad \left(a \neq 0, \frac{1}{2}, -\frac{1}{2}\right).$$

**23** The denominator of this fraction itself involves a complex fraction; so we cannot directly find a clearing factor. Isolating the portion of the denominator that is causing the trouble, we have $\dfrac{1}{1 + \dfrac{1}{2}} \cdot \dfrac{2}{2} = \dfrac{2}{2 + 1} = \dfrac{2}{3}.$

Substituting this in the original fraction produces

$$\frac{1}{1 - \dfrac{1}{1 + \dfrac{1}{2}}} = \frac{1}{1 - \dfrac{2}{3}} = \frac{1}{1 - \dfrac{2}{3}} \cdot \frac{3}{3} = \frac{3}{3 - 2} = 3.$$

**31** Both the numerator and denominator of this one involve complex fractions. The culprits are $\dfrac{3}{1 + \dfrac{a}{b}}$ and $\dfrac{1}{1 + \dfrac{a}{b}}$; so let us first deal with them. Multiplying each of them by $\dfrac{b}{b}$, we obtain $\dfrac{3b}{a + b}$ and $\dfrac{b}{a + b}.$

Substituting, we have

$$\frac{2 - \dfrac{3}{1 + \dfrac{a}{b}}}{1 + \dfrac{1}{1 + \dfrac{a}{b}}} = \frac{2 - \dfrac{3b}{a + b}}{1 + \dfrac{b}{a + b}} = \frac{2 - \dfrac{3b}{a + b}}{1 + \dfrac{b}{a + b}} \cdot \frac{a + b}{a + b}$$

$$= \frac{2(a + b) - 3b}{(a + b) + b} = \frac{2a - b}{a + 2b} \qquad (b \neq 0;\ a + b \neq 0;\ a + 2b \neq 0).$$

# (4)

## First-degree equations and inequalities

### (4.1) (Equations)

It is time to find out how algebra is used to solve equations. Before doing so, however, we had better find out what an equation is, and for that matter what it means to solve an equation.

*equation*

An *equation* is a symbolic statement that two algebraic expressions involving one or more variables are equal. The expressions are called *members* of the equation, and the variables that appear in an equation are sometimes called *unknowns.* The set of possible values of an unknown is its *replacement set,* which, unless otherwise specified, should be taken to be the set of all real numbers. Once complex numbers are introduced at the end of Chapter 5, we will want the replacement set of an unknown to be the set of complex numbers unless we specify differently. When we state no restrictions, the replacement set for $x$ is $R$, but if $x$ denotes the weight of a person in pounds, for example, you would not allow negative numbers in its replacement set.

*members*
*unknowns*
*replacement set*

A typical equation, then, is of the form

(One algebraic expression) = (second algebraic expression).

More concretely, $2x + 3 = 7$ and $2(x + 3) = 2x + 6$ are both examples of equations. But there is a big difference between these two equations; the first one is true only when $x = 2$, whereas the second one holds for all possible values of $x$.

*identity*

An equation is called an *identity* if it is true for all numbers in the replacement sets of its unknowns; otherwise, it is called

*conditional equation*

a *conditional equation*. The equation $x^2 + 5x + 6 = (x + 2)(x + 3)$ is an identity, as are all such laws as

**A.2** $x + y = y + x,$

the commutative law for addition of real numbers.

On the other hand, to show that $5x + 3 = 8$ is a conditional equation, you need only notice that it is false for $x = 0$. To show that an equation is an identity, you must show it to be true for all possible replacements of its variables. To show that an equation is conditional, you need only find one replacement that makes it false.

Given an equation in a single unknown, a number $c$ in its

*root, solution*

replacement set is called a *root,* or a *solution,* of the equation if the equation is true when $c$ is substituted for the unknown. The

*solution set*

set of all solutions of such an equation is called its *solution set,* and to solve an equation is to find the members of its solution set.

*equivalent*

Finally, two equations are called *equivalent* if their solution sets coincide. You know, there are times when this looks more like an English book than an algebra book! The point is that we must establish a common language for communication before we can really get into high gear.

When faced with an equation in one variable, our goal, of course, is to solve it. This is accomplished by systematically replacing the given equation with simpler, equivalent equations until one is reached for which the solution set can be determined.

# Exercise set 4.1

Determine whether or not the specified number is a solution of the given equation.

( ○ )

**1**  $3x + 5 = 8$; 1    **2**  $2x + 7 = 9$; 1

**3**  $4w + 3 = 7$; 0    **4**  $8t - 3 = -2$; 1

**5**  $2a - 3 = 3a + 5$; $-8$    **6**  $6y - 1 = 4y - 5$; $-2$

**7**  $r^2 + 5r = -6$; $-2$    **8**  $s^2 - s = 2$; 2

**9**  $2p^2 - 3p = p^2 + 5$; 4    **10**  $d(d + 1) = -d - 1$; 1

Show that each of the following equations is conditional by finding a number that is not a solution.

( ○ )

**11**  $2x + 5 = 7$    **12**  $5x + 3 = 7$

**13**  $-w + 3 = 2w + 4$    **14**  $5z + 4 = 4z + 5$

**15**  $(u + 3)(u + 2) = u^2 + 5u + 5$

**16**  $(v - 1)^2 = (v - 1)^3$

Simplifying one or both members of an equation will yield an equivalent equation. Thus you can show that the equation $(x + 2)(x + 3) - 5x = x^2 + 6$ is an identity by observing that it is equivalent in turn both to $(x^2 + 5x + 6) - 5x = x^2 + 6$ and to $x^2 + 6 = x^2 + 6$. This last equation is clearly an identity, and consequently, so is the first one.

Show that each of the following is an identity by simplifying one or both members.

**17**  $(x + 2)^2 - 4x = x^2 + 4$

**18**  $(x - 4)(x + 2) + 8 = x^2 - 2x$

( ○ )  **19**  $(w + 2)(w + 1) + (w + 3) = (w + 2)^2 + 1$

**20**  $y(y - 3) - 3y + 9 = (y - 3)^2$

**21**  $(z + 7)(z + 2) - z^2 - 9z = 14$

**22**  $(a - 6)(a + 5) - a^2 + 40 = 10 - a$

**23**  $3(2 - x) + 6(x - 4) = 4(x - 2) - \dfrac{2x + 20}{2}$

**24**  $5(5 - b) - 4(4 - b) = 8(3 + b) - 3(5 + 3b)$

( ○ )

*Sample solutions*
*for exercise set 4.1*

**1**  Substituting $x = 1$ produces the true statement that $3(1) + 5 = 8$; so 1 is a solution.

**11**  Since $2 \cdot 0 + 5 \neq 7$, 0 is not a solution, and the equation is therefore conditional.

**19** Simplifying both members, we have

$$(w^2 + 3w + 2) + (w + 3) = (w^2 + 4w + 4) + 1$$

$$w^2 + 4w + 5 = w^2 + 4w + 5.$$

The last equation is clearly an identity and is equivalent to the original equation.

# (4.2) (Solving equations)

Well, by now you know what equations are, and you know the theory behind their solution. It is time now to get into the actual process of solving an equation. Our methods are based on the following facts about real numbers. Let $a, b, c \in R$. Then

( ○ )  $a = b$ *if and only if* $a + c = b + c$

( ○ )  $a = b$ *if and only if* $ac = bc$ *(provided* $c \neq 0$*)*

Let's phrase these in terms of equations. Suppose that $P$, $Q$, and $R$ represent expressions involving a single variable. Whenever that variable is replaced by a number, then $P$, $Q$, and $R$ each represent real numbers, and the operations $+$ and $\cdot$ on the expressions amount to ordinary addition and multiplication of real numbers. It is easy to see from this that:

( ○ )  EQ.1 *The equation* $P = Q$ *is equivalent to the equation* $P + R = Q + R$

( ○ )  EQ.2 *The equation* $P = Q$ *is equivalent to the equation* $P \cdot R = Q \cdot R$ *(provided* $R$ *is never 0)*

In other words, what EQ.1 means is that if an expression is added to *both* members of an equation, the resulting equation is *equivalent* to the original equation.

EQ.2 states that if both members of an equation are multiplied by an expression that is never zero, the resulting equation is equivalent to the original equation.

Let's apply all of this to the solution of an equation or two. The basic idea is to use EQ.1 and EQ.2 to replace the given equation with an equivalent equation whose solution set can be determined. We begin by looking at an equation of the form

**(1)** $ax + b = 0$ $\quad$ $(a, b \in R; a \neq 0)$.

Such an equation, or more generally one of the form

$ax + b = cx + d$ $\quad$ $(a \neq c)$

is called a *first-degree equation in one unknown*. To solve (1) we multiply both members by $1/a$ to produce the equivalent equation

**(2)** $x + \dfrac{b}{a} = 0$.

Then add $-(b/a)$ to both members of (2) to produce

**(3)** $x = -\dfrac{b}{a}$

The solution set of (3) is clearly $\{-(b/a)\}$; so we deduce that $x = -(b/a)$ is the unique solution of (1). From now on you will no doubt solve an equation like (1) by inspection.

To solve an equation like

**(4)** $2x + 3 = 5x - 8$

we add $-2x$ to both members of (4) to get

**(5)** $3 = 3x - 8$

and then add 8 to both members of (5) to arrive at

**(6)** $11 = 3x$.

The solution set is, therefore, $\{11/3\}$.

Sometimes you will be faced with an equation of a higher degree that simplifies to a first-degree equation. Thus, to solve

**(7)** $(x + 2)(x + 3) = (x - 1)(x + 4)$,

simplify both members to obtain the equivalent equation $x^2 + 5x + 6 = x^2 + 3x - 4$. We then have

$5x + 6 = 3x - 4$ $\quad$ (by adding $-x^2$ to both sides)

$2x + 6 = -4$ $\quad$ (by adding $-3x$ to both sides)

$2x = -10$ $\quad$ (by adding $-6$ to both sides)

$x = -5$ $\quad$ (by multiplying both sides by $1/2$).

The solution set is $\{-5\}$.

When multiplying both members of an equation by an expression that involves a variable, a little caution is called for, since we must be certain that the expression is not allowed to be zero. Thus to solve

$$\textbf{(8)} \quad \frac{x+3}{x} = \frac{3}{x} + 4$$

we might multiply both members by $x$ to get rid of the denominators, obtaining

$$x + 3 = 3 + 4x$$

$$0 = 3x$$

$$0 = x.$$

It appears that $x = 0$ is a solution for (8), but this is wrong! Since division by 0 is not allowed, 0 is not even in the replacement set for $x$. We conclude that (8) has no solution. Whenever you multiply both members of an equation by an expression involving a variable, it is a good idea to check any solutions out by actually substituting them in the original equation. Incidentally, such a check generally helps you to catch any errors in arithmetic.

## Exercise set 4.2

*Solve each equation.*

$( \circ )$    **1**   $2x + 3 = 0$              **2**   $3x + 5 = 0$

     **3**   $5z - 8 = 0$              **4**   $3t - 10 = 0$

     **5**   $4a - 8 = 0$              **6**   $7b - 3 = 0$

$( \circ )$    **7**   $2x - 5 = x + 4$         **8**   $2y - 8 = y + 3$

     **9**   $4x + 7 = -3x - 8$      **10**   $2a - 3 = -5a + 2$

    **11**   $(x + 1)^2 = (x + 2)^2$      **12**   $(x + 2)^2 = (x + 3)^2$

$( \circ )$    **13**   $(w - 1)(w + 2) = (w - 3)^2$

    **14**   $(u - 3)(u + 4) = (u + 1)(u + 3)$

    **15**   $(p + 1)^2 + 2(p - 3) = (p - 2)(p + 4)$

    **16**   $(q - 1)(q - 2) + 2(q + 1) = (q - 3)(q - 4)$

$( \circ )$    **17**   $1 + \dfrac{1}{x} = 2$            **18**   $3 - \dfrac{2}{x} = 5$

**19** $\dfrac{x+1}{x} - 2 = \dfrac{3}{x}$      **20** $\dfrac{3x-2}{x} = 4 + \dfrac{1}{x}$

**21** $\dfrac{4y+1}{y} = 2 + \dfrac{1}{y}$      **22** $\dfrac{6z-4}{4z} = 1 - \dfrac{1}{z}$

( ∘ )   **23**   $5(3x-2) - 4[2 - (3x-1)] = [x - (2x-3)] - 1$

     **24**   $6[3 - (1 - 2x)] - 5[(5 - 2x) - 1] = (x+1) - (3x-2)$

     **25**   $a[a - (a-3)] + (6a-5) - (4a-3) = (a+1)(a+2) - (a+1)^2$

     **26**   $\{b - [(2b+3) - (b-4)]\} - (2b-3) = (b-1)^2 - (2-b)^2$

**( ∘ )**

*Sample solutions*
*for exercise set 4.2*

**1**   The solution set is $\left\{-\dfrac{3}{2}\right\}$.

**7**   $2x - 5 = x + 4$

     $x - 5 = 4$      (by adding $-x$ to both sides)

     $x = 9$      (by adding 5 to both sides)

The solution set is $\{9\}$.

**13**   Simplifying both members of the equation $(w-1)(w+2) = (w-3)^2$, we obtain

$$w^2 + w - 2 = w^2 - 6w + 9$$

$$w - 2 = -6w + 9$$

$$7w = 11$$

$$w = \dfrac{11}{7}.$$

The solution set is $\left\{\dfrac{11}{7}\right\}$.

**17**   Our equation is $1 + \dfrac{1}{x} = 2$. Multiplying both members by $x$ will produce $x + 1 = 2x$, or $1 = x$. Substituting 1 for $x$ in the original equation shows that $1 + \dfrac{1}{1} = 2$; so $x = 1$ is the solution.

**23**   The first step in this one must be the simplification of both members, and here we must be cautious about grouping devices and order of operations. We obtain, in turn, the following equations, all equivalent to the original one:

$$15x - 10 - 4(2 - 3x + 1) = (x - 2x + 3) - 1$$

$$15x - 10 - 4(-3x + 3) = -x + 2$$

$$15x - 10 + 12x - 12 = -x + 2$$

$$27x - 22 = -x + 2$$

$$28x = 24,$$

$$x = \frac{6}{7}.$$

The solution set is $\left\{ \dfrac{6}{7} \right\}$.

# (4.3) (Equations in more than one variable)

You will frequently be faced with an equation that involves more than one variable. For example, the formula for simple interest $I$ on a loan is

$$I = PRT.$$

Here $P$ represents the amount of the loan, $R$ the interest rate, and $T$ the time. If you wanted to solve for $T$, you would simply pretend that $I$, $P$, and $R$ all represented constants, and then divide through by $PR$ to obtain

$$T = \frac{I}{PR}.$$

To solve the equation $y = mx + b$ $(m \neq 0)$ for $x$, you would treat $y$, $m$, and $b$ as constants and proceed as follows:

$$y - b = mx \qquad \text{(subtracting } b \text{ from both sides)}$$

$$x = \frac{y - b}{m} \qquad \text{(dividing by } m \text{ and transposing).}$$

## Exercise set 4.3

Solve for x, indicating any restrictions that must be made to avoid division by zeros.

($\circ$)  **1**  $2x + 3y = 8$                    **2**  $y = 3x + 5$

**3**  $\dfrac{y}{4} + 2x = 8$                    **4**  $8y - 5x = 2y - 3x$

($\circ$)  **5**  $ax + by = c$    $(a \neq 0)$     **6**  $y = kx + gt$    $(k \neq 0)$

**7**  $ax + b = 2ax$    $(a \neq 0)$     **8**  $5cx - ac = acx$    $(c \neq 0)$

**9**  $\dfrac{a}{b}x + c = 0$                    **10**  $\dfrac{a}{c}x + \dfrac{c}{b} = 0$

($\circ$)  **11**  $\dfrac{1}{x} + \dfrac{1}{a} = 1$                    **12**  $\dfrac{a}{x} + \dfrac{b}{a} = c$

**13**  $\dfrac{x}{a} + \dfrac{y}{b} = 1$                    **14**  $\dfrac{x}{a} - \dfrac{y}{b} = c$

**15**  $\dfrac{1}{x} + \dfrac{1}{y} = 1$                    **16**  $\dfrac{a}{x} - \dfrac{b}{y} = ay$

In each remaining problem, you will be given an equation that arises in connection with some geometric figure or some physical situation. Solve for the indicated unknown.

**17**  The area $A$ of a rectangle with base $b$ and altitude $h$ is given by $A = bh$. Solve this equation for $h$.

**18**  The perimeter of a parallelogram that has sides of length $l$ and $w$ is $p = 2l + 2w$. Solve for $l$.

**19**  The circumference of a circle with radius $r$ is $C = 2\pi r$. Solve for $r$.

**20**  The circumference of a circle with diameter $d$ is $C = \pi d$. Solve this equation for $d$.

**21**  The volume of a right circular cylinder whose radius is $b$ and altitude is $c$ is given by the formula $V = \pi b^2 c$. Solve for $c$.

**22**  The area of a triangle with base $b$ and altitude $h$ is $A = \dfrac{bh}{2}$. Solve for $b$.

($\circ$)  **23**  The area of a trapezoid with altitude $h$ and parallel sides $b$, $b'$ is $A = \dfrac{h}{2}(b + b')$. Solve for $b$.

**24**  The distance $s$ travelled by an automobile at constant velocity $r$ in time $t$ is $s = rt$. Solve for $t$.

**25** The total surface area of a right circular cylinder with radius $r$ and altitude $h$ is $A = 2\pi r(r + h)$. Solve for $h$.

**26** The volume of a right circular cylinder with diameter $d$ and altitude $h$ is $V = \dfrac{\pi d^2 h}{4}$. Solve for $h$.

**27** The volume of a rectangular box whose dimensions are $l$, $w$, $h$ is $V = lwh$. Solve for $w$.

**28** If an object is dropped from a height of $s_0$ feet, then at time $t$ its height will be $s = s_0 - \dfrac{1}{2}gt^2$. Solve for $g$.

**29** The velocity of an object undergoing constant acceleration $a$ at time $t$ is $v = k + at$. Solve for $t$.

**30** The kinetic energy $E$ of a body with mass $m$ moving at velocity $v$ is $E = \dfrac{1}{2}mv^2$. Solve for $m$.

**( ∘ )**

*Sample solutions for exercise set 4.3*

**1**   $2x + 3y = 8$

$\quad\quad 2x = 8 - 3y \quad$ (adding $-3y$ to both sides)

$\quad\quad x = \dfrac{1}{2}(8 - 3y) \quad$ (dividing both sides by $2$)

**5**   Given the equation $ax + by = c$, we treat $a$, $b$, $c$, and $y$ as constants, solving for $x$:

$ax = c - by$

$x = \dfrac{c - by}{a} \quad (a \neq 0)$.

**11**   The given equation is $\dfrac{1}{x} + \dfrac{1}{a} = 1$. We see right away that $x \neq 0$ and $a \neq 0$. We treat $a$ as a constant and solve for $x$. Multiplication of both members by $ax$ produces

$a + x = ax$

$\quad\quad a = ax - x \quad$ (subtracting $x$ from both sides)

$\quad\quad a = x(a - 1)$

$\quad\quad x = \dfrac{a}{a - 1} \quad$ (dividing by $a - 1$).

Since we have divided by $a - 1$, we cannot have $a = 1$. If $a \neq 0$, then automatically, $x \neq 0$; so the solution is $x = \dfrac{a}{a - 1}$ $(a \neq 0, 1)$.

23   We are to solve $A = \dfrac{h}{2}(b + b')$ for $b$. Multiplying through by $2$ produces

$2A = h(b + b')$

$2A = hb + hb'$

$hb = 2A - hb'$

$b = \dfrac{2A - hb'}{h}.$

# (4.4) (Word problems)

A word problem, as its name implies, formulates a mathematical problem verbally. To solve it, you must somehow or other reformulate the problem in terms of an equation you can solve. There is no one universal technique that will always work, but at least we can provide a rule-of-thumb method to serve as a general guide.

Step 1: *What do you want to know?* Make a thorough list of all unknown quantities. Read the problem carefully to determine what you are trying to find out. Assign a variable to one of the unknown quantities, and express any other unknowns in terms of this variable. Later in the book you will encounter word problems that lead to equations in more than one unknown, and at that point you may want to assign a variable to each unknown quantity.

Step 2: *What are the facts you have?* Read the problem again, and carefully list all known quantities. If it will help, draw a picture or make a table to see what is happening. (For simple word problems, Step 2 is often omitted.)

Step 3: *Set up an equation.* Find an equation that expresses the unknown quantity in terms of known quantities. This process will often involve expressing some fact given to you in two different ways and setting the resulting expressions equal.

Step 4: *Solve the equation.* Once you have solved the equation, go back and have another look at the original problem. Determine any quantities that the problem asks you to determine.

**Step 5:** *Check your answer.* This is as important as any of the other steps. Be certain that your answer really satisfies the conditions of the problem. This step will help you catch any mistake you have made in setting up or solving your equation. You should also be certain that your answer is physically meaningful. For example, a negative answer for a man's height would have to be thrown out.

It may sound silly to say, but you learn to do word problems by doing word problems. For that reason, the rest of this section is devoted to some examples. The general format to expect is an example illustrating the solution of a class of word problems, followed by two sample problems for you to work on. At the end of the section, detailed solutions of the odd-numbered sample problems will be provided as well as answers for the even-numbered problems.

### Number problems

**Problem:** If 8 is added to a number, the result is one less than tripling the number. Find the number.

**Solution:** The only unknown quantity is the number. Represent it by $x$. The problem then yields the following equation that we solve for $x$:

$$x + 8 = 3x - 1$$

$$9 = 2x$$

$$\frac{9}{2} = x.$$

*Check:* Is it true that $\frac{9}{2} + 8 = 3 \cdot \frac{9}{2} - 1$? We find that $\frac{9}{2} + 8 = \frac{9}{2} + \frac{16}{2} = \frac{25}{2}$ and $3 \cdot \frac{9}{2} - 1 = \frac{27}{2} - 1 = \frac{25}{2}$; so the solution is valid.

( ∘ )    **Sample problem 1:** Find a number such that three times the number decreased by 5 equals twice the number.

**Sample problem 2:** What number, when added to 20, is seven times the number?

## Consecutive integer problems

The idea behind these problems is that consecutive integers differ by 1, and consecutive even or odd integers differ by 2. Thus consecutive integers can be represented by $x$, $x + 1$, $x + 2$,..., and so on, and consecutive even or odd integers by $x$, $x + 2$, $x + 4$,..., and so on.

**Problem:** Find three consecutive integers such that the sum of the first and third is 90.

**Solution:** Let $x$ denote the smallest of the integers. The next two are then given by $x + 1$ and $x + 2$. The problem states that the sum of $x$ and $x + 2$ is 90; so we have the equation $x + (x + 2) = 90$. Solving for $x$, we see that

$$2x + 2 = 90$$

$$2x = 88$$

$$x = 44.$$

The integers are 44, 45, and 46.

*Check:* Is $44 + 46 = 90$? Yes.

( ○ )  **Sample problem 3:** Find three consecutive even integers such that the sum of the first and twice the third equals 68.

**Sample problem 4:** Find two consecutive odd integers such that nine times the first equals seven times the second.

## Age problems

In problems of this kind you will frequently have to represent someone's age at a past or future time in terms of his or her present age. The guidelines for these problems are:

**1** To obtain a person's age $k$ years from now, add $k$ to his present age.

**2** To obtain a person's age $k$ years ago, subtract $k$ from his present age.

**Problem:** Joe is 20 years older than Josephine, and in 3 years he will be twice her age. How old is Josephine now?

**Solution:** Since it is Josephine's present age we are after, we begin by setting it equal to $x$. We then have

|  | Josephine | Joe |
|---|---|---|
| *Present age* | $x$ | $x + 20$ |
| *3 years from now* | $x + 3$ | $x + 23$ |

The problem states that $x + 23$ (Joe's age in 3 years) must equal $2(x + 3)$ (twice Josephine's age in 3 years). This produces the equation

$$x + 23 = 2(x + 3)$$

$$x + 23 = 2x + 6$$

$$17 = x;$$

so Josephine is now 17.

*Check:* If Josephine is now 17, Joe must be 37. In 3 years, their ages will be 20 and 40. Does $40 = 2(20)$? Yes.

( ○ ) **Sample problem 5:** Nine years ago, Jack was three times as old as Jill. Twelve years from now, he will be twice as old as Jill. How old is Jill now?

**Sample problem 6:** If you add the ages of George and his dog, the answer you get will be 12. In 2 years George will be three times as old as the dog. What is the dog's present age?

## Coin problems

The solution of these problems is based on the fact that the value of a certain number of coins of the same kind is the number of coins times the value of each coin:

(Total value) = (number of coins of same kind)(value per coin).

**Problem:** Here is a little riddle for you to unravel. My piggy bank has nickels, dimes, and quarters in it. Their value is $3.90. I have three times as many dimes as nickels and two fewer quarters than dimes. How many of each do I have?

**Solution:** We are given a collection of nickels, dimes, and quarters. The total value = $3.90 and the unknowns = number of coins of each kind. Let $x$ = the number of nickels. There are $3x$ dimes and $3x - 2$ quarters. The following chart may help you to set up the equation.

| Type of coin | Number of coins | Value of coins |
|---|---|---|
| Nickel | x | 5x |
| Dime | 3x | 10(3x) = 30x |
| Quarter | 3x − 2 | 25(3x − 2) = 75x − 50 |

We know that the values must add up to 390; so this leads to the equation

$$5x + 30x + (75x - 50) = 390$$

$$110x - 50 = 390$$

$$110x = 440$$

$$x = 4.$$

Thus there are four nickels, twelve dimes, and ten quarters.

*Check:* Is the value of four nickels, twelve dimes, and ten quarters $3.90? The answer is yes, since $.20 + $1.20 + $2.50 = $3.90.

(∘) **Sample problem 7:** Is it possible to change a $10 bill into half-dollars, quarters, and dimes so that there are five more quarters than half-dollars and five times as many dimes as quarters? If so, state the number of half-dollars, quarters, and nickels needed.

**Sample problem 8:** Is it possible to have $1.90 in nickels and dimes with a total of thirty coins?

(∘)
*Answers to the sample problems*

As we promised you, here are complete solutions to the odd-numbered sample problems and answers to the even-numbered problems.

**Sample problem 1:** Let x be the number we are seeking. The problem then states that 3x decreased by 5 must equal 2x. This produces the equation

$$3x - 5 = 2x$$

$$3x = 2x + 5$$

$$x = 5.$$

*Check:* Is it true that 3·5 − 5 = 2·5? Yes.

**Sample problem 2:** $\dfrac{10}{3}$

Sample problem 3: Let $x$ represent the lowest of the three even integers. The remaining ones are $x + 2$ and $x + 4$. The problem states that the sum of $x$ and twice $x + 4$ must equal 68. The equation stating this is

$$x + 2(x + 4) = 68$$
$$x + 2x + 8 = 68$$
$$3x = 60$$
$$x = 20.$$

The numbers are 20, 22, and 24.

*Check:* Does $20 + 2 \cdot 24 = 68$? Yes.

**Sample problem 4:** The numbers are 7 and 9.

Sample problem 5: We want to find Jill's present age; so let us represent it by the letter $x$. The following table will help you set the problem up.

| Time | Jill | Jack |
|------|------|------|
| Now | $x$ | |
| 9 years ago | $x - 9$ | $3(x - 9)$ |
| 12 years hence | $x + 12$ | $3(x - 9) + (9 + 12)$ |

The problem states that Jack's age 12 years from now equals twice Jill's age 12 years from now. This yields

$$3(x - 9) + (9 + 12) = 2(x + 12)$$
$$3x - 27 + 21 = 2x + 24$$
$$3x - 6 = 2x + 24$$
$$x = 30.$$

Thus Jill is now 30 years old.

*Check:* Nine years ago, Jill was 21 and Jack 63. Twelve years from now, Jill will be 42 and Jack $63 + 21 = 84$. Does $84 = 2(42)$? Yes.

**Sample problem 6:** The dog is now 2 years old.

Sample problem 7: Here are the basic facts of the problem. We are given half-dollars, quarters, and dimes: total value = $10; number of half-dollars + 5 = number of quarters; number of dimes = 5(number of quarters). Let $n$ be the number of half-dollars. To see what is happening, consider the following table.

| Type of coin | Number | Value |
|---|---|---|
| Half-dollar | $n$ | $50n$ |
| Quarter | $n + 5$ | $25(n + 5)$ |
| Dime | $5(n + 5)$ | $50(n + 5)$ |

The total value must be $10 = 1,000¢; so we have

$$50n + 25(n + 5) + 50(n + 5) = 1,000$$

$$50n + 25n + 125 + 50n + 250 = 1,000$$

$$125n + 375 = 1,000$$

$$125n = 625$$

$$n = 5.$$

There must be five half-dollars, ten quarters, and fifty dimes; so it is indeed possible to change a $10 bill in this manner.

*Check:* Does $(50)(5) + (25)(10) + (50)(10) = 1,000$? Yes.

**Sample problem 8:** Yes. It would take twenty-two nickels and eight dimes.

## Exercise set 4.4

Because we included so many sample problems in the text material, there will be no sample solutions at the end of this exercise.

### Number problems

**1** When 40 is added to a number, the result is 8 less than three times the number. What is the number?

**2** If a number is decreased by 1 and then multiplied by 6, the result is 9 more than you would get by multiplying the original number by 4. What was the original number?

**3** If John lost 10 pounds, he would weigh nine times as much as his dog. Together they weigh 200 pounds. How much does each weigh?

**4** Mary is 8 inches shorter than John. If you multiply her height by 4, the result is 40 more than three times John's height. How tall are they? *Note:* All heights are to be measured in inches.

### Consecutive integer problems

**5** Find three consecutive integers that have the property that six times the sum of the first two equals ten times the third.

**6**  Find four consecutive odd integers whose sum is 56.

### Age problems

**7**  My house is three times as old as my car. In another 3 years it will be twice as old as my car. How old is my car?

**8**  In 4 years, Mary's age will be 5 less than twice that of her daughter. Eleven years ago, she was three times as old as her daughter. How old is Mary now?

**9**  George's boss earns $10 per week more than twice George's salary. If they each receive a $20-per-week pay raise, the boss would earn $90 more per week than George. How much will George earn after his pay raise?

### Coin problems

**10**  Suppose you are given three times as many dimes as nickels, and the dimes are worth $3 more than the nickels. How much money were you given?

**11**  Is it possible to change a $10 bill into an equal number of half-dollars, quarters, and nickels? Explain.

**12**  I just bought some 8¢ and 5¢ stamps at the post office. I have sixteen more 8¢ stamps than 5¢ stamps. If the total value of the stamps is $2.97, how many of each did I buy?

# (4.5) (More word problems)

In the last section we had a good look at word problems whose solution involved some kind of a first-degree equation. Now we'll continue our discussion, using the same general approach.

## Cost and mixture problems

Coin problems are a special kind of cost problem. Here, the solution depends on observing that the cost or value of $n$ objects of a certain kind is the number of objects times the cost or value per object.

(Total value) = (number of objects)(value of each)

**Problem:** The Squirrel Nut Co. markets peanuts at 69¢ per pound and a blend of fancy mixed nuts at $1.09 per pound. The mixed nuts aren't selling too well, so the sales manager decides to try blending some peanuts in with the fancy mixed nuts to obtain a blend that can be marketed at 99¢ per pound. How many pounds of each should be put into 10 pounds of the new blend?

**Solution:** It is usually a big help to construct a table for this kind of problem. Let $x$ = the number of pounds of peanuts. We then have:

| Type | Weight | Value (in cents) |
|------|--------|------------------|
| Peanuts | $x$ | $69x$ |
| Fancy mixed nuts | $10 - x$ | $109(10 - x)$ |
| New blend | $10$ | $990$ |

The value of the new blend must be the value of what goes into it; so we are led to the equation

$$69x + 109(10 - x) = 990$$

$$69x + 1090 - 109x = 990$$

$$40x = 100$$

$$x = 2\frac{1}{2}.$$

We therefore must take 2-1/2 pounds of peanuts and 7-1/2 pounds of fancy mixed nuts.

*Check:* Two-and-a-half pounds of peanuts should sell for $1.72 plus 1/2 cent; 7-1/2 pounds of fancy mixed nuts should sell for $8.17 plus 1/2 cent. Do these two figures add up to $9.90? Yes.

( ○ )  **Sample problem 9:** Daisy's Dairy Den sells milk at 36¢ per quart and 64¢ per half-gallon. The gross receipts on a certain day were $164. Daisy knows that she sold 100 more half-gallon containers than quarts but does not know how many of each were sold. Can you help her out?

**Sample problem 10:** A movie theater charges $2.25 admission for adults and 90¢ for children. The receipts for a certain performance were $307.80, and there were twice as many adults as children attending. How many tickets of each kind were sold?

## Interest problems

The key item here is that the annual interest $I$ is the amount $P$ invested times the rate $R$ of interest; in other words, $I = PR$. Remember that unless otherwise specified, $R$ is given as the rate per year, so that $100 invested at 5% would yield $5 per year. Actually, most interest problems may be regarded as a special kind of cost problem.

**Problem:** You have $5,000 invested, part of it at 4% and part at 7%. If the yearly dividend totals $284, how much was invested at each amount?

**Solution:** Represent each amount in symbols. Let $x$ denote the amount invested at 4%. The amount at 7% is then $5,000 - x$. Summarize the known facts now in tabular form:

| Amount | Rate | Interest |
|---|---|---|
| $x$ | 4% | $0.04x$ |
| $5,000 - x$ | 7% | $0.07(5,000 - x)$ |

The total amount of interest must be the sum of $0.04x$ and $0.07(5,000 - x)$; so we have the equation

$$0.04x + 0.07(5000 - x) = 284$$

$$0.04x + 350 - 0.07x = 284$$

$$0.03x = 66$$

$$x = \frac{(100)(66)}{3}$$

$$= 2,200.$$

Thus, you have $2,200 at 4% and $2,800 at 7%.

*Check:* $(0.04)(2,200) = \$\ 88$
$(0.07)(2,800) = \$196$
Total interest $= \$284$

( $\circ$ )　**Sample problem 11:** If a man has $7,000 invested at 3%, how much more money should he invest at 7-1/2% to yield a composite rate of 4%?

**Sample problem 12:** Mr. I. M. Rich has money invested at 4% and 7%. If he has $1,200 more invested at 7% than he has at 4%, and if his yearly income from his investments is $744, how much does he have invested at each rate?

*Travel problems*

There are a number of variations on this general theme, but they are all based on the fact that if you are traveling at a constant rate of speed, then the distance traveled is the speed multiplied by the time spent traveling.

(Distance) = (rate) (time)

$d = rt$

You may, of course, want to use the formula in one of the forms $r = d/t$ or $t = d/r$. A word of caution, however. *The units of measurement must agree.* If the speed is to be measured in miles per hour, then the distance must be measured in miles and the time in hours.

**Problem:** The highway distance from Detroit to St. Louis is 515 miles. An automobile leaves Detroit heading for St. Louis at 9 A.M. A second car leaves St. Louis for Detroit at 10 A.M. If the speed of the first car is 40 miles per hour and that of the second 60 miles per hour, when do they meet, and how far will each have traveled?

**Solution:** First we list the facts that we know:

Car 1 leaves Detroit at 9 A.M.

Car 2 leaves St. Louis at 10 A.M.

Speed of Car 1 = 40 mph.

Speed of Car 2 = 60 mph.

Distance between cities = 515 miles.

At this point, a picture may help:

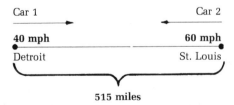

Next, we decide what we are after. We want the time the cars meet. Once this is determined, we can easily compute the distance each car travels. All right, let $t$ be the time in hours that Car 1 travels. The time for Car 2 will be $t - 1$ since it leaves an hour later. The following little table may help set the problem up:

|       | Rate | Time    | Distance    |
|-------|------|---------|-------------|
| Car 1 | 40   | $t$     | $40t$       |
| Car 2 | 60   | $t - 1$ | $60(t - 1)$ |

At the point where the two cars meet, they will have covered a total distance of 515 miles. This leads to the equation $40t + 60(t - 1) = 515$. Solving for $t$, we get

$$40t + 60t - 60 = 515$$

$$100t = 575$$

$$t = 5.75 \text{ hours.}$$

It follows that the two cars will meet 5 hours and 45 minutes after 9 A.M.; that is, at 2:45 P.M. At that time, Car 1 will have traveled $(40)(5.75) = 230$ miles, and Car 2 will have traveled $(60)(4.75) = 285$ miles.

*Check:* **Does** $285 + 230 = 515$? Yes.

$(\circ)$ **Sample problem 13:** Bob's wife takes him to the airport. She starts driving home at the time his plane leaves. She lives 30 miles from the airport, and the plane travels 240 mph faster than she drives. At the time she gets home, the plane has traveled a total of 210 miles. How fast did she drive on the way home?

**Sample problem 14:** Bill is on an all-day bicycle trip but has managed to leave his lunch at home. His father discovers this 2 hours after Bill's departure and decides to catch up with him by car. Bill's speed is about 10 mph, and his father is driving at 50 mph. How long will it take dear old dad to deliver the lunch?

## Doing-the-job problems

This kind of problem involves someone or something performing a task in a certain length of time, then getting some help. The solution is based on the fact that if it takes $x$ hours to do a job, then in 1 hour's time, the portion completed is $1/x$. Thus if you can weed a garden in 3 hours, you will have weeded one-third of it in 1 hour's time; if you can paint a barn in 4 days, you paint one-quarter of it in 1 day, and so on.

**Problem:** Mr. Hand E. Mann can paint his house in 56 hours. His son, working alone, can do the same job in 42 hours. Mr. Mann starts painting. He works for 7 hours by himself on the first day. Starting on the second day, his son works with him until they complete the job. How long does the son work?

**Solution:** Let $t$ denote the number of hours worked by the son. Mr. Mann then works $t + 7$ hours. In each hour, the son paints $1/42$ of the house and the father $1/56$ of the house. We therefore have

| Portion painted by son | | Portion painted by Mr. Mann | | Portion painted by both |
|---|---|---|---|---|
| $\dfrac{t}{42}$ | $+$ | $\dfrac{t + 7}{56}$ | $=$ | $1$ |

Solving this equation, we have

$$56t + 42(t + 7) = 42 \cdot 56$$

$$56t + 42t + 294 = 2{,}352$$

$$98t = 2{,}058$$

$$t = 21 \text{ hours.}$$

Thus we have established that the son worked a total of 21 hours.

*Check:* If the son worked 21 hours, his father worked $21 + 7 = 28$ hours. In 21 hours the son paints $21/42 = 1/2$ of the house, and in 28 hours his father paints $28/56 = 1/2$; so together they paint the whole house.

( ∘ )    **Sample problem 15:** If a lawn can be mowed in 3 hours using a power mower and 7 hours using a hand mower, how long would it take using both?

**Sample problem 16:** A firm's computer can do the weekly payroll in 6 hours' time. A second computer is added; working together, the two machines can do the payroll in 2 hours' time. How long would it take the second machine by itself?

*Answers to the sample problems*

**Sample problem 9:** Let $x$ = number of quarts sold. There were $x + 100$ half-gallons sold.

| Type | Number | Value |
| --- | --- | --- |
| Quarts | $x$ | $.36x$ |
| Half-gallons | $x + 100$ | $.64(x + 100)$ |
| Total | | 164 |

This gives us the equation

$$.36x + .64(x + 100) = 164$$

$$.36x + .64x + 64 = 164$$

$$x = 100.$$

Thus, there were 100 quarts and 200 half-gallons sold.

*Check:*

| | |
| --- | --- |
| 100 quarts at 36¢ yields | $ 36 |
| 200 half-gallons at 64¢ yields | $128 |
| | $164 |

**Sample problem 10:** There were 57 children's tickets and 114 adult's tickets sold.

**Sample problem 11:** We want to determine the extra amount to be invested at $7\frac{1}{2}$%; so let $x$ denote this amount in dollars. The total amount invested is then $x + 7,000$. As usual, we construct a little table to aid the setup of the problem:

| Amount | Rate | Income |
| --- | --- | --- |
| $7,000 | 3% | $210 |
| $x$ | $7\frac{1}{2}$% | $.075x$ |
| $x + 7,000$ | 4% | $.04(x + 7,000)$ |

The table reflects the fact that we want the composite rate on the total investment to be 4%. The total income of $.04(x + 7,000)$ must equal the sum of the portions which make it up; so we have the equation

$$.04(x + 7,000) = .075x + 210$$

$$.04x + 280 = .075x + 210$$

$$.035x = 70$$

$$x = \frac{70}{.035} = 2,000.$$

This says that \$2,000 must be invested at $7\frac{1}{2}\%$.

*Check:*

$2,000$ at $7\frac{1}{2}\%$ = \$150

$7,000$ at $3\%$ = \$210

$9,000$ at $4\%$ = \$360 = \$150 + \$210

**Sample problem 12:** Mr. Rich has \$6,000 invested at 4% and \$7,200 at 7%.

Sample problem 13: The following sketch will help you visualize the situation:

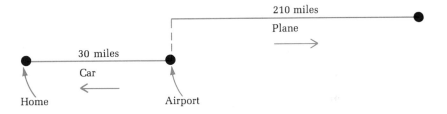

The plane travels 210 miles in the same time as the car travels 30 miles. Let $r =$ speed of car; so the plane's speed $= r + 240$.

$$\text{Time for car} = \frac{30}{r}$$

$$\text{Time for plane} = \frac{210}{r + 240}$$

These times must be equal; so our equation is

$$\frac{30}{r} = \frac{210}{r + 240}$$

$$210r = 30(r + 240)$$

$$210r = 30r + 7{,}200$$

$$180r = 7{,}200$$

$$r = \frac{7200}{180} = 40;$$

so her speed was 40 mph.

*Check:* At 40 mph it will take her three-quarters of an hour to get home. In that length of time the plane will travel $\frac{3}{4}(40 + 240)$ miles. Does $\frac{3}{4}(280)$ = 210? Yes, it does.

**Sample problem 14:** 30 minutes

**Sample problem 15:** In each hour's time, one-third of the lawn is mowed by the power mower and one-seventh by the hand mower. If it takes $t$ hours to mow the lawn using both mowers, then

$$\frac{1}{3}t + \frac{1}{7}t = 1$$

$$7t + 3t = 21$$

$$10t = 21$$

$$t = 2.1 = 2 \text{ hours and 6 minutes.}$$

**Sample problem 16:** It would take the second machine 3 hours.

# Exercise set 4.5

Once again, there will be no sample solutions worked out at the end of the exercise.

## Cost and mixture problems

**1** A man went to a nursery and purchased a total of twenty honeysuckle plants for a hedge he was planting. Some of the plants were 3-year-old plants that were sold at $3.95 each, and some were 4-year-old plants priced at $5.25 each. If the total bill was $89.40, how many of each type did he buy?

**2**  A certain gas station sells gasoline at two kinds of pumps. The first kind has an attendent, and here the gas sells for 63.9¢ per gallon. There is also a self-service pump, at which gas is priced at 61.9¢ per gallon. One day last week the station sold 6,000 gallons of gas and took in a total of $3,790. How much money was taken in at the self-service pump?

**3**  The toll on a certain bridge is 25¢ for a truck and 10¢ for a car. If there are five times more cars than trucks passing over the bridge and a total of $46.50 in tolls is taken in, how many vehicles passed over the bridge?

### Interest problems

**4**  Mr. U. R. Poore has $3,000 invested at 4% and $4,000 at 5%. How much should he invest at $7 \frac{1}{4}$% to get a composite return of 6% on his entire investment?

**5**  Suppose a friend tells you that he invested a total of $6,000. Part of the investment brought a return of 4% and part a loss of $1 \frac{1}{2}$% last year. If he made a total of $163, how much did he have invested in the losing venture?

**6**  A man has $1,750 invested, part of it at 3% and part at 4%. If the annual return on both investments is the same, how much does he have invested at each rate?

### Travel problems

**7**  If I drive to Boston at a rate of 45 mph, the trip takes 30 minutes longer than if I drive at 50 mph. How far do I live from Boston?

**8**  At 9 A.M. an eastbound train leaves Chicago. At 10 A.M. a westbound train leaves, traveling 10 mph faster than the eastbound train. At 2 P.M. they are 445 miles apart. What is the average speed of the eastbound train?

**9**  Harry the hiker starts out at 7 A.M. for an all-day hike. His brother Bob the biker decides at 10:30 A.M. to try to catch up with him by bicycle. If Harry averages 4 mph and Bob 14 mph, at what time will Bob catch up, and how far will he have traveled to reach the meeting point?

### Doing-the-job problems

**10**  During a recent rainstorm the basement of the post office was flooded. The postmaster had a pump that would get rid of the water in 8 hours' time. He sent an employee out to borrow a second pump that, working by itself, would clear the water in 6 hours' time. The first pump was started, and after 2 hours the second pump arrived on the scene. The job was completed using both pumps. How long did it take to clear the basement of water?

**11**  With a certain hose it takes 7 hours to fill a swimming pool with water. A larger hose would take only 5 hours. How long would it take using the two hoses together?

**12**  A pipe takes $22\frac{1}{2}$ hours to fill a tank with water. The process is started, but after 10 hours' time a drainpipe is accidentally opened. It takes another 20 hours to fill the tank. Given a full tank, how long would it take for the drainpipe to empty the tank? (Assume a constant rate of flow.)

# (4.6) (Inequalities)

*inequality*

An equation is an assertion that two expressions (called members) are equal. Well, an *inequality* is a statement to the effect that one expression is "greater than" or "greater than or equal to" a second expression. The expressions are still called *members* of the inequality. Typical inequalities are

$$x + 2 \;<\; x + 3,$$

$$x^2 + 1 \;>\; 0,$$

$$3x - 5 \;<\; 7,$$

$$7x + 6 \;\geq\; 9x - 5.$$

*absolute*

*conditional*

*same sense*

*opposite sense*

*solution*

Your task in an inequality is to find all possible values of the variables that will make the inequality true. Sometimes, as in the first two examples, the inequality is an *absolute inequality* in the sense that it always holds. Other inequalities are sometimes true and sometimes false. These are called *conditional inequalities* and are illustrated in the third and fourth examples that appear above.
  Inequalities like $1 < 2$ and $5 < 7$ are said to be of the *same sense* because in each case the assertion is that the left-hand member is less than the right-hand member. On the other hand, the inequalities $1 < 4$ and $9 > 7$ are of the *opposite sense* in that the first one states that the left-hand member is less than the right-hand member, whereas the second one makes the opposite assertion.
  Suppose we are given an inequality whose members involve a single variable $x$. As in the case of an equation, a *solution* of the

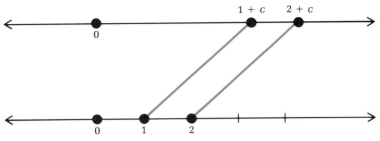

Fig. 4.1

inequality is a value of $x$ that makes its assertion true; the *solution set* is the set of all possible solutions, and to *solve an inequality* is to find the members of its solution set. Two such inequalities are called *equivalent* if they have the same solution set. Our basic method of solving an inequality will be to replace it with an equivalent inequality whose solution set can be determined. Some basic properties of inequalities will, of course, be needed.

Let 3 be added to the inequality $1 < 2$. The resulting inequality states that $4 < 5$. Similarly, addition of $-3$ produces the inequality $-2 < -1$. In fact, addition of *any* number $c$ shows that $1 + c < 2 + c$. The reason for this is that the graph of 2 lies one unit to the right of the graph of 1. Addition of $c$ to both 1 and 2 shifts their graphs by the same amount and in the same direction. The situation is illustrated in Figure 4.1. The same sort of reasoning shows that if $a$ and $b$ are any real numbers for which $a < b$, then $a + c < b + c$. Since any variable in an expression represents a real number, we can make the following general statement:

( ∘ )    I.1   *If an expression is added to both members of an inequality, the resulting inequality (taken in the same sense) is equivalent to the original one.*

Thus addition of $-3x$ to both members of the inequality $4x < 3x + 1$ produces the equivalent inequality $x < 1$ whose solution set is $\{x \mid x < 1\}$.

Let us now multiply both members of the inequality $1 < 2$ by the positive number $c$. The graph of $2c$ lies $c$ units to the right of the graph of $c$; so the resulting inequality is $c < 2c$. Similar reasoning shows that if $a$ and $b$ are any real numbers then $a < b$

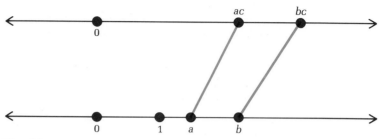

Fig. 4.2

implies $ac < bc$ for any positive number $c$. (See Figure 4.2.) This leads to another generalization:

( ○ )   I.2  *If both members of an inquality are multiplied by an expression that is always positive, the resulting inequality (taken in the same sense) is equivalent to the original one.*

For example, multiplication of $2x < 6$ by $1/2$ shows that $x < 3$; so its solution set is $\{x \mid x < 3\}$.

If both members of $1 < 2$ are multiplied by $-3$, the resulting inequality states that $-3 > -6$. Indeed, if $c > 0$, then multiplication of $1 < 2$ by $-c$ can be thought of as a two-step process:

**1** Multiplication by $c$ to produce $c < 2c$.

**2** Multiplication of $c < 2c$ by $-1$. This second step reflects the graphs of $c$ and $2c$ through the origin, hence puts the graph of $-2c$ to the *left* of the graph of $-c$

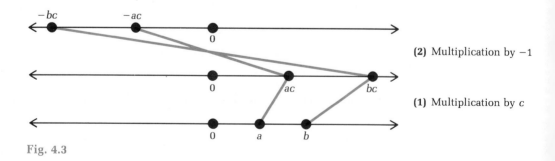

**(2)** Multiplication by $-1$

**(1)** Multiplication by $c$

Fig. 4.3

Thus $1 < 2$ implies $-c > -2c$ for $c > 0$. In fact, if $a < b$ we can argue that $-ac > -bc$ for any positive number $c$ (see Figure 4.3). This leads to the following observation:

$(\circ)$    I.3  *If both members of an inequality are multiplied by an expression that is always negative, the resulting inequality (taken in the opposite sense) is equivalent to the original one.*

We illustrate this by observing that $-3x < 9$ is equivalent to $x > -3$, whose solution set is $\{x \mid x > -3\}$.

If $P$, $Q$, and $R$ each represent algebraic expressions, we may state what we have just learned in symbolic form.

$(\circ)$    I.1   $P < Q$ is equivalent to $P + R < Q + R$.

$(\circ)$    I.2   $P < Q$ is equivalent to $P \cdot R < Q \cdot R$ provided $R$ is always positive.

$(\circ)$    I.3   $P < Q$ is equivalent to $P \cdot R > Q \cdot R$ provided $R$ is always negative.

The foregoing statements remain true even if (1) we replace $<$ by $\leq$ and $>$ by $\geq$; (2) we replace $<$ by $>$ and $>$ by $<$; or (3) we replace $<$ by $\geq$ and $>$ by $\leq$.

These rules are what you will need to solve inequalities. We illustrate their use by considering the solution of $2x + 3 < -7x + 9$. Adding $-3$ to both members, we establish by I.1 that $2x < -7x + 6$. We now add $7x$ to both sides and obtain $9x < 6$. Multiplying through by $1/9$, we use I.2 to write $x < 2/3$. The solution set is $\left\{x \mid x < \dfrac{2}{3}\right\}$

and its graph is

## Exercise set 4.6

Solve each inequality and plot the graph of its solution set.

$(\circ)$    1   $6x - 9 < 0$          2   $5x + 3 \geq 0$

       3   $6y + 8 > 2$          4   $3w - 10 < 13$

       5   $-5a + 3 \leq 18$       6   $-3d - 1 \leq -7$

**7** $3x - 5 < 7x + 6$                    **8** $3z + 2 \geq 8z - 5$

(○) **9** $5 - (4a - 6) > (2a - 3) - (8a + 2)$

**10** $(3b + 1) - (4b + 2) \leq (4b - 2) + (6 - 3b)$

(○) **11** $(x - 2)(x - 3) < (x + 4)^2$

**12** $(y + 1)^2 > (y + 2)^2$

**13** $(w + 1)(w - 2) \leq (w + 3)(w - 1)$

**14** $(k + 2)(k - 3) \geq (k - 2)(k - 4)$

You will sometimes be given an inequality in the form

$$6 < 5 - 2x \leq 8.$$

This really amounts to solving the pair of inequalities $6 < 5 - 2x$ and $5 - 2x \leq 8$, but the problem can be treated as a single inequality using the methods of this section. Let us illustrate the method by solving the given inequality. We know that $6 < 5 - 2x \leq 8$. Adding $-5$ to all three members produces the equivalent inequality $1 < -2x \leq 3$. Multiply through by $-\dfrac{1}{2}$ to obtain $-\dfrac{1}{2} > x \geq -\dfrac{3}{2}$; so the solution

set is $\left[-\dfrac{3}{2}, -\dfrac{1}{2}\right) = \left\{x \ \middle| \ -\dfrac{3}{2} \leq x < -\dfrac{1}{2}, x \in R\right\}$. Its graph is

*Use this technique for the next group of problems.*

(○) **15** $2 < 5x + 3 < 6$                    **16** $-1 < 3x - 4 < 8$

**17** $-6 \geq 2(x + 6) \geq -8$              **18** $-4 > 3y + 5 > -10$

**19** $3 \leq 6 - 2x < 6$                    **20** $4 < 3 + 5x \leq 7$

**21** $-3 < 2t + 5 < 3$                      **22** $-4 < 3 - \dfrac{1}{2}u < 4$

**23** $2z + 1 \leq 5z - 3 < 2z + 3$

**24** $6w + 2 > 4w + 1 > 6w - 4$

Many word problems also lead to inequalities. Consider the following example.

**Problem:** I have twenty coins. Some of them are quarters and some are dimes. I don't remember exactly how much money I have, but I recall that it is at least \$3.25 and no more than \$3.65. What are the possibilities?

**Solution:** Let $n$ denote the number of quarters. I then have $20 - n$ dimes. The value of the quarters is $25n$ and that of the dimes $10(20 - n)$. The problem states that

$$325 \leq 25n + 10(20 - n) \leq 365$$

$$325 \leq 25n + 200 - 10n \leq 365$$

$$325 \leq 15n + 200 \leq 365$$

$$125 \leq 15n \leq 165$$

$$8\frac{1}{3} \leq n \leq 11.$$

Since $n$ must be an integer, I must have $n = 9$, 10, or 11. Therefore, I have either nine quarters and eleven dimes with a value of \$3.35, ten of each with a value of \$3.50, or eleven quarters and nine dimes for \$3.65.

( ○ ) **25** A man has \$3,000 invested at 2%. How much should he invest at 8% to yield a composite return of at least 4% but not more than 5%?

**26** A cheap grade of coffee sells for 89¢ per pound and an expensive grade at \$1.29. How many pounds of each may be used to make up 10 pounds of a mixture to sell at a price between 99¢ and \$1.05 per pound?

**27** Bob the biker has talked his brother into driving him out in the country so that he can bicycle back home. They leave at 8 A.M., traveling at 50 mph. How far can his brother take him so that he can be back home by 5 P.M. if he averages 10 mph on his bicycle?

**28** Hank is picking up a few odd jobs during 5 weeks of his summer vacation. He wants to average at least \$65 per week but doesn't want to work so hard that he will earn more than \$80 per week. The first 4 weeks he earned \$62, \$78, \$93, and \$51. How much can he earn the fifth week and still meet his goal?

**( ○ )**
*Sample solutions*
*for exercise set 4.6*

**1** $6x - 9 < 0$

$$6x < 9$$

$$x < \frac{3}{2}$$

The solution set is $\left\{ x \mid x < \dfrac{3}{2} \right\}$, and its graph is

**9**   Here we must simplify both members of the inequality before doing anything else. Accordingly, we have

$$5 - (4a - 6) > (2a - 3) - (8a + 2)$$

$$5 - 4a + 6 > 2a - 3 - 8a - 2$$

$$11 - 4a > -6a - 5$$

$$-4a > -6a - 16$$

$$2a > -16$$

$$a > -8.$$

The solution set is $\{a \mid a > -8\}$, and its graph is

**11**   The first step again is to simplify $(x - 2)(x - 3) < (x + 4)^2$; so we have $x^2 - 5x + 6 < x^2 + 8x + 16$. Subtracting $x^2$ from both sides produces the equivalent inequality $-5x + 6 < 8x + 16$, which we solve by first subtracting $6$ from both members and then adding $-8x$ to both members of the result:

$$-5x < 8x + 10$$

$$-13x < 10$$

$$x > -\frac{10}{13} \qquad \text{(dividing by} - 13 \text{)}.$$

The solution set is $\left\{ x \;\middle|\; x > -\dfrac{10}{13} \right\}$, with graph as follows:

**15**   $2 < 5x + 3 < 6$

$$-1 < 5x < 3$$

$$-\frac{1}{5} < x < \frac{3}{5};$$

so the solution set is $\left\{-\dfrac{1}{5}, \dfrac{3}{5}\right\}$. Its graph is

**25** Let $x$ denote the amount to be invested at 8%. The total investment is then $3,000 + x$.

| Type | Amount | Income |
|------|--------|--------|
| 2% | $3,000 | 60 |
| 8% | $x$ | $.08x$ |
| Composite | $x + 3,000$ | $.08x + 60$ |

At 4% the composite income would be $.04\,(x + 3,000)$ and at 5%, $.05(x + 3,000)$. We therefore want

$$.04x(x + 3,000) \le .08x + 60 \le .05(x + 3,000)$$

$$.04x + 120 \le .08x + 60 \le .05x + 150$$

$$.04x + 60 \le .08x \le .05x + 90.$$

Now $.04x + 60 \le .08x$ forces $60 \le .04x$ and $1,500 \le x$, whereas $.08x \le .05x + 90$ produces $.03x \le 90$ and $x \le 3,000$. The answer is therefore $1,500 \le x \le 3,000$; so he may invest any amount between $1,500 and $3,000 to achieve his goal.

# (4.7) (Absolute value)

*absolute value*

Back in section 1.3 we defined the *absolute value* of a number $a$ by the rule

$$|a| = \begin{cases} a & \text{if } a \ge 0 \\ -a & \text{if } a < 0 \end{cases}$$

and noted that $|a|$ represented the distance of the graph of $a$ from the origin on a number line. The distance between 1 and 3 on a number line is $3 - 1 = 2$ units. The distance between $-3$ and 2 is $2 - (-3) = 2 + 3 = 5$. In general, the distance between $a$ and $b$ is $b - a$ if $b \ge a$ and $a - b$ if $a > b$.

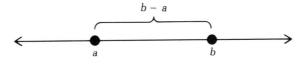

But

$$|b - a| = \begin{cases} b - a \text{ if } b - a \geq 0 & \text{(that is, if } b \geq a) \\ a - b \text{ if } b - a < 0 & \text{(that is, if } a > b); \end{cases}$$

so the distance between $a$ and $b$ is also given by $|b - a|$. You will frequently be called upon to find numbers whose distance from a given number satisfies some specific property. This sort of consideration yields equations and inequalities that involve absolute value. In this section we shall deal with some of the more common types.

Consider the equation $|x - 3| = 5$. To say that $|c| = 5$ is to say that $c = 5$ or $c = -5$; so a number is a solution of the given equation if and only if it is a solution of one of the equations $x - 3 = 5$ or $x - 3 = -5$. This says that $x = 8$ or $x = -2$; so the solution set of $|x - 3| = 5$ is $\{8\} \cup \{-2\} = \{-2, 8\}$. We took the *union* of the two solution sets because we were interested in solutions of one equation or the other. The graph of the solution set is

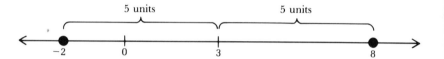

Notice that $-2$ and $8$ are the only numbers that are five units distant from 3. In general, the solution set of the equation $|x - a| = b$ ($b \geq 0$) is the union of the solution sets of the equations $x - a = b$ and $x - a = -b$.

Next, consider the inequality $|x + 2| < 4$. Let's begin by observing that $|c| < 4$ is equivalent to saying that $c$ is strictly between $-4$ and $+4$, or that $-4 < c < 4$. Thus, $|x + 2| < 4$ is equivalent to the statement that $-4 < x + 2 < 4$. Adding $-2$ to all members of this inequality produces $-6 < x < 2$; so the solution set is $(-6, 2) = \{x \mid -6 < x < 2, x \in R\}$. A picture may help you visualize the solution set.

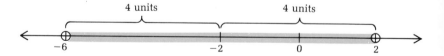

The solution set consists of all those real numbers whose distance from $-2$ is less than four units. More generally, the inequality

$|x - a| < b\,(b > 0)$ is equivalent to the inequality $-b < x - a < b$.

Finally we consider $|x - 4| > 2$. Noting that $|c| > 2$ holds only if $c > 2$ or $c < -2$, we see that a number is a solution of the given inequality if and only if it is a solution of one of the inequalities $x - 4 > 2$ or $x - 4 < -2$. The solution set then is $\{x \mid x > 6\} \cup \{x \mid x < 2\}$. The union of the two solution sets was taken because we wanted solutions of one inequality or the other. Notice that the solution set consists precisely of those real numbers whose distance from 4 is greater than two, as indicated in the following graph:

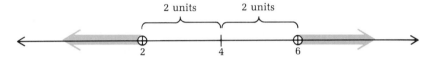

In a similar manner one sees that the solution set of the inequality $|x - a| > b\,(b > 0)$ is the union of the solution sets of the inequalities $x - a > b, x - a < -b$.

We summarize the situation as follows:

| | Form that involves absolute value | Equivalent form that does not involve absolute value | Solution set |
|---|---|---|---|
| AV.1 | $|x - a| = b$ $(b \geq 0)$ | $x - a = b$ or $x - a = -b$ | $\{a - b,\ a + b\}$ |
| AV.2 | $|x - a| < b$ $(b > 0)$ | $-b < x - a < b$ | $\{x \mid a - b < x < a + b\}$ |
| AV.3 | $|x - a| > b$ $(b > 0)$ | $x - a > b$ or $x - a < b$ | $\{x \mid x > a + b\} \cup \{x \mid x < a - b\}$ |

The graph of the respective solution sets would look like this:

AV.1    $a - b$    $a$    $a + b$    $|x - a| = b$

AV.2    $a - b$    $a$    $a + b$    $|x - a| < b$

AV.3    $a - b$    $a$    $a + b$    $|x - a| > b$

Inequalities involving the symbols $\leq$ or $\geq$ are dealt with in a similar manner. For example, it is easy to see that for $b > 0$, $|x - a| \leq b$ is equivalent to the inequality $-b \leq x - a \leq b$; and the solution set of $|x - a| \geq b$ is the union of the solution sets of $x - a \geq b$, $x - a \geq -b$.

## Exercise set 4.7

*Solve each equation or inequality and then plot the graph of its solution set.*

$(\circ)$  **1** $|x + 1| = 1$ 　　　　　　　　　　　**2** $|x + 3| = 2$

**3** $|2x - 3| = 4$ 　　　　　　　　　　　**4** $|3x + 5| = 4$

**5** $|5 - 2y| = 6$ 　　　　　　　　　　　**6** $|3 - 2w| = 4$

**7** $\left| a + \dfrac{1}{2} \right| = \dfrac{3}{2}$ 　　　　　　　　　**8** $\left| z + \dfrac{5}{3} \right| = \dfrac{1}{3}$

**9** $\left| \dfrac{1}{3} - \dfrac{a}{2} \right| = \dfrac{2}{3}$ 　　　　　　　**10** $\left| -\dfrac{x}{3} - \dfrac{1}{4} \right| = \dfrac{5}{4}$

**11** $|x| < 3$ 　　　　　　　　　　　　　**12** $|2w| < 6$

$(\circ)$  **13** $|x - 2| < 3$ 　　　　　　　　　　**14** $|z + 4| < 1$

**15** $|2a + 3| \leq 9$ 　　　　　　　　　**16** $\left| \dfrac{a}{2} - 3 \right| \leq 2$

**17** $\left| \dfrac{x - 5}{-3} \right| \leq 7$ 　　　　　　　　**18** $\left| -\dfrac{x}{2} - \dfrac{5}{3} \right| \leq \dfrac{1}{3}$

**19** $|x| \geq 2$ 　　　　　　　　　　　　**20** $|x| > 4$

$(\circ)$  **21** $|x - 2| > 1$ 　　　　　　　　　　**22** $|2z + 3| \geq 2$

**23** $|2v + 5| > 4$ 　　　　　　　　　**24** $\left| \dfrac{u}{3} + 1 \right| > 1$

**25** $|3 - 2s| \geq 2$ 　　　　　　　　　**26** $\left| 5 - \dfrac{p}{2} \right| \geq 3$

**27** $|x + 3| = 0$ 　　　　　　　　　　**28** $\left| y - \dfrac{3}{2} \right| = 0$

**29** $|2w - 5| \leq 0$ 　　　　　　　　　**30** $|4z + 2| \leq 0$

*Sample solutions*
*for exercise set 4.7*

**1** If $|x + 1| = 1$, then $x + 1 = 1$ or $x + 1 = -1$. If $x + 1 = 1$, then $x = 0$, and if $x + 1 = -1$, we have $x = -2$. The solution set is $\{-2, 0\}$, and its graph is

**13** To say that $|x - 2| < 3$ is equivalent to saying that $-3 < x - 2 < 3$ and $-1 < x < 5$; so the solution set is the open interval $(-1, 5)$, and its graph is

**21** $|x - 2| > 1$ is equivalent to the pair of inequalities $x - 2 > 1$ or $x - 2 < -1$. The solution set is $\{x \mid x > 3\} \cup \{x \mid x < 1\}$ and its graph is

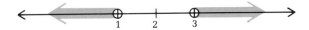

# (5)

## Exponents and radicals

### (5.1) (Integer exponents)

Back in Chapter 2 we defined $b^n =$ (the product of) $n$ factors of $b$ for any real number $b$ and any natural number $n$. We also developed a number of rules (called laws of exponents) to help deal with expressions of this type. Here they are again. Let $a$ and $b \in R$ and $m$ and $n \in N$. Then

$(\circ)$   E.1   $b^m \cdot b^n = b^{m+n}$

$(\circ)$   E.2   $(b^m)^n = b^{mn}$

$(\circ)$   E.3   $(ab)^n = a^n b^n$

$(\circ)$   E.4   $\dfrac{b^m}{b^n} = b^{m-n}$     $(b \neq 0,\ m > n)$

You should know what these laws say. If you feel shaky about exponents, this would be a good time to review Section 2.1.

All right, we know what $b^n$ means for $n$ a positive integer. Our present goal is to give a meaning to $b^n$ for $b \neq 0$ and $n$ an *arbitrary* integer. Hopefully, things can be fixed up so that the laws of exponents will still work, and this provides the clue as to what we must do. Let's start with $n = 0$. If law E.1 is to hold, we must have

$$b^m \cdot b^0 = b^{m+0} = b^m = b^m \cdot 1.$$

But this tells us that $b^0$ must equal 1. Thus, $(1/2)^0 = 1$, $(23)^0 = 1$, $(-631)^0 = 1,\ldots$, and so on. In fact, we have

$$b^0 = 1 \qquad (b \neq 0).$$

What about $b^{-n}$ for $n \in N$? Again, law E.1 tells us what to do. We want

$$b^n \cdot b^{-n} = b^{n-n} = b^0 = 1;$$

so we must take $b^{-n} = 1/b^n$. Thus $2^{-3} = 1/2^3 = 1/8$, $3^{-4} = 1/3^4 = 1/81$, $x^{-6} = 1/x^6,\ldots$, and so on. Therefore, we define

$$b^{-n} = \frac{1}{b^n} \qquad (b \neq 0, n \in N).$$

At this point we have defined $b^n$ for all $b \neq 0$ and all $n \in I$. It turns out that the laws of exponents still hold, and law E.4 even works in the following strengthened form:

---

$(\circ)$  E.4  $\dfrac{b^m}{b^n} = b^{m-n} \qquad (b \neq 0; m, n \in I).$

---

Here are a few examples to illustrate all of this.

**(E.1)** $3^3 \cdot 3^{-4} = 3^{3-4} = 3^{-1} = \dfrac{1}{3}$

**(E.2)** $[(-2)^{-3}]^2 = (-2)^{-6} = \dfrac{1}{(-2)^6} = \dfrac{1}{64}$

**(E.3)** $(3y)^4 = 3^4 y^4 = 81y^4$

**(E.4)** $\dfrac{x^3}{x^{-4}} = x^{3-(-4)} = x^{3+4} = x^7 \qquad (x \neq 0)$

**(E.4)** $\dfrac{2^{-2}}{2^3} = 2^{-2-3} = 2^{-5} = \dfrac{1}{2^5} = \dfrac{1}{32}$

Here is a final item for you to consider. If $P(x) = ax^m$ and $R(x) = bx^n$ $(b \neq 0)$, then law E.4 states that

$$\frac{P(x)}{R(x)} = \frac{a}{b}x^{n-m} \qquad (x \neq 0).$$

This is the same result predicted by the fundamental principle of fractions. Thus, by the fundamental principle,

$$\frac{4x^4}{2x^6} = \frac{2 \cdot 2x^4}{x^2 \cdot 2x^4} = \frac{2}{x^2} \qquad (x \neq 0).$$

By law E.4,

$$\frac{4x^4}{2x^6} = \frac{4}{2}x^{4-6} = 2x^{-2} = \frac{2}{x^2} \qquad (x \neq 0).$$

## Exercise set 5.1

*Express each of the following as an integer or a fraction in lowest terms.*

$(\circ)$   **1**   $5^7 \cdot 5^{-4}$      **2**   $3^5 \cdot 3^{-4}$      **3**   $(2^2)^{-3}$

**4**   $(3^3)^{-1}$      **5**   $\dfrac{4^2}{4^{-1}}$      **6**   $\dfrac{3^2}{3^{-2}}$

**7**   $\dfrac{2^{-2}}{2^{-1}}$      **8**   $\dfrac{3^{-3}}{3^{-2}}$      **9**   $\dfrac{2^{-1}}{2}$

$(\circ)$   **10**   $\dfrac{3^{-3}}{3^2}$      **11**   $4^3 \cdot 16^{-2}$      **12**   $25^{-2} \cdot 5^4$

**13**   $2 \cdot 4^{-2}$      **14**   $3 \cdot 15^{-2}$      **15**   $\dfrac{2^2 \cdot 3^3 \cdot 6^{-1}}{6^2}$

**16**   $\dfrac{2 \cdot 6^{-1}}{4^{-2} \cdot 3^2}$      **17**   $(2^2 + 3^2) \cdot 5^{-2}$

**18**   $(3 + 3^3)(1 + 3^2)^{-1}$      **19**   $\dfrac{2^{-3} \cdot 5^{-2}}{10^{-1} \cdot 8}$

$(\circ)$   **20**   $\dfrac{3^{-4} \cdot 4^{-3} \cdot 2}{2^3 \cdot 3^{-2}}$      **21**   $\left(\dfrac{3^2 \cdot 3^3 \cdot 9^{-2}}{4^2 \cdot 3^{-1} \cdot 7^0}\right)^{-2}$

**22**   $\left(\dfrac{4^{-3} \cdot 6}{2^{-2} \cdot 4 \cdot 5^0}\right)^{-1}$

The expression $\dfrac{6^{-3} \cdot 9^2}{12^{-2} \cdot 4^3}$ can be written as a product of powers of primes as follows:

$$\frac{6^{-3} \cdot 9^2}{12^{-2} \cdot 4^3} = \frac{(2 \cdot 3)^{-3} \cdot (3^2)^2}{(2^2 \cdot 3)^{-2} \cdot (2^2)^3} = \frac{2^{-3} \cdot 3^{-3} \cdot 3^4}{2^{-4} \cdot 3^{-2} \cdot 2^6}$$

$$= 2^{-3} \cdot 3^{-3} \cdot 3^4 \cdot 2^4 \cdot 3^2 \cdot 2^{-6}$$

$$= 2^{-3+4-6} \cdot 3^{-3+4+2} = 2^{-5} \cdot 3^3.$$

*Now, express each of the following as a product of positive or negative powers of distinct primes.*

$(\circ)$  **23**  $\dfrac{6^{-2} \cdot 10^2}{5^{-2} \cdot 12^3}$

**24**  $\dfrac{14^{-1} \cdot 42}{18^{-2} \cdot 9^2}$

**25**  $10^{-2} \cdot 15^2 \cdot 6^{-3} \cdot 8^2$

**26**  $100^{-1} \cdot 20^2 \cdot 5^{-1} \cdot 4^3$

**27**  $\dfrac{24^{-2} \cdot 16^{-3}}{4^{-4} \cdot 12^2} \div \dfrac{36^3 \cdot 60^{-1}}{8^{-3} \cdot 48^2}$

**28**  $\dfrac{64^{-3} \cdot 144^{-2}}{12^{-2} \cdot 18^2} \div \dfrac{72^4}{(-6)^3}$

*Use negative exponents to express each of the following so that the denominator is 1 and so that each variable appears only once.*

$(\circ)$  **29**  $\dfrac{2x^2 yz^3}{4xyz^2}$

**30**  $\dfrac{x^2 y}{x^2 y^2}$

**31**  $\dfrac{6a^2 b}{18bc^2}$

**32**  $\dfrac{14h^2 k^0}{21h^5 k^{-1}}$

**33**  $\dfrac{10s^2 tu^3}{(5su)(30su^2)}$

**34**  $\dfrac{1}{5w^2 x^{-2} y^{-3}}$

**35**  $\left(\dfrac{25xy^3}{5x^3 yz^0}\right)^{-1}$

**36**  $\left(\dfrac{a^{-4} \cdot b^2 \cdot c^{-3}}{a^3 (bc)^{-2}}\right)^{-3}$

*Express the following without negative or zero coefficients in a form in which each variable appears only once.*

$(\circ)$  **37**  $a^2 b^{-3} c^0$

**38**  $a^{-1} b^{-2} c^3$

**39**  $\dfrac{x^2 y}{x^{-1} z^2}$

**40**  $\dfrac{x^{-4} y^2 z^{-6}}{(xy)^{-2} (yz)^{-3}}$

**41**  $\left(\dfrac{(gh)^{-2} (hk)^{-3}}{(gk)^3 (hk)^{-1}}\right)^{-2}$

**42**  $\left(\dfrac{(stu)^2 (tuv)^{-3} (uvw)^{-4}}{(su)^{-1} (tv)^2 (uv)^{-3}}\right)^{-1}$

*Express each of the following as an integer or fraction times a product of positive or negative powers of the variables that are involved, with no variable appearing twice and no variable in the denominator. Here's an example to show you how it's done.*

$$\frac{2x^{-1}yz^4}{6xy^2(yz)^{-1}} = \frac{1}{3}x^{-1}yz^4 x^{-1}y^{-2}(yz) = \frac{1}{3}x^{-2}y^{1-2+1}z^{4-1}$$

$$= \frac{1}{3}x^{-2}z^3 \qquad (x,\ y,\ z \neq 0).$$

(○) 43 $\quad \dfrac{16x^2\, yz^{-2}}{12xy^{-1}z}$

44 $\quad \dfrac{(5x)^2(-y)^{-2}(2z)^{-1}}{20(xyz)^{-2}}$

45 $\quad \dfrac{18xy^0\, z^2}{9x^{-1}\, yz^0}$

46 $\quad \dfrac{10abc^2}{5a^2 b^0 c^{-1}}$

47 $\quad \dfrac{15gh^2}{25h^{-1}k} \div \dfrac{35h^2}{10k^{-3}}$

48 $\quad \dfrac{(3p)^{-1}(2p)^2}{(5p)^{-3}(25p^2)} \div \dfrac{3p}{5p^{-2}}$

*Express each of the following as a fraction in lowest terms involving no negative exponents. Thus* $\dfrac{x^{-1} + y^{-1}}{x^{-1} - y^{-1}}$ *would be*

$$\frac{\dfrac{1}{x} + \dfrac{1}{y}}{\dfrac{1}{x} - \dfrac{1}{y}} \cdot \frac{xy}{xy} = \frac{y+x}{y-x} \qquad (x,\ y \neq 0;\ y \neq x)$$

(○) 49 $\quad \dfrac{(2a)^{-1} - (3ab)^{-1}}{2b + (b^2 - 1)(b^2 + b)^{-1}}$

50 $\quad \dfrac{x^{-1} - y^{-1}}{(x+y)^{-1}}$

51 $\quad \dfrac{a^{-1} + b^{-1} + c^{-1}}{abc}$

52 $\quad x^{-1}y + xy^{-1}$

53 $\quad \dfrac{s}{t^{-2}} + \dfrac{t}{s^2}$

54 $\quad a + (1 - a)^{-2}$

55 $\quad (x^{-1} + 2xy^{-1} - 3xyz^{-1})^{-1}$

56 $\quad (1 + 2x^{-2} + x^{-4})^{-1}$

*In the remaining problems n denotes an integer. Express each problem in a form so that the denominator is 1 and so that each variable appears only once.*

(○) 57 $\quad \dfrac{x^{2n-1}}{x^{n-3}}$

58 $\quad \dfrac{x^{3n+5}}{x^{7-2n}}$

**59** $(a^{2n} \cdot a^{1-n})^{1-n}$

**60** $\left( \dfrac{s^{2n} t^{n-1}}{s^{-n} t^{1-n}} \right)^{n-2}$

(∘)

*Sample solutions*
*for exercise set 5.1*

**1** $5^7 \cdot 5^{-4} = 5^{7-4} = 5^3 = 125$. Here we have used law E.1.

**11** $4^3 \cdot 16^{-2} = 4^3 \cdot (4^2)^{-2} = 4^3 \cdot 4^{-4} = 4^{3-4} = 4^{-1} = \dfrac{1}{4}$. Notice how we tried to use the laws of exponents to avoid computations with large numbers. We could also have written $4^3 \cdot 16^{-2} = 64 \cdot \dfrac{1}{256} = \dfrac{1}{4}$, but this involves more computational work.

**21**
$$\left[ \frac{3^2 \cdot 3^3 \cdot 9^{-2}}{4^2 \cdot 3^{-1} \cdot 7^0} \right]^{-2} = \left[ \frac{3^2 \cdot 3^3 \cdot 3^{-4}}{4^2 \cdot 3^{-1}} \right]^{-2}$$
$$= (3^2 \cdot 3^3 \cdot 3^{-4} \cdot 4^{-2} \cdot 3^1)^{-2}$$
$$= (3^2 \cdot 4^{-2})^{-2} = 3^{-4} \cdot 4^4 = \frac{256}{81}$$

**23**
$$\frac{6^{-2} \cdot 10^2}{5^{-2} \cdot 12^3} = \frac{(2 \cdot 3)^{-2}(2 \cdot 5)^2}{5^{-2}(2^2 \cdot 3)^3} = \frac{2^{-2} \cdot 3^{-2} \cdot 2^2 \cdot 5^2}{5^{-2} \cdot 2^6 \cdot 3^3}$$
$$= \frac{3^{-2} \cdot 5^2}{2^6 \cdot 3^3 \cdot 5^{-2}} = 2^{-6} \cdot 3^{-5} \cdot 5^4$$

**29** $\dfrac{2x^2 yz^3}{4xyz^2} = 2^{-1} xz \qquad (x, y, z \neq 0)$

**37** $a^2 b^{-3} c^0 = \dfrac{a^2}{b^3} \qquad (b, c \neq 0)$

**43** $\dfrac{16x^2 yz^{-2}}{12xy^{-1}z} = \dfrac{16}{12} xy^2 z^{-3} = \dfrac{4}{3} xy^2 z^{-3} \qquad (x, y, z \neq 0)$

**49**
$$\frac{(2a)^{-1} - (3ab)^{-1}}{2b + (b^2 - 1)(b^2 + b)^{-1}} = \frac{\dfrac{1}{2a} - \dfrac{1}{3ab}}{2b + \dfrac{b^2 - 1}{b^2 + b}}$$
$$= \frac{\dfrac{1}{2a} - \dfrac{1}{3ab}}{2b + \dfrac{(b - 1)(b + 1)}{(b)(b + 1)}}$$

$$= \cfrac{\dfrac{1}{2a} - \dfrac{1}{3ab}}{2b + \dfrac{b-1}{b}}$$

Having simplified the denominator as much as possible, we see that a clearing factor is $6ab$. Multiplying numerator and denominator by $6ab$ produces

$$\cfrac{\dfrac{1}{2a} - \dfrac{1}{3ab}}{2b + \dfrac{b-1}{b}} \cdot \dfrac{6ab}{6ab} = \dfrac{3b-2}{12ab^2 + 6ab - 6a}$$

$$= \dfrac{3b-2}{6a(2b^2 + b - 1)}$$

$$= \dfrac{3b-2}{6a(2b-1)(b+1)} \qquad \left(a \neq 0; \, b \neq 0, \dfrac{1}{2}, -1\right).$$

**57** $\quad \dfrac{x^{2n-1}}{x^{n-3}} = x^{2n-1-n+3} = x^{n+2} \qquad (x \neq 0)$

## (5.2) (Rational exponents)

You have just learned what $b^n$ means for $b \neq 0$ and $n \in I$. The goal of this section is to define $b^r$ for $r$ an arbitrary rational number. Let's start with rational numbers of the form $1/n$ with $n \in N, (n > 2)$. If the laws of exponents are to remain valid, we may apply law E.2 in order to obtain

$$(b^{1/n})^n = b^{(1/n)n} = b^1 = b;$$

*nth root*

*square root*

*cube root*

so if $a = b^{1/n}$ then $a^n = b$. A number $a$ that has the property that $a^n = b$, is called an n*th root* of $b$; we point out that for $n = 2$, an nth root is usually called a *square root*, and for $n = 3$, a *cube root*. The table on page 145 indicates the conditions under which a real number $b$ has an nth root. Thus $-3$ and $+3$ are both square roots of 9, since $(-3)^2 = (+3)^2 = 9$; $-2$ is the unique cube root of $-8$, since $(-2)^3 = -8$; and $-4$ cannot have a square root, since the square of any nonzero real number is positive.

We are now ready to define $b^{1/n}$. If $b$ has only one nth root,

| $b$ | $n$ | nth root of $b$ |
|-----|-----|-----------------|
| Positive | even | two nth roots, a positive one and its additive inverse |
| Positive | odd | unique positive nth root |
| Negative | even | no nth root of $b$ |
| Negative | odd | unique negative nth root |

we define $b^{1/n}$ to be this unique $n$th root of $b$. In case $b$ has two $n$th roots, we agree to let $b^{1/n}$ represent its positive $n$th root. Naturally, if $b$ does not have an $n$th root, we leave $b^{1/n}$ undefined. Thus $4^{1/2} = 2$, $27^{1/3} = 3$, $0^{1/5} = 0$, $(-64)^{1/3} = -4$, and $(-4)^{1/2}$ is not defined.

You should become familiar with another type of notation. We often write $\sqrt{b}$ in place of $b^{1/2}$ and for $n > 2$, $\sqrt[n]{b}$ in place of $b^{1/n}$. The symbol $\sqrt{b}$ is read "square root of $b$," $\sqrt[3]{b}$ is read "cube root of $b$," and for $n > 3$, $\sqrt[n]{b}$ is read "nth root of $b$." However, a little caution is called for. In those cases where $b$ has two $n$th roots, the symbol $\sqrt[n]{b}$ denotes the *positive* $n$th root of $b$. Thus $+5$ and $-5$ are both square roots of 25, but the symbol $\sqrt{25}$ denotes $+5$. If we wanted to specify $-5$, we would write $-\sqrt{25}$. Along these lines, it is important that you understand the difference between the symbols $-b^{1/n}$ and $(-b)^{1/n}$. We have that

$$-b^{1/n} = -\sqrt[n]{b} = \text{the negative of the } n\text{th root of } b$$

whereas

$$(-b)^{1/n} = \sqrt[n]{-b} = \text{the } n\text{th root of negative } b.$$

To illustrate the distinction, we point out that $-4^{1/2} = -2$, whereas *radical notation* $(-4)^{1/2}$ is not defined because there is no real number whose square is $-4$. The notation $\sqrt[n]{b}$ is called *radical notation*, and we shall have more to say about it in the next few sections. Both radical and exponential notation have their uses, and you should be prepared to switch back and forth between them.

In order to define $b^r$ for $r \in Q^+$, we write $r = m/n$ ($m, n \in N$) as a fraction in lowest terms. If $b^{1/n}$ is defined, we let $b^r = (b^{1/n})^m$, and if $b^{1/n}$ does not exist we leave $b^r$ undefined. For example, $4^{3/2} = (4^{1/2})^3 = 2^3 = 8$ and $(-64)^{8/6} = (-64)^{4/3} = [(-64)^{1/3}]^4 = (-4)^4 = 256$. It is easy to check that when $b^{m/n}$ is defined it equals $(b^m)^{1/n}$; in other words, that

$$b^{m/n} = (b^{1/n})^m = (b^m)^{1/n}$$

or, in radical notation

$$b^{m/n} = (\sqrt[n]{b})^m = \sqrt[n]{b^m}.$$

Consequently, $4^{3/2} = \sqrt{4^3} = \sqrt{64} = 8$ and $(-64)^{4/3} = \sqrt[3]{(-64)^4}$
$= \sqrt[3]{(-2^6)^4} = (2^{24})^{1/3} = 2^8 = 256$.

What about negative rational exponents? For $r \in Q^+$ we simply define

$$b^{-r} = 1/b^r \qquad (b \neq 0),$$

thereby ensuring that $b^r \cdot b^{-r} = b^{r-r} = b^0 = 1$. By way of illustration we point out that

$$9^{-3/2} = \frac{1}{9^{3/2}} = \frac{1}{(9^{1/2})^3} = \frac{1}{3^3} = \frac{1}{27}.$$

We have now accomplished our goal and have defined $b^r$ for every rational number $r$. Incidentally, we continue with the terminology we introduced earlier and call $b$ the *base* and $r$ the *exponent* of the expression $b^r$ (the $r$th power of $b$). It is certainly worth mentioning that powers involving rational exponents obey the laws of exponents we mentioned earlier. In the chapter on logarithms we shall discuss powers of positive numbers that have an arbitrary real number exponent, and we shall see that the laws of exponents are valid even in that context.

## Exercise set 5.2

In this exercise you are to assume (unless otherwise indicated) that all variables and all bases are positive.

In problems 1 to 14, perform the indicated operations. Express your answer as an integer or fraction in lowest terms.

(∘)  1  $4^{5/2}$       2  $9^{3/2}$       3  $(-27)^{1/3}$

4  $(-64)^{2/3}$       5  $32^{2/5}$       6  $125^{4/3}$

(∘)  7  $(1{,}000)^{-2/3}$       8  $4^{-7/2}$       9  $(-216)^{-1/3}$

10  $(-125)^{-2/3}$       11  $\sqrt{\dfrac{4}{9}}$       12  $\sqrt[3]{\dfrac{8}{27}}$

13  $\sqrt[5]{-32}$       14  $\dfrac{\sqrt{81}}{\sqrt{36}}$

*Express the following in radical form.*

$(\circ)$  15  $x^{3/2}$     16  $y^{7/3}$     17  $a^{-2/3}$

   18  $a^{-3/4}$     19  $5x^{4/3}$     20  $7w^{9/10}$

   21  $(a-2b)^{5/3}$     22  $(a+5b)^{-7/3}$

*State problems 23 to 28 in exponential form with no denominators involved.*

$(\circ)$  23  $\sqrt[3]{x^4}$     24  $\sqrt[4]{x^3}$     25  $\dfrac{1}{\sqrt[3]{x^2}}$

   26  $\dfrac{1}{\sqrt{a}}$     27  $\sqrt{4xy^6}$     28  $\sqrt[3]{27x^2\,y^9}$

*Use the laws of exponents to write each of the next expressions as a term or sum of terms in which each variable appears only once in a given term, and so that the denominator is always 1.*

$(\circ)$  29  $x^{1/2}x^{2/3}$        30  $y^{3/4}\,y^{5/6}$

   31  $(w^{4/5}\,w^{3/5})^{1/7}$     32  $(a^{1/4}\,a^{1/2})^2$

   33  $x^{2/3}\cdot x^{-1/2}\cdot x^{1/6}$    34  $(g^{7/9}\cdot g^{-1/3})^3$

$(\circ)$  35  $\dfrac{x^{1/2}\cdot x^{3/4}}{x^{1/6}}$     36  $\dfrac{y^{1/3}\,y^{3/8}}{y^{-5/2}}$

   37  $(a^{2/3})^{3/4}$      38  $(a^3\,b^{3/5})^{4/3}$

$(\circ)$  39  $(x^{1/2}+y^{1/2})^2$    40  $(a^{3/2}+1)^2$

   41  $(x^{1/2}+3)(x^{1/2}+1)$   42  $(x^{1/2}-5)(x^{3/2}+x^{1/2})$

   43  $x^{1/2}(x^{3/2}-5x^{1/2}+2x^{-1/2})$   44  $x^{1/3}(3x^{4/3}-x^{2/3})$

   45  $(x^{1/3}-3)(x^{1/3}+2)$   46  $(2x^{1/4}+1)(2x^{1/4}-3)$

Until now all factoring problems have used positive integral powers of the variables involved. We can also factor out rational powers of a variable. For example,

$$x^{3/2}+3x^{-1/2}=x^{-1/2}(x^2+3) \qquad (x>0).$$

*Factor each of the following in the indicated fashion.*

$(\circ)$  47  $x^{5/4}+x^{3/4}=x^{3/4}(\quad)$

   48  $y^{7/2}-y^{3/4}=y^{3/4}(\quad)$

   49  $6x^{7/9}y^{1/3}+9x^{2/3}y^{2/3}=3x^{2/3}y^{1/3}(\quad)$

   50  $4a^{3/2}b^{5/4}+6a^{1/5}b^{5/8}=2a^{1/8}b^{5/8}(\quad)$

**51** $x^{1/2} - x^{-1/2} = x^{-1/2}(\quad)$

**52** $2x^{5/3} + 4x^{-2/3} = 2x^{-1}(\quad)$

**53** $4a^{3/4} + 10a^{-1/4} = 4a^{1/2}(\quad)$

**54** $16b^{7/3} - 6b^{2/9} = 2b^2(\quad)$

$(\circ)$   **55** $x - 9 = (x^{1/2} + 3)(\quad - \quad)$

**56** $x + 3x^{1/2} + 2 = (x^{1/2} + 2)(\quad + \quad)$

*Given $n \in N$, simplify problems 57 to 60.*

$(\circ)$   **57** $(x^{2n} \cdot x^{n/2})^{1/n}$
                                       **58** $\left(\dfrac{x^{n+3}}{x^{n-1}}\right)^{1/2}$

**59** $\left(\dfrac{x^{n+3} \cdot y^{n-2}}{x^3 y^{-2}}\right)^{2/n}$
                              **60** $\dfrac{(x^{n^2} \cdot x^{2n} \cdot x)^{1/(n+1)}}{x^n}$

$(\circ)$

**Sample solutions**
**for exercise set 5.2**

**1** $4^{5/2} = (4^{1/2})^5 = 2^5 = 32$

**7** $(1{,}000)^{-2/3} = \dfrac{1}{(1{,}000)^{2/3}} = \dfrac{1}{(1{,}000^{1/3})^2} = \dfrac{1}{10^2} = \dfrac{1}{100}$

**15** $x^{3/2} = (\sqrt{x})^3$. An equally correct answer is $x^{3/2} = \sqrt{x^3}$.

**23** $\sqrt[3]{x^4} = x^{4/3}$

**29** $x^{1/2} x^{2/3} = x^{1/2 + 2/3} = x^{7/6}$

**35** $\dfrac{x^{1/2} \cdot x^{3/4}}{x^{1/6}} = \dfrac{x^{5/4}}{x^{1/6}} = x^{5/4 - 1/6} = x^{15/12 - 2/12} = x^{13/12}$

**39** $(x^{1/2} + y^{1/2})^2 = x^{1/2} x^{1/2} + 2x^{1/2} y^{1/2} + y^{1/2} y^{1/2}$

                          $= x + 2x^{1/2} y^{1/2} + y$

**47** $x^{5/4} + x^{3/4} = x^{3/4}(x^{1/2} + 1)$

**55** $x - 9 = (x^{1/2} + 3)(x^{1/2} - 3)$

**57** $(x^{2n} \cdot x^{n/2})^{1/n} = (x^{5n/2})^{1/n} = x^{(5n/2)(1/n)} = x^{5/2}$

# (5.3) (Radical notation)

*radical sign of order*
*index, radicand*

In the last section we pointed out that for $n \geq 2$, the notation $\sqrt[n]{b}$ is often used to denote $b^{1/n}$. The symbol $\sqrt{\phantom{x}}$ is called a *radical sign of order $n$; $n$ is called the *index* and $b$ the *radicand* of $\sqrt[n]{b}$.

If no index appears, it is understood that the index is 2. There are times when radical notation is more natural than exponential notation; so these next few sections are devoted to developing the basic properties of radicals.

*In what follows, it will be assumed that when the index of a radical is even, all variables and all radicands are non-negative.* Much of what we need to know about radicals will turn out to be a translation of the laws of exponents into radical notation. For notational convenience we'll number the various laws as we have other important ones.

The second law of exponents, E.2, says, among other things, that for $m$ and $n \in I$ with $n \geq 2$, $(b^m)^{1/n} = (b^{1/n})^m$. In radical notation this same idea is the basis for the following statement.

( ∘ )   **RN.1**  $\sqrt[n]{b^m} = (\sqrt[n]{b})^m$      *(m, n ∈ I; n ≥ 2)*

This can be used to compute $\sqrt[4]{16^5} = (\sqrt[4]{16})^5 = 2^5 = 32$. Notice how much easier this is than raising 16 to the fifth power and trying to find the fourth root of the result!

So long as we do not allow $b$ to become negative with $n$ even (a condition we have carefully avoided) this also tells us that $\sqrt[n]{b^n} = (\sqrt[n]{b})^n = b$. Thus for $x \geq 0$, $\sqrt{x^2} = x$; for arbitrary $y$, $\sqrt[3]{y^3} = y$, and so on.

The third law of exponents, E.3, tells us that $(ab)^{1/n} = a^{1/n}b^{1/n}$. In radical notation this becomes the next statement.

( ∘ )   **RN.2**  $\sqrt[n]{ab} = \sqrt[n]{a}\,\sqrt[n]{b}$      *(n ∈ N; n ≥ 2)*

Thus we can multiply radicals of the same index as follows:

$$\sqrt{6} \cdot \sqrt{3} = \sqrt{18},$$

$$\sqrt[3]{x} \cdot \sqrt[3]{20} = \sqrt[3]{20x}.$$

We can also insert expressions into a radicand:

$$4x\sqrt[3]{2x^5} = \sqrt[3]{64x^3}\,\sqrt[3]{2x^5} = \sqrt[3]{128x^8}.$$

Finally, law RN.2 enables us to remove from a radicand any factor that appears raised to a power greater than or equal to the index of the radical. Thus

$$\sqrt{32} = \sqrt{16 \cdot 2} = \sqrt{16}\,\sqrt{2} = 4\sqrt{2},$$

$$\sqrt[3]{54x^4} = \sqrt[3]{(27x^3)(2x)} = \sqrt[3]{27x^3}\,\sqrt[3]{2x} = 3x\sqrt[3]{2x}.$$

Another consequence of the third law of exponents is the fact that $\left(\dfrac{a}{b}\right)^{1/n} = \dfrac{a^{1/n}}{b^{1/n}}$ $(b \neq 0)$. In radical form this becomes the following statement.

$(\circ)$ RN.3 $\quad \sqrt[n]{\dfrac{a}{b}} = \dfrac{\sqrt[n]{a}}{\sqrt[n]{b}} \qquad (b \neq 0;\ n \in N;\ n \geq 2)$

This procedure is used to remove any fraction from the radicand. For example,

$$\sqrt{\dfrac{3}{4}} = \dfrac{\sqrt{3}}{\sqrt{4}} = \dfrac{\sqrt{3}}{2},$$

$$\sqrt[3]{\dfrac{5}{27x^3}} = \dfrac{\sqrt[3]{5}}{\sqrt[3]{27x^3}} = \dfrac{\sqrt[3]{5}}{3x} \qquad (x \neq 0).$$

Sometimes the denominator of the radicand of a radical of order $n$ is not an expression raised to the $n$th power. One must then multiply the numerator and denominator by a factor that will put the denominator in this form. The process is called *rationalizing the denominator*, and here is how it goes:

*rationalizing the denominator*

$$\sqrt{\dfrac{5}{2x}} = \sqrt{\dfrac{5 \cdot 2x}{2x \cdot 2x}} = \dfrac{\sqrt{10x}}{2x} \qquad (x \neq 0),$$

$$\sqrt[3]{\dfrac{7}{3y^2}} = \sqrt[3]{\dfrac{7 \cdot 9y}{3y^2 \cdot 9y}} = \sqrt[3]{\dfrac{63y}{27y^3}} = \dfrac{\sqrt[3]{63y}}{3y} \qquad (y \neq 0),$$

$$\dfrac{2}{\sqrt[3]{5y}} = \dfrac{2 \cdot \sqrt[3]{5y} \cdot \sqrt[3]{5y}}{\sqrt[3]{5y} \cdot \sqrt[3]{5y} \cdot \sqrt[3]{5y}} = \dfrac{2\sqrt[3]{25y^2}}{5y} \qquad (y \neq 0).$$

Let's break off our discussion here to give you a chance to apply these ideas to some actual problems.

## Exercise set 5.3

In this exercise you are to assume that all variables are positive.

Using the fact that $b^{m/n} = \sqrt[n]{b^m} = (\sqrt[n]{b})^m$, one can write $6^{3/2} = \sqrt{216} = (\sqrt{6})^3$. *Do the corresponding thing in each of problems 1 to 8.*

**1** $5^{2/3}$   **2** $3^{4/3}$   **3** $2^{4/6}$   **4** $2^{6/8}$

**5** $x^{3/10}$   **6** $y^{7/5}$   **7** $a^{-3/5}$   **8** $b^{-23/9}$

*Find the roots in each of the following problems.*

**9** $\sqrt{25}$   **10** $\sqrt[3]{-64}$   **11** $\sqrt[4]{\dfrac{81}{16}}$   **12** $-\sqrt{\dfrac{9}{4}}$

**13** $-\sqrt{9x^2}$   **14** $\sqrt[3]{27w^9}$   **15** $\sqrt[4]{16a^4b^8}$   **16** $\sqrt{49s^2t^6}$

*Express each of problems 17 to 28 as the radical of a single expression.*

**17** $\sqrt[3]{8}\,\sqrt[3]{4}$   **18** $\sqrt[4]{17}\,\sqrt[4]{3}$   **19** $\sqrt{3x}\,\sqrt{6x}$

**20** $\sqrt[4]{5ab^2c^2}\,\sqrt[4]{5bc^3}$   **21** $3\sqrt{2}$   **22** $5\sqrt{3}$

**23** $2\cdot\sqrt[3]{15}$   **24** $3\sqrt[3]{2}$   **25** $5x\sqrt{3xy}$

**26** $2ab^2\sqrt[4]{a^3b}$   **27** $(x+y)\sqrt{x}$   **28** $(x+1)\sqrt[3]{x}$

*Remove all possible factors from each of the following radicands.*

**29** $\sqrt{18}$   **30** $\sqrt[3]{16}$   **31** $\sqrt[4]{405}$   **32** $\sqrt{200}$

**33** $\sqrt{25x^3}$   **34** $\sqrt{27a^4b^5}$   **35** $\sqrt[4]{32x^5}$   **36** $\sqrt[3]{250x^4y^7}$

*Express problems 37 to 44 as the radical of a single expression.*

**37** $\dfrac{\sqrt{5}}{\sqrt{2}}$   **38** $\dfrac{\sqrt[3]{9}}{\sqrt[3]{7}}$   **39** $\dfrac{\sqrt{7}}{\sqrt{x}}$   **40** $\dfrac{\sqrt[4]{8}}{\sqrt[4]{2w}}$

**41** $\dfrac{\sqrt[3]{6x}}{\sqrt[3]{x}}$   **42** $\dfrac{\sqrt{8w^2}}{\sqrt{4w}}$   **43** $\dfrac{\sqrt{5}\,\sqrt{2x}}{\sqrt{10}}$   **44** $\dfrac{\sqrt{3}\,\sqrt{6x}}{\sqrt{5}\,\sqrt{3x^2}}$

*In problems 45 to 60, rationalize each denominator.*

**45** $\sqrt{\dfrac{2}{9}}$   **46** $\sqrt{\dfrac{5}{16}}$   **47** $\sqrt{\dfrac{2}{5}}$   **48** $\sqrt{\dfrac{7}{10}}$

**49** $\sqrt[3]{\dfrac{3}{5}}$   **50** $\sqrt[4]{\dfrac{7}{2}}$   **51** $\sqrt{\dfrac{5}{4x^2}}$   **52** $\sqrt{\dfrac{23}{9x^4}}$

**53** $\sqrt[3]{\dfrac{2}{3x}}$   **54** $\sqrt[4]{\dfrac{5}{2x^3}}$   **55** $\sqrt[4]{\dfrac{3x}{2}}$   **56** $\sqrt[3]{\dfrac{3y}{2x}}$

**57** $\dfrac{1}{\sqrt{x}}$   **58** $\dfrac{1}{\sqrt[3]{y^2}}$   **59** $\dfrac{\sqrt[3]{5}}{3\cdot\sqrt[3]{2x}}$   **60** $\dfrac{\sqrt{6x}}{2\sqrt{2x^5}}$

1 $5^{2/3} = \sqrt[3]{25} = (\sqrt[3]{5})^2$

7 $a^{-3/5} = \sqrt[5]{a^{-3}} = (\sqrt[5]{a})^{-3}$. An equally correct form of the answer is provided by

$$a^{-3/5} = \sqrt[5]{\dfrac{1}{a^3}} = \dfrac{1}{(\sqrt[5]{a})^3}.$$

9 $\sqrt{25} = 5$

13 $-\sqrt{9x^2} = -3x$

17 $\sqrt[3]{8}\,\sqrt[3]{4} = \sqrt[3]{32}$

21 $3\sqrt{2} = \sqrt{9}\,\sqrt{2} = \sqrt{18}$

25 $5x\sqrt{3xy} = \sqrt{25x^2}\,\sqrt{3xy} = \sqrt{75x^3\,y}$

29 $\sqrt{18} = \sqrt{9 \cdot 2} = 3\sqrt{2}$

35 $\sqrt[4]{32x^5} = \sqrt[4]{16x^4 \cdot 2x} = 2x\sqrt[4]{2x}$

37 $\dfrac{\sqrt{5}}{\sqrt{2}} = \sqrt{\dfrac{5}{2}}$

41 $\dfrac{\sqrt[3]{6x}}{\sqrt[3]{x}} = \sqrt[3]{\dfrac{6x}{x}} = \sqrt[3]{6}$

47 $\sqrt{\dfrac{2}{5}} = \sqrt{\dfrac{2 \cdot 5}{5 \cdot 5}} = \dfrac{\sqrt{10}}{5}$

53 $\sqrt[3]{\dfrac{2}{3x}} = \dfrac{\sqrt[3]{2}\,\sqrt[3]{3x}\,\sqrt[3]{3x}}{\sqrt[3]{3x}\,\sqrt[3]{3x}\,\sqrt[3]{3x}} = \dfrac{\sqrt[3]{18x^2}}{\phantom{xxxxx}}$

# (5.4) (More on radical notation)

This section is a continuation of Section 5.3, and all assumptions made in that section remain in force. In particular, if the index of a radical is even, we specifically assume that the radicand and all variables are non-negative. We developed three laws of radicals in Section 5.3, and that brings us to the fourth law.

The second law of exponents, E.2, may be used to develop a

rule for changing the index of a radical. Given integers $c$ and $n$, both greater than 1, we have

$$b^{m/n} = b^{cm/cn} \qquad (m \in I).$$

Translated to radical form this becomes our next law.

$(\circ)$   RN.4  $\sqrt[n]{b^m} = \sqrt[cn]{b^{cm}}$ $\qquad$ ($m \in I$; $c$, $n \in N$ and $> 1$)

This law tells us that we can raise the index of a radical by multiplying both its index and the power of its radicand by the same natural number. We illustrate by observing some examples:

$$\sqrt{2} = \sqrt[2 \cdot 2]{2^{2 \cdot 1}} = \sqrt[4]{4},$$

$$\sqrt[3]{5x^2} = \sqrt[4 \cdot 3]{(5x^2)^{4 \cdot 1}} = \sqrt[12]{(5x^2)^4} = \sqrt[12]{625x^8},$$

$$\sqrt[4]{x^3} = \sqrt[2 \cdot 4]{x^{2 \cdot 3}} = \sqrt[8]{x^6}.$$

The same sort of idea may be used to lower the index of a radical, such as $\sqrt[n]{b^m}$. To see how to do this, let $k > 1$ be a factor of both $m$ and $n$. Then, since $\dfrac{m}{n} = \dfrac{m/k}{n/k}$, we have:

$(\circ)$   RN.4a  $\sqrt[n]{b^m} = \sqrt[n/k]{b^{m/k}}$

In other words, the index of a radical may be lowered by dividing both the index and the power of its radicand by a common positive factor. Thus we find

$$\sqrt[4]{x^2} = \sqrt{x},$$

$$\sqrt[9]{64a^6} = \sqrt[9]{(4a^2)^3} = \sqrt[3]{4a^2}.$$

The same sort of thing can be done using exponential notation. To see this, consider the following:

$$\sqrt[9]{64a^6} = (64a^6)^{1/9} = [(4a^2)^3]^{1/9} = (4a^2)^{3/9}$$

$$= (4a^2)^{1/3} = \sqrt[3]{4a^2}.$$

The table at the top of page 154 summarizes the rules we have developed and where they came from.

By systematically applying the laws set forth in this table, we can put a radical in a form in which:

| Rule | Law of radicals | | Rule | Law of exponents |
|------|-----------------|---|------|------------------|
| **RN.1** | $\sqrt[n]{b^m} = (\sqrt[n]{b})^m$ | | **E.2** | $(b^m)^{1/n} = (b^{1/n})^m$ |
| **RN.2** | $\sqrt[n]{ab} = \sqrt[n]{a}\,\sqrt[n]{b}$ | | **E.3** | $(ab)^{1/n} = a^{1/n}b^{1/n}$ |
| **RN.3** | $\sqrt[n]{\dfrac{a}{b}} = \dfrac{\sqrt[n]{a}}{\sqrt[n]{b}}$ | $(b \neq 0)$ | **E.3** | $\left(\dfrac{a}{b}\right)^{1/n} = \dfrac{a^{1/n}}{b^{1/n}}$ |
| **RN.4** | $\sqrt[n]{b^m} = \sqrt[cn]{b^{cm}}$ $(c,k \in N;\ c,\ k > 1)$ | | **E.2** | $b^{m/n} = b^{cm/cn}$ |
| **RN.4a** | $\sqrt[n]{b^m} = \sqrt[n/k]{b^{m/k}}$ | $(k|m,\ n)$ | **E.2** | $b^{m/n} = b^{cm/cn}$ |

**1** All denominators have been rationalized

**2** No zero or negative exponents appear

**3** The laws of exponents have been used to simplify products and quotients

**4** No integer or polynomial factor can be removed from the radicand

**5** The index of the radical cannot be reduced

When all these things have been done, we say that we have *simplified* the radical. Thus

$$\sqrt[6]{\frac{4x^8}{9y^4}} = \sqrt[6]{\frac{(2x^4)^2}{(3y^2)^2}} = \sqrt[3]{\frac{2x^4}{3y^2}} = \sqrt[3]{\frac{2x^4 \cdot 9y}{3y^2 \cdot 9y}}$$

$$= \frac{\sqrt[3]{18x^4 y}}{3y} = \frac{x\sqrt[3]{18xy}}{3y}$$

is in simplified form.

Until now, our standing assumption has been that in the presence of an even index, all radicands and all variables are non-negative. Let us now see what happens when this assumption is dropped. To preview the sort of problem that can arise, we consider $\sqrt[6]{x^2}$. For $x \geq 0$ we have that $\sqrt[6]{x^2} = (\sqrt[6]{x})^2 = \sqrt[3]{x}$. This statement is just not true for $x$ negative. Indeed, for $x = -1$ we have that $\sqrt[3]{-1} = -1$, $\sqrt[6]{(-1)^2} = \sqrt[6]{+1} = +1$, whereas $(\sqrt[6]{-1})^2$ is not even defined! It is time to take a closer look at $\sqrt[n]{x^m}$ for $n$ even and $x$ possibly negative. If $m$ is odd and $x < 0$, then $x^m < 0$; so $\sqrt[n]{x^m}$ is not defined. On the other hand, if $m$ is even it is clear that $x^m = |x|^m$; so $\sqrt[n]{x^m} = \sqrt[n]{|x|^m}$ for all possible values of $x$. In summary, we have that

$$\sqrt[n]{x^m} = \begin{cases} \sqrt[n]{|x|^m} & \text{if } m \text{ is even} \\ \text{is undefined if } m \text{ is odd and } x < 0 \\ \text{is defined if } m \text{ is odd and } x \geq 0 \end{cases}$$

In particular, if this is applied to $\sqrt[n]{x^n}$, we see that

$$\sqrt[n]{x^n} = \begin{cases} x & n \text{ odd} \\ |x| & n \text{ even} \end{cases}$$

so that $\sqrt{(-2)^2} = |-2| = 2$, and so on.

## Exercise set 5.4

In problems 1 to 50, you are to assume that all variables represent positive real numbers.

*Reduce the order of each radical in problems 1 to 10 as far as possible.*

(∘)　1　$\sqrt[4]{25}$  　　2　$\sqrt[6]{4}$　　　3　$\sqrt[6]{8}$

(∘)　4　$\sqrt[12]{64}$  　　5　$\sqrt[10]{16x^2}$　　6　$\sqrt[9]{8y^3}$

　　7　$\sqrt[15]{32x^{10}}$　　8　$\sqrt[15]{8x^{12}}$　　9　$\sqrt[6]{64x^6 y^3}$

　10　$\sqrt[8]{256a^{16} b^8 c^4}$

In order to use the laws of radicals to express the product $\sqrt{3} \cdot \sqrt[3]{4}$ as a single radical, both radicals must be expressed in an equivalent form that has the same order. The LCM of the indices of the radicals is 6; so we may express them as radicals of order 6. Thus, using the laws of radicals we have $\sqrt{3} = \sqrt[6]{3^3} = \sqrt[6]{27}$, $\sqrt[3]{4} = \sqrt[6]{4^2} = \sqrt[6]{16}$; so $\sqrt{3}\sqrt[3]{4} = \sqrt[6]{27}\sqrt[6]{16} = \sqrt[6]{27 \cdot 16} = \sqrt[6]{432}$.

*Express each of the following as a single radical of suitable order.*

(∘)　11　$\sqrt[4]{3}\,\sqrt[6]{5}$　　　　　12　$\sqrt[5]{7}\,\sqrt{2}$

　13　$\sqrt[3]{8}\,\sqrt[4]{6}$　　　　　14　$\sqrt{3}\,\sqrt[3]{6}$

(∘)　15　$\sqrt[6]{5x}\,\sqrt{3x}$　　　　16　$\sqrt[7]{4y}\,\sqrt[3]{2y}$

　17　$\sqrt{3w}\,\sqrt[6]{5w^4}$　　　18　$\sqrt[4]{5a^3 b}\,\sqrt[6]{2ab^2}$

　19　$\sqrt{3x}\,\sqrt[3]{xy^2}\,\sqrt[5]{2y}$　　20　$\sqrt[6]{2a}\,\sqrt[3]{a}\,\sqrt{3a^2}$

*Simplify each of the following radicals.*

(∘)　21　$\sqrt{18}$　　　　　22　$\sqrt[3]{-250}$

　23　$\sqrt[4]{80x^5}$　　　　24　$\sqrt{52x^4}$

　25　$\sqrt[3]{108a^2 b^3}$　　　26　$\sqrt[4]{81g^4 h^5}$

( ∘ ) **27** $\sqrt{12}\ \sqrt{15}$      **28** $\sqrt{75}\ \sqrt{21}$

**29** $\sqrt[3]{54}\ \sqrt[3]{250}$      **30** $\sqrt[4]{1,250}\ \sqrt[4]{24}$

( ∘ ) **31** $\sqrt{12s^2t}\ \sqrt{21st^3}$      **32** $\sqrt{10uv}\ \sqrt{20u^3}$

**33** $\sqrt[3]{36a^4b^3c}\ \sqrt[3]{6ab^2}$      **34** $\sqrt[4]{49a^3b}\ \sqrt[4]{7ab^2}\ \sqrt[4]{14ab}$

( ∘ ) **35** $\dfrac{\sqrt{27}}{\sqrt{6}}$      **36** $\dfrac{\sqrt{12}}{\sqrt{21}}$

**37** $\dfrac{\sqrt[3]{16}}{\sqrt[3]{54}}$      **38** $\dfrac{\sqrt[4]{243}}{\sqrt[4]{375}}$

**39** $\dfrac{\sqrt[3]{32x^3y}}{\sqrt[3]{500xy^2}}$      **40** $\dfrac{\sqrt[3]{54c^2}}{\sqrt[3]{24cd^2}}$

**41** $\dfrac{\sqrt{16ab^4}}{\sqrt{28a^2b^7}}$      **42** $\dfrac{\sqrt{250hk^3}}{\sqrt{45h^3k^2}}$

( ∘ ) **43** $\dfrac{\sqrt{9x^3y}\ \sqrt{2xy^2}}{\sqrt{150x^{-1}y}}$      **44** $\dfrac{\sqrt{8xy^{-1}}\ \sqrt{2x^{-1}y^0}}{\sqrt{18xy^{-2}}}$

**45** $\dfrac{\sqrt[3]{9ab^4}\ \sqrt[3]{48ab}}{\sqrt[3]{6a^{-1}b^{-2}}}$      **46** $\dfrac{\sqrt[3]{a^{-1}b^3}\ \sqrt[3]{25a^{-2}b^{-1}}}{\sqrt[3]{4ab^{-4}}}$

**47** $\dfrac{\sqrt[4]{3x^3}}{\sqrt[4]{48x}}$      **48** $\dfrac{\sqrt[6]{64x^2}}{\sqrt[6]{36x^2y^4}}$

**49** $\dfrac{\sqrt[9]{16x^4y}}{\sqrt[9]{54x^4y^7}}$      **50** $\dfrac{\sqrt[12]{250a^6b^2}}{\sqrt[12]{2b^5}}$

*In the remaining problems, consider any variable to be an arbitrary real number and simplify. State any restrictions you must make on the variables.*

( ∘ ) **51** $\sqrt{4x^2}$      **52** $\sqrt{9y^2}$

**53** $\sqrt{x^2-2x+1}$      **54** $\sqrt{x^2-6x+9}$

( ∘ ) **55** $\sqrt{-6x^3}$      **56** $\sqrt{(6-x)^3}$

**57** $\dfrac{1}{\sqrt{9x^2-6x+1}}$      **58** $\dfrac{1}{\sqrt{25y^2-10xy+x^2}}$

**59** $\sqrt{16x^4}$      **60** $\sqrt{25(x^2+y^2)^2}$

**1** $\sqrt[4]{25} = \sqrt[4]{5^2} = \sqrt{5}$

**5** $\sqrt[10]{16x^2} = \sqrt[10]{(4x)^2} = \sqrt[5]{4x}$

**11** The LCM of 4 and 6 is 12; so both radicals must be expressed as radicals of order 12. Thus $\sqrt[4]{3}\ \sqrt[6]{5} = \sqrt[12]{3^3}\ \sqrt[12]{5^2} = \sqrt[12]{3^3 \cdot 5^2} = \sqrt[12]{675}$.

**15** $\sqrt[6]{5x}\ \sqrt{3x} = \sqrt[6]{5x}\ \sqrt[6]{(3x)^3} = \sqrt[6]{(5x)(3x)^3}$

$\qquad\qquad = \sqrt[6]{135x^4}$

**21** $\sqrt{18} = \sqrt{9 \cdot 2} = 3\sqrt{2}$

**27** $\sqrt{12}\ \sqrt{15} = \sqrt{12 \cdot 15} = \sqrt{2^2 \cdot 3 \cdot 3 \cdot 5} = \sqrt{2^2 \cdot 3^2}\sqrt{5}$

$\qquad\qquad = 2 \cdot 3\sqrt{5} = 6\sqrt{5}$

**31** $\sqrt{12s^2 t}\ \sqrt{21st^3} = \sqrt{12 \cdot 21 \cdot s^3 \cdot t^4} = \sqrt{2^2 \cdot 3 \cdot 3 \cdot 7 \cdot s^3 \cdot t^4}$

$\qquad\qquad = \sqrt{2^2 \cdot 3^2 \cdot s^2 \cdot t^4}\ \sqrt{7s} = 6st^2 \sqrt{7s}$

**35** $\dfrac{\sqrt{27}}{\sqrt{6}} = \sqrt{\dfrac{27}{6}}$ by RN.3 and $\sqrt{\dfrac{27}{6}} = \sqrt{\dfrac{3 \cdot 3 \cdot 3}{2 \cdot 3}} = \sqrt{\dfrac{3 \cdot 3}{2}}$

$= 3\sqrt{\dfrac{1}{2}}$ . We must still rationalize the denominator; so we write $3\sqrt{\dfrac{1}{2}}$

$= 3\sqrt{\dfrac{1 \cdot 2}{2 \cdot 2}} = 3\dfrac{\sqrt{2}}{\sqrt{4}} = \dfrac{3}{2}\sqrt{2}$. Alternately, we could write $\dfrac{\sqrt{27}}{\sqrt{6}}$

$= \dfrac{\sqrt{27}}{\sqrt{6}} \cdot \dfrac{\sqrt{6}}{\sqrt{6}} = \dfrac{\sqrt{162}}{6} = \dfrac{\sqrt{81 \cdot 2}}{6} = \dfrac{9\sqrt{2}}{6} = \dfrac{3\sqrt{2}}{2} = \dfrac{3}{2}\sqrt{2}$ .

**43** $\dfrac{\sqrt{9x^3 y}\ \sqrt{2xy^2}}{\sqrt{150x^{-1}y}} = \sqrt{\dfrac{(9x^3 y)(2xy^2)}{150x^{-1}y}} = \sqrt{\dfrac{2 \cdot 3^2 \cdot x^4 \cdot y^3}{2 \cdot 3 \cdot 5^2 \cdot x^{-1} \cdot y}}$

$\qquad\qquad = \sqrt{\dfrac{3x^4 y^3}{25x^{-1} y}} = \sqrt{\dfrac{3x^5 y^2}{25}}$

$\qquad\qquad = \sqrt{\dfrac{x^4 y^2}{25} \cdot 3x} = \dfrac{x^2 y}{5}\sqrt{3x}$

**51** $\sqrt{4x^2} = 2|x|$

**55**  In order for $\sqrt{-6x^3}$ to be defined, we must have $-6x^3 \geq 0$. This happens only when $x \leq 0$, and then $\sqrt{-6x^3} = \sqrt{-6x \cdot x^2} = |x|\sqrt{-6x} = -x\sqrt{-6x}$. This used the fact that $|x| = -x$ for $x \leq 0$.

## (5.5) (Arithmetic of radicals)

The distributive law

$$(a + b)c = ac + bc \qquad (a, b, c \in R)$$

provides the means of writing sums and differences containing radicands of the same order as a single term. Thus $2\sqrt{3} + 3\sqrt{3} = (2 + 3)\sqrt{3} = 5\sqrt{3}$, whereas $8\sqrt[3]{x} - 4\sqrt[3]{x} = (8 - 4)\sqrt[3]{x} = 4\sqrt[3]{x}$. Of course, there are times when you will first want to simplify the radicals before making any attempt to combine terms. To illustrate this, let's consider

$$
\begin{aligned}
\sqrt{12} + \sqrt[3]{24} + \sqrt{75} + \sqrt[3]{81} \ &= \ 2\sqrt{3} + 2\sqrt[3]{3} - 5\sqrt{3} + 3\sqrt[3]{3} \\
&= \ (2 - 5)\sqrt{3} + (2 + 3)\sqrt[3]{3} \\
&= \ -3\sqrt{3} + 5\sqrt[3]{3}.
\end{aligned}
$$

Expressions involving radicals are multiplied in the same way that any algebraic expressions are multiplied, except that you must use your knowledge of radicals to simplify the answer. Thus

$$(\sqrt{5} + \sqrt{3})(\sqrt{5} - \sqrt{3}) = (\sqrt{5})^2 - (\sqrt{3})^2 = 5 - 3 = 2,$$

$$(\sqrt{2} + \sqrt{3x})^2 = (\sqrt{2})^2 + 2\sqrt{2}\sqrt{3x} + (\sqrt{3x})^2 = 2 + 2\sqrt{6x} + 3x,$$

and as a final example,

$$
\begin{aligned}
(\sqrt{2} + 3\sqrt{5})(\sqrt{2} - 2\sqrt{5}) &= (\sqrt{2})^2 + (3 - 2)\sqrt{2}\sqrt{5} - (3\sqrt{5})(2\sqrt{5}) \\
&= 2 + \sqrt{10} - 30 \\
&= -28 + \sqrt{10}.
\end{aligned}
$$

This type of computation also provides the clue for rationalizing denominators that have two terms. What you need is the fact that

$$(a + b)(a - b) = a^2 - b^2;$$

so there is no term involving $ab$. Each of the expressions $a + b$, *conjugate*  $a - b$ is said to be the *conjugate* of the other. If you multiply $a + \sqrt{b}$ by its conjugate, the result,

$$(a + \sqrt{b})(a - \sqrt{b}) = a^2 - b,$$

does not involve a radical. We may apply this to observe that the fraction $\dfrac{c}{a + \sqrt{b}}$ may be rationalized as follows:

$$\frac{c}{a + \sqrt{b}} = \frac{c}{a + \sqrt{b}} \cdot \frac{a - \sqrt{b}}{a - \sqrt{b}} = \frac{c(a - \sqrt{b})}{a^2 - b}.$$

Similarly,

$$\frac{c}{\sqrt{a} + \sqrt{b}} = \frac{c}{\sqrt{a} + \sqrt{b}} \cdot \frac{\sqrt{a} - \sqrt{b}}{\sqrt{a} - \sqrt{b}} = \frac{c(\sqrt{a} - \sqrt{b})}{a - b}$$

takes care of an expression of the form $\dfrac{c}{\sqrt{a} + \sqrt{b}}$. Here's an example to illustrate the process:

$$\frac{\sqrt{5} + \sqrt{3}}{\sqrt{5} - \sqrt{3}} = \frac{\sqrt{5} + \sqrt{3}}{\sqrt{5} - \sqrt{3}} \cdot \frac{\sqrt{5} + \sqrt{3}}{\sqrt{5} + \sqrt{3}} = \frac{5 + 2\sqrt{5}\sqrt{3} + 3}{5 - 3}$$

$$= \frac{8 + 2\sqrt{15}}{2} = 4 + \sqrt{15}.$$

## Exercise set 5.5

*In each of the following problems, simplify each radical, and then combine coefficients of equal radicands that have the same indices. You may assume any variable appearing in a radicand is positive.*

**1** $2\sqrt{3} + 5\sqrt{3}$         **2** $8\sqrt{2} + \sqrt{2}$

**3** $4\sqrt{7} - 2\sqrt{7}$         **4** $3\sqrt{10} - 5\sqrt{10}$

(∘) **5** $\sqrt{8} + \sqrt{2}$         **6** $\sqrt{18} + 2\sqrt{2}$

**7** $\sqrt{12} + \sqrt{27}$         **8** $\sqrt{20} + \sqrt{245}$

**9** $\sqrt[3]{16} + \sqrt[3]{54}$         **10** $\sqrt[3]{40} + \sqrt[3]{320}$

**11** $\sqrt{8} - \sqrt{98}$         **12** $\sqrt{28} - \sqrt{175}$

**13** $\sqrt{5} + \sqrt[3]{16} + \sqrt{45} - \sqrt[3]{432}$

**14** $\sqrt[4]{162} - \sqrt{24} - \sqrt[4]{32} + \sqrt{150}$

(∘) **15** $\sqrt{ab^2 c} + \sqrt[3]{a^4 b^7 c^2} + \sqrt{a^3 c} - \sqrt[3]{a^4 bc^5}$

**16** $\sqrt{x^3 z} - \sqrt{xy^2 z} + 3\sqrt{xz^5}$

**17** $\sqrt[3]{s^4 t^2} - t\sqrt[3]{st^2} + \sqrt[6]{s^2 t^4} - \sqrt[3]{s^7 t^5}$

**18** $\sqrt[3]{xz^2} - \sqrt[3]{xy^3 z^5} + y\sqrt[3]{xz^5}$

*Find the product in problems 19 to 34.*

$(\circ)$   **19**   $(\sqrt{2} + \sqrt{5})(\sqrt{2} - \sqrt{5})$     **20**   $(\sqrt{7} + \sqrt{3})(\sqrt{7} - \sqrt{3})$

$(\circ)$   **21**   $(\sqrt{7} + 2\sqrt{5})(\sqrt{7} - 3\sqrt{5})$     **22**   $(\sqrt{3} + 4\sqrt{2})(\sqrt{3} + 3\sqrt{2})$

    **23**   $(\sqrt{2} - \sqrt{3})^2$     **24**   $(\sqrt{5} - \sqrt{3})^2$

    **25**   $(1 + \sqrt{2})^2$     **26**   $(1 - \sqrt{3})^2$

    **27**   $(5 - 4\sqrt{3})(8 + 2\sqrt{3})$     **28**   $(3 - \sqrt{2})(5 + 2\sqrt{2})$

    **29**   $(\sqrt{2x} + \sqrt{y})(\sqrt{2x} - \sqrt{y})$     **30**   $(\sqrt{5w} + \sqrt{3z})(\sqrt{5w} - \sqrt{3z})$

$(\circ)$   **31**   $(3 - 2\sqrt{x})^2$     **32**   $(5 + 2\sqrt{t})^2$

    **33**   $(3\sqrt{x} + 1)(2\sqrt{x} - 1)$     **34**   $(\sqrt{3y} + \sqrt{2z})(\sqrt{8y} - \sqrt{27z})$

*Rationalize each of the following denominators.*

$(\circ)$   **35**   $\dfrac{1}{1 + \sqrt{2}}$        **36**   $\dfrac{2}{3 - 2\sqrt{2}}$

    **37**   $\dfrac{\sqrt{3}}{1 - \sqrt{3}}$        **38**   $\dfrac{\sqrt{5}}{\sqrt{5} - 1}$

    **39**   $\dfrac{1 + \sqrt{7}}{1 - \sqrt{7}}$        **40**   $\dfrac{2 + \sqrt{3}}{2 - \sqrt{3}}$

    **41**   $\dfrac{\sqrt{3} + \sqrt{5}}{\sqrt{3} - \sqrt{5}}$        **42**   $\dfrac{\sqrt{2} - \sqrt{3}}{\sqrt{2} + \sqrt{3}}$

$(\circ)$   **43**   $\dfrac{\sqrt{6} + \sqrt{3}}{\sqrt{2} - \sqrt{3}}$        **44**   $\dfrac{\sqrt{12} - \sqrt{3}}{2 + \sqrt{5}}$

    **45**   $\dfrac{x}{\sqrt{x} - 2}$        **46**   $\dfrac{1}{\sqrt{3} - y}$

    **47**   $\dfrac{\sqrt{a} + \sqrt{b}}{\sqrt{a} - \sqrt{b}}$        **48**   $\dfrac{\sqrt{a} - \sqrt{b}}{\sqrt{a} + \sqrt{b}}$

$(\circ)$   **49**   $\dfrac{x - y}{\sqrt{x} - \sqrt{y}}$        **50**   $\dfrac{x - 4y^2}{\sqrt{x} + 2y}$

*Perform the indicated additions or subtractions, and then rationalize the denominators.*

$(\circ)$   **51**   $\dfrac{1}{\sqrt{x - 1}} + \sqrt{x - 1}$        **52**   $\dfrac{1}{\sqrt{y + 3}} + \sqrt{y + 3}$

    **53**   $\dfrac{1}{3 + \sqrt{x}} + \dfrac{1}{3 - \sqrt{x}}$        **54**   $\dfrac{1}{\sqrt{5} + \sqrt{x}} - \dfrac{1}{\sqrt{5} - \sqrt{x}}$

$(\circ)$    55   $\dfrac{1}{\sqrt{2} + \sqrt{3}} - \dfrac{1}{\sqrt{2} + 2\sqrt{3}}$     56   $\dfrac{1}{\sqrt{5} + \sqrt{2}} + \dfrac{1}{3\sqrt{5} - \sqrt{2}}$

57   $\dfrac{1}{\sqrt{x} - 1} + \dfrac{1}{2\sqrt{x} + 1}$     58   $\dfrac{3}{3\sqrt{y} - 2} - \dfrac{2}{2\sqrt{y} - 3}$

59   $\dfrac{x}{\sqrt{x^2 + 2}} + \dfrac{\sqrt{x^2 + 2}}{x}$     60   $\dfrac{x}{\sqrt{x^2 + 2}} - \dfrac{\sqrt{x^2 + 2}}{x}$

$(\circ)$

*Sample solutions*
*for exercise set 5.5*

5   $\sqrt{8} + \sqrt{2} = 2\sqrt{2} + \sqrt{2} = 3\sqrt{2}$

15   $\sqrt{ab^2 c} + \sqrt[3]{a^4 b^7 c^2} + \sqrt{a^3 c} - \sqrt[3]{a^4 bc^5}$

$= b\sqrt{ac} + ab^2\sqrt[3]{abc^2} + a\sqrt{ac} - ac\sqrt[3]{abc^2}$

$= (a + b)\sqrt{ac} + (ab^2 - ac)\sqrt[3]{abc^2}$

19   $(\sqrt{2} + \sqrt{5})(\sqrt{2} - \sqrt{5}) = 2 - 5 = -3$

21   $(\sqrt{7} + 2\sqrt{5})(\sqrt{7} - 3\sqrt{5}) = 7 + (2 - 3)\sqrt{35} - 30$

$= -23 - \sqrt{35}$

31   $(3 - 2\sqrt{x})^2 = 9 - 12\sqrt{x} + 4x$

35   $\dfrac{1}{1 + \sqrt{2}} \cdot \dfrac{1 - \sqrt{2}}{1 - \sqrt{2}} = \dfrac{1 - \sqrt{2}}{1 - 2} = \dfrac{1 - \sqrt{2}}{-1} = \sqrt{2} - 1$

43   $\dfrac{\sqrt{6} + \sqrt{3}}{\sqrt{2} - \sqrt{3}} \cdot \dfrac{\sqrt{2} + \sqrt{3}}{\sqrt{2} + \sqrt{3}} = \dfrac{\sqrt{12} + \sqrt{6} + \sqrt{18} + 3}{2 - 3}$

$= -(2\sqrt{3} + \sqrt{6} + 3\sqrt{2} + 3)$

49   $\dfrac{x - y}{\sqrt{x} - \sqrt{y}} \cdot \dfrac{\sqrt{x} + \sqrt{y}}{\sqrt{x} + \sqrt{y}} = \dfrac{(x - y)(\sqrt{x} + \sqrt{y})}{x - y}$

$= \sqrt{x} + \sqrt{y} \qquad (x \neq y)$

Alternately, you can factor the numerator and write

$\dfrac{x - y}{\sqrt{x} - \sqrt{y}} = \dfrac{(\sqrt{x} + \sqrt{y})(\sqrt{x} - \sqrt{y})}{\sqrt{x} - \sqrt{y}} = \sqrt{x} + \sqrt{y} \qquad (x \neq y).$

51   The LCD of the two fractions is clearly $\sqrt{x - 1}$. We must put both fractions over this common denominator, add, and then rationalize:

$$\frac{1}{\sqrt{x-1}} + \sqrt{x-1} = \frac{1}{\sqrt{x-1}} + \sqrt{x-1} \cdot \frac{\sqrt{x-1}}{\sqrt{x-1}}$$

$$= \frac{1}{\sqrt{x-1}} + \frac{x-1}{\sqrt{x-1}} = \frac{x}{\sqrt{x-1}}$$

$$= \frac{x\sqrt{x-1}}{x-1} \qquad (x \neq 1).$$

Of course, you can first rationalize, and then add:

$$\frac{1}{\sqrt{x-1}} + \sqrt{x-1} = \frac{\sqrt{x-1}}{x-1} + \sqrt{x-1}$$

$$= \frac{\sqrt{x-1}}{x-1} + \frac{\sqrt{x-1}\,(x-1)}{x-1}$$

$$= \frac{x\sqrt{x-1}}{x-1} \qquad (x \neq 1).$$

55
$$\frac{1}{\sqrt{2}+\sqrt{3}} - \frac{1}{\sqrt{2}+2\sqrt{3}} = \frac{(\sqrt{2}+2\sqrt{3}) - (\sqrt{2}+\sqrt{3})}{(\sqrt{2}+\sqrt{3})(\sqrt{2}+2\sqrt{3})}$$

$$= \frac{\sqrt{3}}{2+3\sqrt{6}+6} = \frac{\sqrt{3}}{8+3\sqrt{6}}$$

$$= \frac{\sqrt{3}\,(8-3\sqrt{6})}{(8+3\sqrt{6})(8-3\sqrt{6})}$$

$$= \frac{8\sqrt{3}-3\sqrt{18}}{64-54} = \frac{8\sqrt{3}-9\sqrt{2}}{10}$$

# (5.6) (Sums and differences of complex numbers)

*complex numbers*

*C*

In our discussion of radicals we were very careful to point out that negative numbers do not have square roots in the real number system. It is possible, however, to enlarge the set of real numbers to a new set of numbers that *does* contain square roots of negative numbers. This new set is called the set of *complex numbers* and will be denoted by the letter *C*.

We assume the existence of a complex number $i = \sqrt{-1}$ having the property that $i^2 = -1$, and then define $C$ to be the set of all expressions of the form $a + bi$, with $a$ and $b$ real numbers. If multiplication of complex numbers is to be commutative and associative, then for $b > 0$,

$$(i\sqrt{b})(i\sqrt{b}) = (i^2)(\sqrt{b})^2 = (-1)(b) = -b,$$

so that $\sqrt{-b} = i\sqrt{b}$ does exist in the complex number system. Thus $\sqrt{-4} = 2i$, and $\sqrt{-3} = i\sqrt{3}$, and so on.

Since any real number $a$ may be written in the form $a + 0i$, the real numbers may be regarded as a subset of $C$. It should be noted that to avoid notational confusion there are times when it will be convenient to write a complex number as $a + ib$ instead of $a + bi$. If $b \neq 0$, the number $a + bi$ is sometimes called an *imaginary* number. This terminology dates back to a time when people did not really understand complex numbers. There is nothing "imaginary" about $C$. Mathematicians are able to construct the complex number system, and it is just as real as the real number system. Since the word "imaginary" is misleading, we simply won't use it.

*imaginary*

Two complex numbers $a + bi$, $c + di$ are called *equal* when both $a = c$ and $b = d$. Thus $2 + 3i = (5 - 3) + (1 + 2)i$, but $2 + 3i \neq 3 + 2i$. Addition of complex numbers is defined by the following statement.

---

CN.1 $(a + bi) + (c + di) = (a + c) + (b + d)i$

---

Thus $(2 + 3i) + (4 + 5i) = (2 + 4) + (3 + 5)i = 6 + 8i$, whereas $(6 - 4i) + (-3 + 4i) = (6 - 3) + (-4 + 4)i = 3$. It should be noted, of course, that $6 - 4i$ denotes the complex number $6 + (-4)i$.

Subtraction of complex numbers is defined, as you might expect.

---

CN.2 $(a + bi) - (c + di) = (a - c) + (b - d)i$

---

We illustrate the process by noting that $(2 + 3i) - (4 + 5i) = (2 - 4) + (3 - 5)i = -2 - 2i$, and $(6 - 4i) - (-3 + 4i) = [6 - (-3)] + (-4 - 4)i = 9 - 8i$.

In closing, we must also mention that addition of complex numbers is both commutative and associative. Furthermore, $0 + 0i$ acts as an additive identity element, and it is easy to show that $-a - bi$ is the additive inverse of $a + bi$, in the sense that $(a + bi)$

$+ (-a - bi) = 0 + 0i$. Thus addition of complex numbers has all the properties of addition of real numbers.

## Exercise set 5.6

*Put each of the following in the form a + bi or a − bi, with a, b ∈ R.*

( ∘ )  **1**  $3 + \sqrt{-10}$            **2**  $4 + \sqrt{-7}$

  **3**  $2 + \sqrt{-9}$            **4**  $6 + \sqrt{-16}$

  **5**  $12 - 3\sqrt{-12}$            **6**  $3 - \sqrt{-8}$

*Perform the indicated operations, and express your answer in the form a + bi or a − bi, (a, b ∈ R).*

( ∘ )  **7**  $(3 + 4i) + (8 + 2i)$            **8**  $(7 + i) + (3 + 2i)$

  **9**  $i + (5 + 2i)$            **10**  $3i + (7 + i)$

  **11**  $(1 + 2i) + (1 - 2i)$            **12**  $(3 + 4i) + (3 - 4i)$

  **13**  $(3 + 4i) - (6 - 2i)$            **14**  $(1 + 2i) - (-2 + 3i)$

  **15**  $(\sqrt{2} + 3i) - (1 + \sqrt{2} + i)$     **16**  $(1 + i\sqrt{2}) - (1 + \sqrt{2} + i\sqrt{2})$

( ∘ )  **17**  $(3 - i) - [(4 + 2i) - (1 + 3i)]$   **18**  $(2 + i) - [(3 + 7i) - (2 + 3i)]$

  **19**  $\left[ 3 - \left( \dfrac{5}{2} - \dfrac{3}{4} i \right) \right] - \left[ (7 - 2i) + \left( \dfrac{3}{2} + \dfrac{5}{4} i \right) \right]$

  **20**  $\left[ 4 + \left( \dfrac{3}{7} - \dfrac{1}{3} i \right) \right] - \left[ (2 + 3i) - \left( 1 + \dfrac{13}{3} i \right) \right]$

( ∘ )
*Sample solutions*
*for exercise set 5.6*

  **1**  $3 + \sqrt{-10} = 3 + i\sqrt{10}$

  **7**  $(3 + 4i) + (8 + 2i) = (3 + 8) + (4 + 2)i = 11 + 6i$

  **17**  $(3 - i) - [(4 + 2i) - (1 + 3i)] = (3 - i) - [(4 - 1) + (2 - 3)i]$

  $= (3 - i) - (3 - i) = 0 + 0i$

## (5.7) (Products and quotients of complex numbers)

Among other things, a complex number is an algebraic expression. The rules for forming products of complex numbers are simply those that apply to any product of algebraic expressions, except that one must use the fact that $i^2 = -1$. Thus

$$(a + bi)(c + di) = ac + (ad + bc)i + bdi^2$$

$$= ac + (ad + bc)i - bd$$

$$= (ac - bd) + (ad + bc)i,$$

which leads us to our next statement.

( ○ )    **CN.3** $(a + bi)(c + di) = (ac - bd) + (ad + bc)i.$

Direct application of this definition shows that

$$(2 + 3i)(5 - 2i) = [2 \cdot 5 - 3(-2)] + [2(-2) + (3 \cdot 5)]i$$

$$= (10 + 6) + (-4 + 15)i$$

$$= 16 + 11i.$$

But there is no point in memorizing the definition of the product of complex numbers, for it is just as easy to treat it as a product of algebraic expressions and write

$$(2 + 3i)(5 - 2i) = 10 - 4i + 15i - 6i^2$$

$$= (10 + 6) + (15 - 4)i$$

$$= 16 + 11i.$$

For much the same reason, it is not necessary to learn a formula for the quotient of two complex numbers. The key items here are the fundamental principle of fractions (which is true in $C$) and the fact that $(c + di)(c - di) = c^2 + d^2$. For example,

$$\frac{2 + 3i}{5 - 2i} = \frac{2 + 3i}{5 - 2i} \cdot \frac{5 + 2i}{5 + 2i} = \frac{10 + 4i + 15i - 6}{25 + 4}$$

$$= \frac{4 + 19i}{29} = \frac{4}{29} + \frac{19}{29}i.$$

The general situation is that for $c + di \neq 0 + 0i,$

$$\frac{a + bi}{c + di} = \frac{a + bi}{c + di} \cdot \frac{c - di}{c - di} = \frac{(ac - bd) + (ad + bc)i}{c^2 + d^2},$$

so that we have the following.

---

$(\circ)$ CN.4 $\quad \dfrac{a + bi}{c + di} = \dfrac{ac - bd}{c^2 + d^2} + \dfrac{ad + bc}{c^2 + d^2} i$

---

It is fairly easy to show that multiplication of complex numbers is both associative and commutative, that 1 serves as a multiplicative identity, and that every complex number other than $0 + 0i$ has a multiplicative inverse. Since multiplication distributes over addition we see from this (and a similar comment made in Section 5.6) that the laws of arithmetic for the reals carry over to the complex number system.

It is important to notice, though, that there *are* properties of the real numbers that do not carry over to the complex numbers. For example, if $a$ and $b \in R^+$, then $\sqrt{a}\,\sqrt{b} = \sqrt{ab}$, whereas $\sqrt{-a}$ $\cdot \sqrt{-b} = -\sqrt{ab}$ $[\neq \sqrt{(-a)(-b)}]$. Thus $\sqrt{3}\,\sqrt{4} = \sqrt{12}$, whereas $\sqrt{-3}\,\sqrt{-4} = i\sqrt{3}\,i\sqrt{4} = i^2\sqrt{12} = -\sqrt{12}$. Any confusion along these lines can be avoided by simply writing $\sqrt{-a}$ in the form $i\sqrt{a}$ for $a \in R^+$.

## Exercise set 5.7

Perform the indicated operations. Express each answer in the form $a + bi$ or $a - bi$ $(a, b \in R)$.

$(\circ)$    **1**   $3(4 - 2i)$               **2**   $5(5 + 3i)$

**3**   $-6\left(\dfrac{1}{2} + 2i\right)$         **4**   $-4\left(3 + \dfrac{1}{2}i\right)$

**5**   $i(2 + 3i)$             **6**   $i(7 - 3i)$

**7**   $-2i(1 + 5i)$         **8**   $-6i(3 - 2i)$

$(\circ)$    **9**   $(1 + 2i)(7 - i)$      **10**   $(3 - 4i)(2 + 3i)$

**11**   $(\sqrt{2} - 3i)(\sqrt{2} + i)$      **12**   $(\sqrt{3} - 4i)(2\sqrt{3} + i)$

**13**   $(5 - i\sqrt{5})(3 + i\sqrt{5})$      **14**   $(2 - i\sqrt{3})(4 + i\sqrt{12})$

**15**   $(2 + 3i)^2$           **16**   $(1 - 2i)^2$

$(\circ)$   **17**   $(3 + 4i)(3 - 4i)$      **18**   $(1 - 2i)(1 + 2i)$

**19**   $(\sqrt{2} - i\sqrt{3})(\sqrt{2} + i\sqrt{3})$      **20**   $(2 + 3i)(2 - 3i)$

*Now for some practice in division of complex numbers. As before, express each answer in the form a + bi or a − bi (a, b ∈ R).*

(∘)  21  $\dfrac{1}{i}$

22  $\dfrac{3}{2i}$

23  $\dfrac{-5}{3i}$

24  $\dfrac{3}{-4i}$

25  $\dfrac{2}{1+i}$

26  $\dfrac{5}{3+4i}$

27  $\dfrac{i}{1-i}$

28  $\dfrac{i}{1+i}$

29  $\dfrac{-3i}{2+i}$

30  $\dfrac{-4i}{3-2i}$

(∘)  31  $\dfrac{1+2i}{1-2i}$

32  $\dfrac{3+4i}{3-4i}$

33  $\dfrac{5-3i}{3+5i}$

34  $\dfrac{1-4i}{4+i}$

35  $\dfrac{6+7i}{3+4i}$

36  $\dfrac{3-i}{3-4i}$

37  $\dfrac{8-i}{2+i}$

38  $\dfrac{8-2i}{3-5i}$

One of the interesting things about complex numbers is that every nonzero complex number has exactly $n$ $n$th roots. Since $1^4 = (-1)^4 = i^4 = (-i)^4 = 1$, we see, for example, that 1, −1, $i$, and −$i$ are each fourth roots of 1. Problems 39 to 41 are based on similar observations.

(∘)  39  Verify that $\left(-\dfrac{1}{2} + i\,\dfrac{\sqrt{3}}{2}\right)^3 = 1$ and $\left(-\dfrac{1}{2} - i\,\dfrac{\sqrt{3}}{2}\right)^3 = 1$. What are the three cube roots of 1?

40  Verify that $\left(\dfrac{1}{2} + i\,\dfrac{\sqrt{3}}{2}\right)^3 = -1$ and $\left(\dfrac{1}{2} - i\,\dfrac{\sqrt{3}}{2}\right)^3 = -1$. What are the three cube roots of −1?

41  Use problems 39 and 40 to find the 6 sixth roots of 1.

1   $3(4 - 2i) = 12 - 6i$

9   $(1 + 2i)(7 - i) = 7 - i + 14i + 2 = 9 + 13i$

17   $(3 + 4i)(3 - 4i) = 9 + 16 = 25$

21   $\dfrac{1}{i} = \dfrac{1}{i} \cdot \dfrac{i}{i} = \dfrac{i}{-1} = 0 - i$

31   $\dfrac{1 + 2i}{1 - 2i} = \dfrac{1 + 2i}{1 - 2i} \cdot \dfrac{1 + 2i}{1 + 2i} = \dfrac{1 + 4i - 4}{1 + 4} = \dfrac{-3 + 4i}{5} = -\dfrac{3}{5} + \dfrac{4}{5} i$

39   $\left( -\dfrac{1}{2} + i\dfrac{\sqrt{3}}{2} \right)^2 = \dfrac{1}{4} - \dfrac{3}{4} - i\dfrac{\sqrt{3}}{2} = -\dfrac{1}{2} - i\dfrac{\sqrt{3}}{2}$

$\left( -\dfrac{1}{2} + i\dfrac{\sqrt{3}}{2} \right)^3 = \left( -\dfrac{1}{2} + i\dfrac{\sqrt{3}}{2} \right)^2 \left( -\dfrac{1}{2} + i\dfrac{\sqrt{3}}{2} \right)$

$= \left( -\dfrac{1}{2} - i\dfrac{\sqrt{3}}{2} \right)\left( -\dfrac{1}{2} + i\dfrac{\sqrt{3}}{2} \right)$

$= \dfrac{1}{4} + \dfrac{3}{4} = 1$

A similar argument shows that $\left( -\dfrac{1}{2} - i\dfrac{\sqrt{3}}{2} \right)^3 = 1$. The three cube roots

of 1 are 1, $-\dfrac{1}{2} + i\dfrac{\sqrt{3}}{2}$, and $-\dfrac{1}{2} - i\dfrac{\sqrt{3}}{2}$.

# (6)

## Quadratic equations and inequalities

### (6.1) (Solution by factoring)

Chapter 4 was devoted to the study of first-degree equations and inequalities in one variable. In Section 4.1 we gave you the definition of an equation and explained what it means to solve an equation. In Section 4.2, certain rules were developed whereby an equation could be replaced with a simpler, but equivalent, equation. We shall use these ideas often in this chapter. Right now, we shall make use of the fact that addition of an expression to both members of an equation produces an equivalent equation and that multiplication of both members by an expression that is never zero will also produce an equivalent equation. These facts will be applied to develop methods for solving an equation equivalent to one in the form

*second-degree equation in one variable*
*quadratic equation*

**QE.1** $ax^2 + bx + c = 0$     $(a, b, c \in R; a \neq 0)$.

Such an equation is called a *second-degree equation in one variable*, or a *quadratic equation*.

Let's begin by trying to solve the quadratic equation $x^2 + 5x + 6 = 0$. The expression $x^2 + 5x + 6$ is $(x + 2)(x + 3)$ in factored form; so the equation is equivalent to $(x + 2)(x + 3) = 0$. To solve this, we note that the product of the real numbers $a$ and $b$ is zero if and only if $a = 0$ or $b = 0$. To see why this is true, we observe that if $a$ or $b$ is 0, then $ab = 0$ since 0 times anything is 0. On the other hand, if $ab = 0$ and $a \neq 0$, then we may multiply both members of $ab = 0$ by $1/a$ to produce

$$\frac{1}{a}(ab) = \frac{1}{a} \cdot 0$$

$$\left(\frac{1}{a} \cdot a\right)b = 0$$

$$b = 0.$$

Thus $(x + 2)(x + 3) = 0$ holds if and only if $x + 2 = 0$ or $x + 3 = 0$. It follows that the solution set of $x^2 + 5x + 6 = 0$ is $\{-2, -3\}$.

In general, if the left member of the quadratic equation

$$ax^2 + bx + c = 0 \qquad (a, b, c \in I)$$

can be expressed in the factored form

$$(ex + f)(gx + h) \qquad (e, f, g, h \in I)$$

then its solution set is

$$\left\{-\frac{f}{e}, -\frac{h}{g}\right\}.$$

solution by factoring

Since this technique involves factoring the left member of $ax^2 + bx + c = 0$, it is called *solution by factoring*.

It is worth mentioning that a quadratic equation can have either two real solutions, one real solution, or two nonreal, complex solutions. We have just witnessed a case with two real solutions. To see a one-element solution set, we consider $9x^2 - 6x + 1 = 0$. This equation is equivalent to $(3x - 1)^2 = 0$, whose solution set is $\left\{\frac{1}{3}\right\}$.

For an example with nonreal complex solutions, we consider the equation $x^2 + 1 = 0$, which is equivalent to $x^2 = -1$, whose solution set is $\{-i, i\}$.

Of course, a little manipulation may be needed before you can solve certain equations by factoring. Thus, to solve $(2x - 4)^2 = 16x + 96$, we must first write the equivalent equations

$$4x^2 - 16x + 16 = 16x + 96$$

$$4x^2 - 32x - 80 = 0 \quad \text{(adding } -16x - 96 \text{ to both members)}$$

$$x^2 - 8x - 20 = 0 \quad \left(\text{multiplying by } \frac{1}{4}\right).$$

The left member of this last equation factors as $(x + 2)(x - 10)$, and the solution set for the resulting equation, $(x + 2)(x - 10) = 0$, is $\{-2, 10\}$.

All this is well and good, but what do you do when you are faced with a quadratic equation $ax^2 + bx + c = 0$ whose left member cannot readily be factored? Sections 6.2 and 6.3 will give you some additional techniques that can be applied to such an equation.

## Exercise set 6.1

*Solve each of the following quadratic equations by the method of factoring.*

$(\circ)$  **1**  $x^2 - 4 = 0$          **2**  $y^2 - 9 = 0$

**3**  $w^2 + 6w = 0$        **4**  $x^2 + \dfrac{5}{2}x = 0$

$(\circ)$  **5**  $x^2 + 6x - 16 = 0$     **6**  $w^2 + w - 12 = 0$

**7**  $a^2 - 10a + 24 = 0$     **8**  $b^2 + 12b + 32 = 0$

**9**  $z^2 + 10z + 25 = 0$     **10**  $x^2 - 18x + 81 = 0$

*In the next few problems a bit of simplification is necessary before the method of factoring can be applied.*

$(\circ)$  **11**  $x^2 + 5x = 24$        **12**  $x^2 - 4x = 21$

**13**  $d^2 + d = 15 - d$      **14**  $k^2 - k = 12 + 3k$

**15**  $2w^2 - 2w = 3w + 3$    **16**  $4x^2 - 4x = 8x - 5$

**17**  $8y(y + 1) = 15 - 6y$    **18**  $2w(9w + 1) = 5(w + 2)$

**19**  $8x^2 + 8x = 15 - 8x^2$    **20**  $x^2 - 4x + 4 = -8x^2 - 24x$

**21**  $2x^2 + 5x + 2 = (x + 2)(x - 2)$    **22**  $2(d - 2)(d - 1) = d^2 - 5$

**23**  $3(y - 1)(y + 1) = y(y - 5)$

**24**  $(3z - 1)^2 + (z + 5)(z + 2) = 13$

$( \circ )$  25  $\dfrac{1}{x + 2} - \dfrac{2}{x - 1} = 2$

26  $\dfrac{2}{x + 3} - \dfrac{2}{x + 4} = 1$

27  $\dfrac{14}{y + 1} - \dfrac{15}{y + 3} = 1$

28  $\dfrac{10}{w + 3} - \dfrac{3}{w + 1} = 1$

To construct a quadratic equation that has the given numbers $r$ and $s$ as its roots, one merely observes that any such equation must be equivalent to

$$(x - r)(x - s) = 0;$$

that is, equivalent to

$$x^2 - (r + s)x + rs = 0.$$

Thus, to construct an equation that has 3 and $-\dfrac{5}{3}$ as its roots, we begin with $(x - 3)\left( x + \dfrac{5}{3} \right) = 0$ and replace it with the equivalent equations $(x - 3)(3x + 5) = 0$ and $3x^2 - 4x - 15 = 0$.

*In the remaining problems, find a quadratic equation $ax^2 + bx + c = 0$*
*($a, b, c \in I$) with the indicated roots.*

$( \circ )$  29  1 and 3

30  2 and $-1$

31  5 and $-7$

$( \circ )$  32  $-2$ and $-4$

33  $-3$

34  $-5$

$( \circ )$  35  0 and $-5$

36  0 and 8

37  $\dfrac{7}{3}$ and $\dfrac{3}{7}$

38  $\dfrac{1}{3}$ and $\dfrac{2}{5}$

39  $\dfrac{7}{6}$ and 5

40  $\dfrac{3}{5}$ and $-\dfrac{2}{7}$

$( \circ )$

*Sample solutions*
*for exercise set 6.1*

1  In factored form the equation is $(x + 2)(x - 2) = 0$; so the solution set is $\{-2, 2\}$.

5  By factoring $x^2 + 6x - 16 = 0$ we get $(x + 8)(x - 2) = 0$; so the solution set is $\{2, -8\}$.

11  The equation $x^2 + 5x = 24$ must first be put in the form $ax^2 + bx + c = 0$. To accomplish this we subtract 24 from both members and obtain

$$x^2 + 5x - 24 = 0$$

$$(x + 8)(x - 3) = 0$$

$$x = -8 \text{ or } x = 3.$$

The solution set is $\{-8, 3\}$.

**25** We begin by multiplying both members of the equation by $(x + 2)(x - 1)$ to produce $(x - 1) - 2(x + 2) = 2(x + 2)(x - 1)$. Simplifying this equation, we get

$$x - 1 - 2x - 4 = 2(x^2 + x - 2)$$
$$-x - 5 = 2x^2 + 2x - 4$$
$$0 = 2x^2 + 3x + 1$$
$$0 = (2x + 1)(x + 1).$$

The solution set is $\left\{-\dfrac{1}{2}, -1\right\}$. It should be noted, however, that we could not allow $x = -2$ or $x = 1$ to be roots (even if we had obtained one of them), for these numbers would yield a zero denominator in the original equation.

**29** $(x - 1)(x - 3) = 0$
$$x^2 - 4x + 3 = 0$$

**33** $(x + 3)^2 = 0$
$$x^2 + 6x + 9 = 0$$

**37** The equation $\left(x - \dfrac{7}{3}\right)\left(x - \dfrac{3}{7}\right) = 0$ has the desired roots. Multiply by $3 \cdot 7 = 21$ to get rid of the fractions. This produces

$$(3x - 7)(7x - 3) = 0$$
$$21x^2 - 58x + 21 = 0.$$

# (6.2) (Extraction of roots and completing the square)

As we promised at the end of Section 6.1, we are going to show you several other approaches to the problem of solving quadratic equations. For example, the equation

$$x^2 = b$$

*extraction of roots*

can be solved by a method called *extraction of roots*. Any solution of the given equation must be a (real or complex) square root of $b$. Hence the solution set of $x^2 = b$ is $\{-\sqrt{b}, \sqrt{b}\}$. To solve $3x^2 = 15$

we divide both members by 3 to obtain the equation $x^2 = 5$, whose solution set is $\{-\sqrt{5}, \sqrt{5}\}$.

This same method may be used to solve an equation of the form

$$(x - a)^2 = b$$

since we can get $x - a = \pm\sqrt{b}$ by extraction of roots; so $x = a + \sqrt{b}$ or $x = a - \sqrt{b}$. For example, the solution set of $(x + 2)^2 = 3$ is $\{-2 + \sqrt{3}, -2 - \sqrt{3}\}$.

*completing the square*

As an introduction to the method of *completing the square,* let's have a look at the quadratic equation $x^2 + 2x - 3 = 0$. We begin by writing the equivalent equation $x^2 + 2x = 3$. We next observe that $(x + 1)^2 = x^2 + 2x + 1$; so if we were to add 1 to both members of the equation $x^2 + 2x = 3$, the left member of the resulting equation would be a perfect square. This leads us to the equations

$$x^2 + 2x + 1 = 3 + 1$$

$$(x + 1)^2 = 4.$$

By extraction of roots, $x + 1 = 2$ or $-2$; so $x = 1$ or $-3$.

When you use this method you will frequently have to add a constant to an expression of the form $x^2 + kx$ in order to produce a perfect square; that is, an expression of the form $(x + a)^2$. The key item here is that

$$\left(x + \frac{k}{2}\right)^2 = x^2 + kx + \frac{k^2}{4};$$

so you will want to take half the coefficient of $x$, square it, and use this for the constant. Thus

$$x^2 + 3x + \left(\frac{3}{2}\right)^2 = \left(x + \frac{3}{2}\right)^2,$$

$$x^2 - \frac{7}{2}x + \left(\frac{7}{4}\right)^2 = \left(x - \frac{7}{4}\right)^2, \text{ and so on.}$$

Now let's solve the equation $3x^2 + x - 1 = 0$ by completing the square. First, we write the equivalent equation, $3x^2 + x = 1$, and then divide through by 3 to obtain $x^2 + (1/3)x = 1/3$. The coefficient of $x$ is $1/3$. Take half of $1/3$ and square it to obtain $(1/6)^2 = 1/36$; so $1/36$ must be added to both members of the equation in order to obtain a perfect square on the left side. We therefore write

$$x^2 + \frac{1}{3}x + \frac{1}{36} = \frac{1}{3} + \frac{1}{36}$$

$$\left(x + \frac{1}{6}\right)^2 = \frac{13}{36}$$

$$x + \frac{1}{6} = \frac{\sqrt{13}}{6} \quad \text{or} \quad -\frac{\sqrt{13}}{6}$$

$$x = -\frac{1}{6} + \frac{\sqrt{13}}{6} \quad \text{or} \quad x = -\frac{1}{6} - \frac{\sqrt{13}}{6}.$$

The solution set is $\left\{ -\frac{1}{6} + \frac{\sqrt{13}}{6}, \ -\frac{1}{6} - \frac{\sqrt{13}}{6} \right\}$.

Here, then, is the general procedure for solving the quadratic equation $ax^2 + bx + c = 0$ by completing the square:

**Step 1:** Write the equation in the form $ax^2 + bx = -c$.

**Step 2:** Divide both sides by $a$ to obtain $x^2 + \frac{b}{a}x = -\frac{c}{a}$.

**Step 3:** Take half of $\frac{b}{a}$, square it to obtain $\left(\frac{b}{2a}\right)^2 = \frac{b^2}{4a^2}$ and add this to both sides to produce $x^2 + \frac{b}{a}x + \left(\frac{b}{2a}\right)^2 = \left(\frac{b}{2a}\right)^2 - \frac{c}{a}$ or $\left(x + \frac{b}{2a}\right)^2 = d$ where $d = \left(\frac{b}{2a}\right)^2 - \frac{c}{a}$.

**Step 4:** Use extraction of roots to see that $x + \frac{b}{2a} = \pm \sqrt{d}$, from which the solution set can be obtained.

## Exercise set 6.2

*Solve each of the following equations by extraction of roots.*

($\circ$)  **1**  $x^2 = 4$      **2**  $x^2 = 9$      **3**  $y^2 = 8$

        **4**  $y^2 = 3$      **5**  $3x^2 = 18$      **6**  $2x^2 = 10$

($\circ$)  **7**  $\frac{w^2}{3} = 5$      **8**  $\frac{v^2}{8} = 3$      **9**  $3x^2 = -18$

( ∘ )   **10**  $2x^2 = -10$   **11**  $(x + 2)^2 = 1$   **12**  $(x - 3)^2 = 4$

**13**  $(x - 1)^2 = 9$   **14**  $(x + 7)^2 = 16$   **15**  $\left(y + \dfrac{1}{2}\right)^2 = 1$

**16**  $\left(w - \dfrac{1}{3}\right)^2 = \dfrac{4}{9}$   **17**  $(3z - 2)^2 = 25$   **18**  $(2x + 5)^2 = 9$

**19**  $(7t + 3)^2 = 49$   **20**  $(5v - 6)^2 = 36$

In each of the following problems you are to determine $k$ so that the expression will be a perfect square, and then express it in the form $(x + a)^2$ or $(x - a)^2$.

( ∘ )   **21**  $x^2 - 2x + k$   **22**  $x^2 + 6x + k$

**23**  $y^2 - 5y + k$   **24**  $z^2 - 9z + k$

**25**  $a^2 - \dfrac{4}{3}a + k$   **26**  $b^2 - \dfrac{6}{5}b + k$

( ∘ )   **27**  $x^2 + kx + 100$   **28**  $w^2 + kw + 49$

**29**  $t^2 + kt + \dfrac{25}{4}$   **30**  $s^2 - ks + \dfrac{36}{49}$

Solve by completing the square.

( ∘ )   **31**  $x^2 + 4x + 3 = 0$   **32**  $x^2 - 6x - 16 = 0$

**33**  $p^2 + 9p + 18 = 0$   **34**  $q^2 - 5q - 14 = 0$

**35**  $x^2 + 3x - 1 = 0$   **36**  $x^2 - 5x - 2 = 0$

**37**  $y^2 + y - 4 = 0$   **38**  $w^2 + 2w - \dfrac{1}{2} = 0$

( ∘ )   **39**  $b^2 - 5b + 2 = 0$   **40**  $a^2 - 3a + 1 = 0$

( ∘ )   **41**  $x^2 + 2x + 2 = 0$   **42**  $w^2 + 6w + 10 = 0$

**43**  $2x^2 + 3x + 1 = 0$   **44**  $2x^2 - 5x + 3 = 0$

( ∘ )   **45**  $2x^2 - 5x - 3 = 0$   **46**  $3x^2 + 2x - 6 = 0$

**47**  $9w^2 - 24w + 16 = 0$   **48**  $4x^2 + 20x + 25 = 0$

**49**  $x^2 + ax - 3a^2 = 0$   $(a \in R^+)$

**50**  $x^2 + 3bx + b^2 = 0$   $(b \in R^+)$

( ∘ )

*Sample solutions for exercise set 6.2*   **1**  If $x^2$ is to equal 4, then $x$ must be a square root of 4. Hence, $x = 2$ or $x = -2$; so the solution set is $\{-2, 2\}$.

**9**  To solve $3x^2 = -18$, we divide through by 3 to obtain $x^2 = -6$. From this we see that the solution set is $\{i\sqrt{6}, -i\sqrt{6}\}$.

**11** If $(x + 2)^2 = 1$, then $x + 2 = 1$ or $-1$; so $x = -1$ or $x = -3$. The solution set is $\{-1, -3\}$.

**21** We must make $x^2 - 2x + k$ into a perfect square. Half of the coefficient of $x$ is $(1/2)(-2) = -1$, and $(-1)^2 = 1$; so we take $k = 1$ and write $x^2 - 2x + 1 = (x - 1)^2$.

**27** The square root of 100 is 10; so the coefficient of $x$ must be $2 \cdot 10 = 20$. We therefore take $k = 20$ and write $x^2 + 20x + 100 = (x + 10)^2$.

**31** We are given the equation $x^2 + 4x + 3 = 0$. Step by step, here is how to solve it by completing the square.

$$x^2 + 4x = -3$$

$$x^2 + 4x + 4 = -3 + 4$$

$$(x + 2)^2 = 1$$

$$x + 2 = 1 \quad \text{or} \quad -1$$

$$x = -1 \quad \text{or} \quad x = -3 \quad \text{(see problem 11)}$$

The solution set is $\{-3, -1\}$.

**39**
$$b^2 - 5b + 2 = 0$$

$$b^2 - 5b = -2$$

$$b^2 - 5b + \left(-\frac{5}{2}\right)^2 = \left(-\frac{5}{2}\right)^2 - 2$$

$$\left(b - \frac{5}{2}\right)^2 = \frac{25}{4} - 2 = \frac{25}{4} - \frac{8}{4} = \frac{17}{4}$$

$$b = \frac{5}{2} + \frac{\sqrt{17}}{2} \quad \text{or} \quad \frac{5}{2} - \frac{\sqrt{17}}{2}$$

The solution set is $\left\{\dfrac{5}{2} - \dfrac{\sqrt{17}}{2}, \dfrac{5}{2} + \dfrac{\sqrt{17}}{2}\right\}$.

**41**
$$x^2 + 2x + 2 = 0$$

$$x^2 + 2x = -2$$

$$x^2 + 2x + 1 = -2 + 1$$

$$(x + 1)^2 = -1$$

$$x + 1 = i \quad \text{or} \quad -i$$

$$x = -1 + i \quad \text{or} \quad x = -1 - i$$

The solution set is $\{-1 + i, -1 - i\}$.

45 $\quad 2x^2 - 5x - 3 = 0$

$$2x^2 - 5x = 3$$

$$x^2 - \frac{5}{2}x = \frac{3}{2}$$

$$x^2 - \frac{5}{2}x + \left(-\frac{5}{4}\right)^2 = \frac{3}{2} + \left(-\frac{5}{4}\right)^2$$

$$\left(x - \frac{5}{4}\right)^2 = \frac{3}{2} + \frac{25}{16} = \frac{24}{16} + \frac{25}{16} = \frac{49}{16}$$

$$x - \frac{5}{4} = \frac{7}{4} \quad \text{or} \quad -\frac{7}{4}$$

$$x = \frac{12}{4} \quad \text{or} \quad -\frac{2}{4}$$

$$x = 3 \quad \text{or} \quad x = -\frac{1}{2}$$

The solution set is $\left\{3, -\dfrac{1}{2}\right\}$.

# (6.3) (The quadratic formula)

Our third method of solving quadratic equations involves a formula for the roots. This formula is derived by completing the square on the general quadratic equation

**QE.1** $ax^2 + bx + c = 0 \quad (a \neq 0)$

in the usual way: if $ax^2 + bx + c = 0$, then $ax^2 + bx = -c$. We now divide through by $a$ (noting that $a \neq 0$) to obtain $x^2 + (b/a)x = -c/a$. The left member of this equation becomes a perfect square if $(b/2a)^2 = b^2/4a^2$ is added to both members:

$$x^2 + \frac{b}{a}x + \frac{b^2}{4a^2} = \frac{b^2}{4a^2} - \frac{c}{a}$$

$$\left(x + \frac{b}{2a}\right)^2 = \frac{b^2 - 4ac}{4a^2}$$

$$x + \frac{b}{2a} = \frac{\pm\sqrt{b^2 - 4ac}}{2a}$$

$$x = \frac{-b \pm \sqrt{b^2 - 4ac}}{2a}$$

Equation QE.2 is known as the *quadratic formula*. It tells us that the solution set of the general quadratic equation QE.1 is

$$\left\{ \frac{-b + \sqrt{b^2 - 4ac}}{2a}, \frac{-b - \sqrt{b^2 - 4ac}}{2a} \right\}.$$

The quantity $b^2 - 4ac$ that occurs in the quadratic formula is called the *discriminant* of the equation $ax^2 + bx + c = 0$. The nature of the solutions of a quadratic equation can be determined from the value of its discriminant as follows:

**1** When the discriminant is negative, the equation has two nonreal, complex solutions.

**2** When the discriminant is zero, there is a unique, real solution that takes the form $x = -(b/2a)$.

**3** When the discriminant is positive, there are two real solutions.

To illustrate the use of the quadratic formula, consider the equation $3x^2 + x - 2 = 0$. Here $a = 3$, $b = 1$, $c = -2$, and we have

$$x = \frac{-1 \pm \sqrt{1 - 4(3)(-2)}}{2 \cdot 3}$$

$$= \frac{-1 \pm \sqrt{1 + 24}}{6}$$

$$= \frac{-1 \pm 5}{6}.$$

Thus

$$x = \frac{-1 + 5}{6} = \frac{4}{6} = \frac{2}{3} \quad \text{or} \quad x = \frac{-1 - 5}{6} = \frac{-6}{6} = -1;$$

so the solution set is $\left\{ -1, \frac{2}{3} \right\}$.

# Exercise set 6.3

*Determine the nature of the solutions of each of the following quadratic equations by examining its discriminant. Do not solve the equation.*

( ∘ )    **1**    $x^2 - 6x + 9 = 0$            **2**    $x^2 - 6x + 10 = 0$

       **3**    $x^2 - 6x + 8 = 0$            **4**    $y^2 + 5y - 6 = 0$

( ∘ )    **5**    $2w^2 - 3w + 4 = 0$        **6**    $7t^2 - 2t + 1 = 0$

       **7**    $3a^2 - 5 = 0$               **8**    $3b^2 + 5 = 0$

       **9**    $3t^2 + 7t = 0$            **10**    $3s^2 - 5s + \dfrac{25}{12} = 0$

*Solve the next equations by using the quadratic formula.*

( ∘ )    **11**    $x^2 - 5x - 14 = 0$       **12**    $x^2 + 22x + 120 = 0$

      **13**    $p^2 - 8p + 16 = 0$       **14**    $w^2 + 12w + 36 = 0$

      **15**    $x^2 - 4x + 5 = 0$        **16**    $x^2 + 5x + 8 = 0$

      **17**    $4x^2 - x - 3 = 0$        **18**    $2x^2 + 3x - 2 = 0$

( ∘ )    **19**    $3w^2 + \pi w - 2 = 0$      **20**    $4v^2 + \pi^2 v - 3 = 0$

*Solve for x, y, or z. All other letters are to be treated as constants.*

( ∘ )    **21**    $x^2 + 3x - k^2 = 0$       **22**    $y^2 + cy - 1 = 0$

      **23**    $az^2 + 2z - 3a = 0$       **24**    $ax^2 + bx - 2a = 0$

*For what values of k will each of the following have a real solution?*

( ∘ )    **25**    $x^2 - 2x + k = 0$        **26**    $y^2 + 6y + k = 0$

      **27**    $w^2 + 8w - k = 0$       **28**    $3z^2 + 5z - k = 0$

      **29**    $2x^2 + 3x + 2k = 0$      **30**    $2t^2 - 7t - 3k = 0$

*Determine k so that each equation has a unique real root.*

      **31**    $2w^2 - 5w + k = 0$       **32**    $3s^2 - s + k = 0$

      **33**    $2kx^2 - 3x + 2 = 0$       **34**    $ky^2 + 7y + 3 = 0$

*For what values of k will each of the following equations have nonreal, complex solutions?*

( ∘ )    **35**    $2x^2 + 3x + 2k = 0$      **36**    $2t^2 - 7t - 3k = 0$

      **37**    $2kx^2 - 3x + 2 = 0$       **38**    $ky^2 + 7y + 3 = 0$

*If* $r_1$, $r_2$ *are roots of the quadratic equation* $ax^2 + bx + c = 0$ ($a \neq 0$), *then*

$$r_1 + r_2 = \left(\frac{-b}{2a} + \frac{\sqrt{b^2 - 4ac}}{2a}\right) + \left(\frac{-b}{2a} - \frac{\sqrt{b^2 - 4ac}}{2a}\right)$$

$$= \frac{-2b}{2a} = -\frac{b}{a}$$

while

$$r_1 r_2 = \left(\frac{-b}{2a} + \frac{\sqrt{b^2 - 4ac}}{2a}\right)\left(\frac{-b}{2a} - \frac{\sqrt{b^2 - 4ac}}{2a}\right)$$

$$= \frac{b^2}{4a^2} - \frac{b^2 - 4ac}{4a^2}$$

$$= \frac{4ac}{4a^2} = \frac{c}{a}.$$

In other words, the sum of the roots of the equation $ax^2 + bx + c = 0$ $(a \neq 0)$ is $-(b/a)$, and the product of its roots is $c/a$.

*Make use of these facts to find the sum and product of the roots of each of the remaining equations.*

( ∘ )    39   $x^2 - 5x + 8 = 0$          40   $x^2 - 8x + 5 = 0$

41   $x^2 - 3x - 7 = 0$          42   $x^2 + 4x + 6 = 0$

43   $2x^2 - 3x + 5 = 0$        44   $5x^2 + 4x - 3 = 0$

45   $\dfrac{x^2}{9} + \dfrac{3x}{5} - 3 = 0$        46   $-\dfrac{x^2}{2} - \dfrac{3x}{4} + \dfrac{5}{6} = 0$

( ∘ )

*Sample solutions for exercise set 6.3*

1   We have $a = 1$, $b = -6$, and $c = 9$; so the discriminant is $(-6)^2 - 4(1)(9) = 36 - 36 = 0$. Hence, there is exactly one root.

5   The discriminant is $(-3)^2 - 4(2)(4) = 9 - 32 < 0$; so there are two non-real, complex solutions.

11   The discriminant is $(-5)^2 - 4(1)(-14) = 25 + 56 = 81$. By the quadratic formula, the roots are

$$x = \frac{-(-5) + \sqrt{81}}{2} = \frac{5 + 9}{2} = \frac{14}{2} = 7$$

and

$$x = \frac{5 - 9}{2} = \frac{-4}{2} = -2.$$

The solution set is $\{-2, 7\}$.

**19**  Here we have $a = 3$, $b = \pi$, and $c = -2$; so

$$x = \frac{-\pi \pm \sqrt{\pi^2 - 4(3)(-2)}}{6} = \frac{-\pi \pm \sqrt{\pi^2 + 24}}{6}$$

The solution set is $\left\{ \dfrac{-\pi - \sqrt{\pi^2 + 24}}{6}, \dfrac{-\pi + \sqrt{\pi^2 + 24}}{6} \right\}$.

**21**  In this one, $a = 1$, $b = 3$, $c = -k^2$; so the solutions are

$$x = \frac{-3 \pm \sqrt{9 - 4(-k^2)}}{2} = \frac{-3 \pm \sqrt{9 + 4k^2}}{2}.$$

**25**  The discriminant is $(-2)^2 - 4k = 4 - 4k$. We must determine $k$ so that the discriminant is non-negative. This amounts to solving the inequality $4 - 4k \geq 0$. We see from this that $4k \leq 4$, or that $k \leq 1$.

**35**  The discriminant of the equation $2x^2 + 3x + 2k = 0$ is $9 - 4(2)(2k)$ $= 9 - 16k$. For the equation to have nonreal complex solutions, one must have $9 - 16k < 0$, $9 < 16k$, or $k > 9/16$.

**39**  Since $a = 1$, $b = -5$, and $c = 8$, the sum of the roots must be $-(-5/1)$ $= 5$, and their product must be $8/1 = 8$.

# (6.4) (Equations quadratic in form)

Sometimes an equation is not quadratic in the given variable but is quadratic in some expression involving the variable. Thus

**(1)**  $x^4 - 4x^2 - 12 = 0$

is not quadratic, but if we let $u = x^2$, the equation becomes $u^2 - 4u - 12 = 0$, which is quadratic. Similarly,

**(2)**  $2(3x^2 - 1)^2 + 3(3x^2 - 1) + 1 = 0$

is not quadratic in $x$, but if $u = 3x^2 - 1$, the result is the quadratic equation $2u^2 + 3u - 1 = 0$.

How do you solve such an equation? Let's illustrate the technique by solving equations (1) and (2). To solve (1) we look at $u^2 - 4u - 12 = 0$, where $u = x^2$. This factors as $(u - 6)(u + 2) = 0$, to yield the roots $u = 6$, $u = -2$. Thus $x^2 = 6$ or $x^2 = -2$. The equation $x^2 = 6$ has $\pm\sqrt{6}$ as its roots, and $\pm i\sqrt{2}$ are the roots of $x^2 = -2$. It follows that the solution set of $x^4 - 4x^2 - 12 = 0$ is $\{\sqrt{6}, -\sqrt{6}, i\sqrt{2}, -i\sqrt{2}\}$.

To solve $2(3x^2 - 1)^2 + 3(3x^2 - 1) + 1 = 0$, we look at $2u^2 + 3u + 1 = 0$ with $u = 3x^2 - 1$. This equation factors as $(2u + 1)(u + 1) = 0$; so $u = -1/2$ or $u = -1$.

| If $u = -1/2$ | If $u = -1$ |
|---|---|
| $3x^2 - 1 = -\dfrac{1}{2}$ | $3x^2 - 1 = -1$ |
| $3x^2 = \dfrac{1}{2}$ | $3x^2 = 0$ |
| $x^2 = \dfrac{1}{6}$ | $x = 0$. |
| $x = \pm\dfrac{\sqrt{6}}{6}$. | |

We conclude that the solution set of the equation $2(3x^2 - 1)^2 + 3(3x^2 - 1) + 1 = 0$ is $\left\{0, \dfrac{\sqrt{6}}{6}, -\dfrac{\sqrt{6}}{6}\right\}$.

To sum it all up, an equation in the variable $x$ is quadratic in form if it looks like $au^2 + bu + c = 0$ $(a \neq 0)$ where $u$ is an expression in $x$. To solve such an equation you first solve $au^2 + bu + c = 0$ to obtain the roots $u = r$, $u = s$ $(r, s \in R)$. Then regard $u = r$ and $u = s$ as equations in $x$ to determine the roots of the original equation.

## Exercise set 6.4

*Each of the following equations is quadratic in form. Use the methods of this section to find the solution set of each equation.*

($\circ$)  1   $x^4 - 13x^2 + 36 = 0$       2   $x^4 - 17x^2 + 16 = 0$

  3   $x^4 - 7x^2 - 18 = 0$       4   $x^4 - x^2 - 12 = 0$

  5   $(x^2 - 3)^2 - 4(x^2 - 3) = 12$   6   $\left(\dfrac{x^2 - 15}{5}\right)^2 + 3\left(\dfrac{x^2 - 15}{5}\right) = 10$

*Find all real solutions of the following equations.*

  7   $y^8 - 6y^4 = 160$       8   $z^6 - 7z^3 = 8$

  9   $x^{10} - 3x^5 + 2 = 0$       10   $x^{12} - 4x^6 + 4 = 0$

*In problems 11 to 16, find each solution set.*

$(\circ)$ **11** $x - 3\sqrt{x} + 2 = 0$ **12** $x - 5\sqrt{x} + 6 = 0$

$(\circ)$ **13** $w - \sqrt{w} - 6 = 0$ **14** $z + 3\sqrt{z} = 10$

**15** $x^{2/3} - 6x^{1/3} + 9 = 0$ **16** $x^{2/5} - 3x^{1/5} + 2 = 0$

An equation like $2x^3 + 5x^2 - 3x = 0$ may be solved by observing that it factors as $x(2x^2 + 5x - 3) = 0$ with $2x^2 + 5x - 3 = 0$ a quadratic equation. In factored form, our equation becomes $x(2x - 1)(x + 3) = 0$, for which the solution set is $\left\{0, \dfrac{1}{2}, -3\right\}$. Be careful though! Don't forget the solution $x = 0$. To do so is a common error that you must guard against.

*The foregoing technique should enable you to solve problems 17 to 20.*

**17** $x^4 + 2x^3 - 3x^2 = 0$ **18** $2x^3 - 3x^2 + x = 0$

$(\circ)$ **19** $x^2 - x^{3/2} = 6x$ **20** $x^{3/2} - 7x + 10x^{1/2} = 0$

$(\circ)$

*Sample solutions for exercise set 6.4*

**1** The given equation is $x^4 - 13x^2 + 36 = 0$. With $u = x^2$, this becomes

$$u^2 - 13u + 36 = 0$$

$$(u - 9)(u - 4) = 0$$

$$u = 9 \quad \text{or} \quad u = 4.$$

Thus $x^2 = 9$ or $x^2 = 4$, so $x = \pm 3$ or $x = \pm 2$, and the solution set is $\{3, -3, 2, -2\}$.

**11** With $u = \sqrt{x}$, the equation becomes

$$u^2 - 3u + 2 = 0$$

$$(u - 2)(u - 1) = 0$$

$$u = 2 \quad \text{or} \quad u = 1$$

$$x = 4 \quad \text{or} \quad x = 1.$$

The solution set, then, is $\{1, 4\}$.

**13** Setting $u = \sqrt{w}$ produces the equation $u^2 - u - 6 = 0$; so

$$(u - 3)(u + 2) = 0$$

$$u = 3 \quad \text{or} \quad u = -2$$

$$\sqrt{w} = 3 \quad \text{or} \quad \sqrt{w} = -2.$$

But $\sqrt{w}$ cannot be negative so that the solution $\sqrt{w} = -2$ must be thrown out. If $\sqrt{w} = 3$, then $w = 9$ and we have the solution set $\{9\}$.

**19**   We are given the equation $x^2 - x^{3/2} = 6x$; so

$$x^2 - x^{3/2} - 6x = 0$$

$$x(x - x^{1/2} - 6) = 0.$$

With $u = x^{1/2}$, the equation $x - x^{1/2} - 6 = 0$ becomes $u^2 - u - 6 = 0$, which factors as $(u - 3)(u + 2) = 0$ and has $u = 3$, $u = -2$ as its roots. As in problem 13, $u = -2$ must be discarded since $x^{1/2}$ can never equal a negative number. Thus $x^{1/2} = 3$, and $x = 9$. The solution set of $x(x - x^{1/2} - 6) = 0$ is therefore seen to be $\{0, 9\}$.

# (6.5) (Equations involving square roots)

An equation that involves the square root of an expression containing an unknown may be solved by judicious use of the following principle:

(∘)

*The solution set of any equation is a subset of the solution set of the equation formed by squaring both of its members.*

*extraneous roots*

One must be careful though, for the squaring operation does not necessarily lead to an equation equivalent to the original one; indeed, it may introduce extra solutions, called *extraneous roots*. Thus if we square both members of the equation $x - 1 = 2$, the new equation is $x^2 - 2x + 1 = 4$, which may be solved by writing

$$x^2 - 2x - 3 = 0$$

$$(x - 3)(x + 1) = 0$$

$$x = 3 \quad \text{or} \quad x = -1.$$

The root $x = -1$ is a solution of $x^2 - 2x + 1 = 4$, but it is not a solution of the original equation, $x - 1 = 2$. It must therefore be discarded as an extraneous root.

For a more complicated example let us consider $x = 3 + \sqrt{5 - x}$. If we were to square both members immediately, the resulting equation would still involve a radical; so we first form the equivalent equation $x - 3 = \sqrt{5 - x}$. Squaring this equation produces

$$x^2 - 6x + 9 = 5 - x$$

$$x^2 - 5x + 4 = 0$$

$$(x - 4)(x - 1) = 0$$

$$x = 1 \quad \text{or} \quad x = 4.$$

We must now test for extraneous roots by actually substituting any suspected root into the original equation.

| $x = 1$ | $x = 4$ |
|---|---|
| $1 = 3 + \sqrt{5 - 1}$ | $4 = 3 + \sqrt{5 - 4}$ |
| $1 = 3 + 2$ | $4 = 3 + 1$ |
| False! | True! |

We therefore see that $x = 1$ is extraneous, and the solution set is $\{4\}$.

Here is a step-by-step procedure for solving an equation that contains the square root of a single expression:

**Step 1:** Replace the equation (if necessary) with an equivalent equation that has the radical as one of its members.

**Step 2:** Form a new equation by squaring both members.

**Step 3:** Solve the new equation.

**Step 4:** Test for extraneous roots by actual substitution in the original equation.

## *Exercise set 6.5*

*Find all real solutions of each equation and check each answer.*

(∘)  **1** $\sqrt{x - 2} = 3$          **2** $\sqrt{y + 3} = 2$

**3** $\sqrt{2x}\,\sqrt{x - 2} = 4$        **4** $\sqrt{x}\,\sqrt{x - 8} = 3$

**5** $\sqrt{y}\,\sqrt{y + 5} = 6$         **6** $\sqrt{w}\,\sqrt{w - 9} = 6$

(∘)  **7** $\sqrt{2x + 19} = x + 2$       **8** $\sqrt{22 - 3x} = x - 4$

**9** $x = 2 + \sqrt{3x - 6}$        **10** $t = \sqrt{4t + 17} - 5$

Sometimes an equation will involve more than one square root. Consider, for example, the equation $\sqrt{x + 7} - \sqrt{x} = 1$. The technique here is to form an equivalent equation that has one of the radicals as its left-hand member. Thus we have $\sqrt{x + 7} = \sqrt{x} + 1$. Squaring both members we obtain

$$x + 7 = x + 2\sqrt{x} + 1$$

$$6 = 2\sqrt{x}$$

$$\sqrt{x} = 3.$$

Squaring again produces $x = 9$. The next step is to check the answer. With $x = 9$, does $\sqrt{9 + 7} - \sqrt{9} = 1$? Yes. It follows that $x = 9$ is the unique solution.

($\circ$)    11   $\sqrt{x - 4} - \sqrt{x} = 2$        12   $\sqrt{x - 9} - \sqrt{x} = 3$

       13   $\sqrt{x} - \sqrt{x + 9} = 3$        14   $\sqrt{x + 40} - \sqrt{x} = 4$

       15   $\sqrt{x + 4} - \sqrt{x - 3} = 1$        16   $\sqrt{x + 9} - \sqrt{x - 30} = 3$

       17   $\sqrt{4y + 21} - \sqrt{2y + 2} = 3$        18   $\sqrt{3w - 2} - \sqrt{2w - 8} = 2$

($\circ$)    19   $\sqrt{3t} - \sqrt{t + 1} = \sqrt{6t - 3}$        20   $\sqrt{2z} + \sqrt{z - 7} = \sqrt{5z - 13}$

($\circ$)

*Sample solutions for exercise set 6.5*

1   We are given the equation $\sqrt{x - 2} = 3$. Squaring both members yields $x - 2 = 9$ or $x = 11$. The solution set is $\{11\}$. To check, does $\sqrt{11 - 2} = 3$? Yes.

7   Squaring both members of the given equation will produce the equations

$$2x + 19 = x^2 + 4x + 4$$

$$0 = x^2 + 2x - 15$$

$$0 = (x + 5)(x - 3);$$

so $x = -5$ or $x = 3$ are the candidates for solutions. We next test for extraneous roots by actual substitution in the original equation.

| $x = -5$ | $x = 3$ |
|---|---|
| $\sqrt{-10 + 19} = -5 + 2$ | $\sqrt{6 + 19} = 3 + 2$ |
| $3 = -3$ | $5 = 3 + 2$ |

False!                                            True!

This shows that $x = -5$ is an extraneous root, and consequently that the solution set is $\{3\}$.

11   We are given the equation $\sqrt{x - 4} - \sqrt{x} = 2$. We begin by replacing it with the equivalent equation $\sqrt{x - 4} = \sqrt{x} + 2$, whose left member is one of the radicals. Squaring both members will then produce

$$x - 4 = x + 4\sqrt{x} + 4$$

$$0 = 4\sqrt{x} + 8$$

$$\sqrt{x} = -2.$$

But this solution is impossible since $\sqrt{x}$ is always non-negative. We conclude that the solution set is empty.

**19** This one involves three radicals, but the general technique is the same. Square to eliminate the radicals, and then check for extraneous roots. Squaring both members of the given equation yields

$$3t + (t + 1) - 2\sqrt{3t}\sqrt{t + 1} = 6t - 3$$

$$4t + 1 - 2\sqrt{3t}\sqrt{t + 1} = 6t - 3$$

$$-2\sqrt{3t}\sqrt{t + 1} = 2t - 4$$

$$-\sqrt{3t}\sqrt{t + 1} = t - 2.$$

Squaring this last equation now gives us

$$3t(t + 1) = t^2 - 4t + 4$$

$$3t^2 + 3t = t^2 - 4t + 4$$

$$2t^2 + 7t - 4 = 0$$

$$(2t - 1)(t + 4) = 0$$

$$t = \frac{1}{2} \quad \text{or} \quad t = -4.$$

Now we check for extraneous roots. When $t = 1/2$, we have

$$\sqrt{3\left(\frac{1}{2}\right)} - \sqrt{\frac{1}{2} + 1} = 0 = \sqrt{6\left(\frac{1}{2}\right) - 3};$$

so $t = 1/2$ is a solution. For $t = -4$,

$$\sqrt{3(-4)} - \sqrt{-4 + 1} = \sqrt{-12} - \sqrt{-3} = 2i\sqrt{3} - i\sqrt{3} = i\sqrt{3},$$

while

$$\sqrt{6(-4) - 3} = \sqrt{-24 - 3} = \sqrt{-27} = 3i\sqrt{3} \neq i\sqrt{3},$$

and $t = -4$ is extraneous. The solution set is $\left\{\dfrac{1}{2}\right\}$.

# (6.6) (More inequalities)

In Section 4.6 you learned how to solve a first-degree inequality. The present section is devoted to the solution of second-degree inequalities. Such an inequality arises from a second-degree equation by replacing the "equals" sign with one of the four inequality symbols $<$, $\leq$, $>$, or $\geq$. The solution of these inequalities will involve what you did in Section 4.6 (so this might be a good time for a quick review) plus some new techniques that you are about to

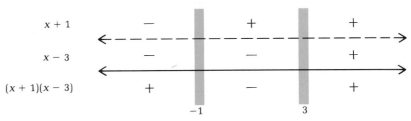

Fig. 6.1

sign graph

learn. Naturally, for this idea to be meaningful, all variables must be restricted to the set $R$ of real numbers.

Let's start by finding the solution set of $x^2 - 3 < 2x$. We want to find all values of $x$ that will make the assertion of the inequality true. We begin by putting it in the form $P(x) < 0$. This is accomplished by adding $-2x$ to both members, which yields the equivalent inequality $x^2 - 2x - 3 < 0$. Factorization now produces $(x - 3)(x + 1) < 0$.

We can now arrive at the solution by indicating on a number line the signs associated with each factor for various number replacements of the variable. Such a picture is called a *sign graph,* and it gives us the solution because $x^2 - 2x - 3$ can be negative only if $x - 3$ and $x + 1$ are opposite in sign. Here then, in Figure 6.1, is a sign graph for the inequality $x^2 - 2x - 3 < 0$.

The horizontal lines in our sign graph each represent number lines, and the two vertical bars are present to remind you that the expression $x^2 - 2x - 3$ can change sign only when $x + 1$ or $x - 3$ changes sign; that is, only when $x = -1$ or $x = 3$. The first row indicates those regions on which the expression $x + 1$ is positive and those on which it is negative. The second row does the same thing for $x - 3$. Using the fact that the product of two numbers is positive if and only if they are both positive or both negative, the third row can be filled in as indicated and the sign of the product determined. We deduce that the solution set is the open interval $(-1,3)$.

To summarize, here is the method for solving a second-degree inequality:

**Step 1:** Replace the inequality with an equivalent one whose right member is 0.

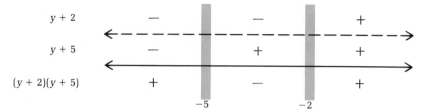

Fig. 6.2

Step 2: Completely factor the left member.

Step 3: Solve the resulting inequality by drawing an appropriate sign graph.

Let's work this procedure through again by considering $y(y + 7) \geq -10$. First we simplify, $y^2 + 7y \geq -10$, then put the equation in a form whose right member is 0, $y^2 + 7y + 10 \geq 0$. Factoring, we obtain $(y + 2)(y + 5) \geq 0$. The expression $(y + 2)(y + 5)$ can change sign only at $y = -2$ or at $y = -5$; so we have the sign graph shown in Figure 6.2. Thus $(y + 2)(y + 5) \geq 0$ when $y \leq -5$ as well as when $y \geq -2$. The solution set is $\{y \mid y \leq -5\} \cup \{y \mid y \geq -2\}$ and its graph looks like this:

As a third example we consider the inequality $x/(x + 1) > 2$. Since $x + 1$ may be either positive or negative we cannot multiply through by $x + 1$ unless we consider $x + 1 > 0$ and $x + 1 < 0$

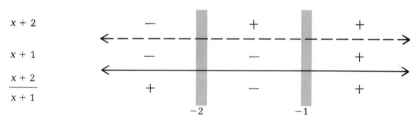

Fig. 6.3

190 (∘) Quadratic equations and inequalities

separately. A good technique for solving this kind of inequality is to replace it with an equivalent inequality whose right member is 0, and then simplify. Proceeding in this manner we have

$$\frac{x}{x + 1} - 2 > 0$$

$$\frac{x}{x + 1} - 2\left(\frac{x + 1}{x + 1}\right) > 0$$

$$\frac{x - 2(x + 1)}{x + 1} > 0$$

$$\frac{x - 2x - 2}{x + 1} > 0$$

$$\frac{-x - 2}{x + 1} > 0$$

$$\frac{x + 2}{x + 1} < 0.$$

Now the quotient of two numbers is negative if and only if they have opposite signs, so that we have the sign graph shown in Figure 6.3.

The solution set is the open interval $(-2, -1)$, whose graph is

Although you should be prepared for variations on the foregoing ideas, you now have the ammunition you will need to do the exercises.

## Exercise set 6.6

*Solve each inequality and sketch a graph of the solution set.*

(∘)  1  $(x - 3)(x + 2) > 0$           2  $(x - 4)(x + 1) < 0$

3  $x(x + 1) \leq 0$           4  $(x - 5)(x + 4) \geq 0$

5  $x^2 - 5x + 6 \geq 0$           6  $x^2 - 4x - 12 \leq 0$

**7** $y^2 + 2 \le 3y$

**8** $w^2 - 5 > 4w$

**9** $x^2 + 1 > 0$

**10** $x^2 > -3$

$(\circ)$ **11** $\dfrac{x}{x-2} > 3$

**12** $\dfrac{y}{y+1} < 5$

**13** $\dfrac{x-2}{x+1} \ge 3$

**14** $\dfrac{z+3}{z-1} < -2$

**15** $\dfrac{3x}{x^2+2} < 1$

**16** $1 \le \dfrac{6x}{x^2+5}$

**17** $x^2(x-3) < 0$

**18** $(y-2)^2(y+1) > 0$

$(\circ)$ **19** $(x+1)(x^2-1) \ge 0$

**20** $w^4(w+1)^3 \le 0$

**21** $x(x-1)(x-2) < 0$

**22** $(x-3)(x^2-1) > 0$

$(\circ)$

*Sample solutions
for exercise set 6.6*

**1** This expression is already in factored form; so we merely note that $(x-3)(x+2)$ can change sign only at $x = 3$ or at $x = -2$. Therefore, we get the following sign graph.

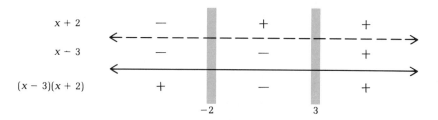

The solution set is seen to be $\{x \mid x < -2\} \cup \{x \mid x > 3\}$, whose graph is

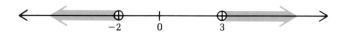

**11** The given inequality is equivalent in turn to each of the following:

$$\frac{x}{x-2} - 3 > 0$$

$$\frac{x - 3(x-2)}{x-2} > 0$$

$$\frac{x - 3x + 6}{x - 2} > 0$$

$$\frac{-2x + 6}{x - 2} > 0.$$

The expression can change sign only at $x = 2$ or at $x = 3$; so we have the situation shown in the following sign graph.

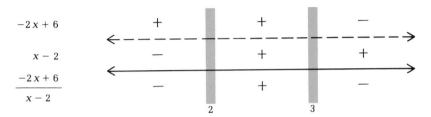

The solution set is $(2,3)$ whose graph is

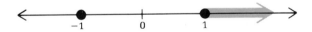

**19** In factored form, we are solving $(x + 1)(x + 1)(x - 1) \geq 0$. The left member equals 0 when $x = 1$ or $x = -1$; so we must ask when it is strictly greater than 0. Since $(x + 1)^2 \geq 0$ for *all* $x$, this is equivalent to saying that $x - 1 > 0$ or that $x > 1$. The solution set is $\{x | x \geq 1\}$ $\cup \{-1\}$ and its graph is

## (6.7) (Word problems)

Back in Sections 4.4 and 4.5 we considered word problems whose solution involved first-degree equations. Now it is time to look at word problems that lead to quadratic equations. Fortunately, there are no extra difficulties involved, and the techniques are essentially those of Section 4.4.

**Problem:** Your neighbor has just installed a swimming pool. You'd like to know how big it is but you do not want to seem nosy.

You manage to overhear a workman say that he has installed a 3-foot walkway around the perimeter of the pool and that the area of this walkway is 264 square feet less than the area of the pool. You also manage to find out that the pool is a rectangular one whose length is 10 feet more than its width. You now have enough information to determine the dimensions of the pool; so go to it.

Solution: Let $x$ be the width of the pool in feet. Then $x + 10$ must be its length. At this point a picture may help. As the diagram

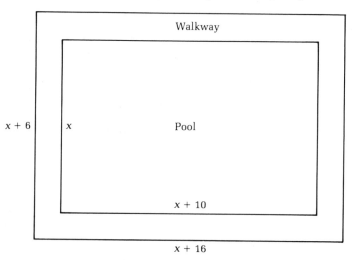

shows, the width of the region enclosed by the pool and the walkway together is $x + 3 + 3 = x + 6$, whereas its length is $(x + 10) + 3 + 3 = x + 16$. Hence the area of the pool plus the walkway is $(x + 6)(x + 16)$. The area of the pool alone is $x(x + 10)$, and the area of the walkway alone is $(x + 6)(x + 16) - x(x + 10)$. The information from the workman tells us that the area of the walkway $+ 264 =$ the area of the pool.

In mathematical notation we have that $[(x + 6)(x + 16) - x(x + 10)] + 264 = x(x + 10)$. Simplifying this, we have

$$(x + 6)(x + 16) + 264 = 2x(x + 10)$$

$$x^2 + 22x + 96 + 264 = 2x^2 + 20x$$

$$2x + 360 = x^2$$

$$x^2 - 2x - 360 = 0$$

$$(x - 20)(x + 18) = 0.$$

This equation has $x = 20$ and $x = -18$ as roots, but the solution $x = -18$ must be rejected on the grounds that a negative width is meaningless. We deduce that the pool is 20 feet wide and 30 feet long.

*Check:* Area of pool plus walkway = $(26)(36) = 936$

$$\text{Area of pool} = (20)(30) = 600$$

$$\text{Area of walkway} = 336$$

Does $336 + 264 = 600$? Yes.

The pool-area problem serves to illustrate a very important fact about word problems in general. When translating a word problem into an equation, the equation acts only as a model for some physical situation. Its roots may or may not be solutions for the original problem. The only way to tell is to check your answer by actual substitution. This fact is also reflected in the next example.

**Problem:** A man drives to work through 30 miles of suburbs followed by 10 miles of city traffic. His speed in the city is 20 mph less than in the suburbs. The trip takes a total of 1 hour and 4 minutes. How fast did he drive in the city?

**Solution:** Let $r$ be his speed through the city. Then $r + 20$ is his speed in the suburbs. We know that *time = distance / rate*, provided that we keep our units in terms of miles and hours. Thus, the time in the suburbs is $30/(r + 20)$, the time in the city is $10/r$, and the total time (in hours) $= 64/60 = 16/15$. It follows that

$$\frac{30}{r + 20} + \frac{10}{r} = \frac{16}{15}$$

$$\frac{15}{r + 20} + \frac{5}{r} = \frac{8}{15}$$

$$15[15r + 5(r + 20)] = 8r(r + 20)$$

$$300r + 1{,}500 = 8r^2 + 160r$$

$$8r^2 - 140r - 1{,}500 = 0$$

$$2r^2 - 35r - 375 = 0$$

$$(2r + 15)(r - 25) = 0$$

$$r = -\frac{15}{2} \text{ or } r = 25.$$

The root $r = -15/2$ must be rejected on the grounds that a negative speed makes no sense in the problem. We are left with $r = 25$.

*Check:* Thirty miles at 45 mph takes 40 minutes and 10 miles at 25 mph takes 24 minutes. Does $40 + 24 = 64$? Yes.


## Exercise set 6.7

### Number problems

( ∘ )  **1**  Find two positive integers whose difference is 9 and whose product is 136.

**2**  Find two positive integers whose difference is 6 and whose product is 91.

**3**  Find two numbers whose sum is 22 and whose product is 120.

**4**  Find two numbers whose sum is 20 and whose product is 96.

### Area problems

( ∘ )  **5**  The perimeter of the basement of a rectangular building is 220 feet, and its floor area is 2,400 square feet. What are its dimensions?

**6**  A rectangular garden whose length exceeds its width by 3 feet is surrounded by a rock border 1 foot wide. If the area of the border is 10 square feet less than that of the garden, what are the dimensions of the garden?

**7**  A man wishes to fence in a rectangular exercise region for his dog. The region borders his house; so he will only have to fence in three sides. If he uses 28 feet of fencing and if the area of the region is 96 square feet, what are its dimensions?

**8**  The length of a rectangular swimming pool is three times its width. The shallow end of the pool is marked by a rope that goes across the pool parallel to one side. If the area of the shallow end of the pool is 630 square feet and if the rope is 69 feet from the deep end of the pool, what are the dimensions of the pool?

### Cost and mixture problems

( ∘ )  **9**  The Tie Rack is a shop that sells two kinds of neckties. The expensive ones sell for $1.50 more per tie than the cheaper line. On a certain day, $240 was taken in on sales of the expensive ties and $225 on sales of the cheaper ones, with a total of 150 ties sold. What was the selling price of each kind of tie?

**10**  A grocer mixes together two kinds of nuts to produce 60 pounds of a blend. If he uses $8 worth of one kind and $60 worth of the other and if the first kind sells for 40¢ per pound less than the other kind, find the price per pound of each kind.

**11** The toll on a bridge is 15¢ more for a truck than it is for a car. On a certain day a total of 1,000 vehicles passed over the bridge. If the truck revenue was $80 and the car revenue $200, determine the toll for a car.

**12** Movie tickets cost $1.10 more for an adult than for a child. On a certain night 200 tickets were sold. The total revenue was $334, of which $54 came from the sale of children's tickets. What is the cost of a children's admission?

## Interest problems

**13** A man has a total of $9,000 invested, some of it in a high-quality stock and some of it in a more speculative issue that yields 1% more per year. If his income from the quality issue is $250 and from the speculative issue $240, what is the percentage rate of return on each investment?

**14** A man has $6,000 invested, some of it in real estate and some in corporate bonds whose yield is 2% less per year than the real-estate investment. If his income is $200 on the real estate and $320 on the bonds, find the rate of return on his real estate holdings.

## Travel problems

( ∘ ) **15** The current in a certain stream is 3 mph. A man rows 10 miles downstream, rests for a half-hour, and then rows upstream back to where he started. If the total trip took 3 hours, how fast did he travel going downstream and how fast upstream?

**16** A plane flies a total of 700 miles with a tail wind of 40 mph. On the return trip it encounters a head wind of 10 mph. If the total air time is 3 hours and 45 minutes, find the air speed of the plane.

**17** To reach a certain waterfall, a man must travel a total of 20 miles; 15 miles by car, and the remainder on foot. If he drives 45 mph faster than he walks and if the trip takes 78 minutes, how fast did he drive?

**18** A man has been injured at a remote mining camp. The camp foreman radios for an ambulance to meet them at the junction of the camp access road and the main highway. The injured man will be taken to this point by jeep, as the access road is not safe for the ambulance. Both vehicles leave at once for the meeting point. The jeep must travel 20 miles and the ambulance 60 miles, but the average speed of the ambulance is 65 mph greater than that of the jeep. When the jeep reaches the intersection it develops that the ambulance had been waiting 35 minutes. What was the average speed of the jeep? At what speed should the ambulance have traveled for them both to arrive at the same time?

*Work problems*

( ∘ ) **19** It takes Painter A 2 hours longer than Painter B to paint a certain room. Working together, they can paint it in 2 hours and 24 minutes. How long would it take Painter A if he worked separately?

**20** It takes 2 hours longer to fill a certain storage tank than it does to drain it. If the intake and drain pipes are both opened, it takes 60 hours to drain a full tank. How long would it take to refill it with the drain closed?

**21** Harry the handyman and his son are painting a room. Harry starts at 8:00 A.M. but must leave at 11:00 A.M. His son comes to help him at 10:00 A.M. They work together for an hour, and the son finishes the job by himself after Harry leaves. The son completes the job at noon. If Harry painted the room by himself, it would take him 2 hours longer than it would his son painting alone. How long would it take Harry to paint the room by himself?

**22** During a recent rainstorm the basement of the police station was flooded. A pump was started, but after 2 hours it was decided that the process was taking too long. At that point a second pump was hooked up, and the job was completed using both pumps in another 3 hours. Working separately, it would have taken the original pump 4 hours longer than the new one. How long would it have taken the original pump?

( ∘ )

*Sample solutions
for exercise set 6.7*

**1** If $x$ is the smaller number, then $x + 9$ is the larger, and the problem states that $x(x + 9) = 136$. Thus

$$x^2 + 9x - 136 = 0$$

$$(x - 8)(x + 17) = 0.$$

Since we are after positive integers, this tells us that $x = 8$; so the two numbers are 8 and 17.

*Check:* $17 - 8 = 9$, and $8 \cdot 17 = 136$.

**5** Let $x$ be the width of the basement. Then $2{,}400/x$ must be its length, and we have

$$2x + 2\left( \frac{2{,}400}{x} \right) = 220$$

$$x + \frac{2{,}400}{x} = 110$$

$$x^2 + 2{,}400 = 110x$$

$$x^2 - 110x + 2{,}400 = 0$$

$$(x - 80)(x - 30) = 0$$

$$x = 80 \text{ or } x = 30.$$

This shows that the building must be 30 feet wide and 80 feet long.

*Check:* $2 \cdot 30 + 2 \cdot 80 = 220$, and $(30)(80) = 2,400$.

**9**  Let $x$ be the price in dollars of an expensive tie. Then $x - (3/2)$ is the price of a cheap tie, and we have that number of *ties sold* = *revenue/ price*. With the number of cheap ties = $225/[x - (3/2)]$ and the number of expensive ties = $240/x$, we are led to the equation

$$\frac{225}{x - \dfrac{3}{2}} + \frac{240}{x} = 150.$$

Dividing through by 15 and multiplying by $x[x - (3/2)]$ will produce

$$15x + 16x - 24 = 10x\left(x - \frac{3}{2}\right)$$

$$31x - 24 = 10x^2 - 15x$$

$$10x^2 - 46x + 24 = 0$$

$$5x^2 - 23x + 12 = 0$$

$$(5x - 3)(x - 4) = 0$$

$$x = \frac{3}{5}, \text{ or } x = 4.$$

The solution $x = 3/5$ would force the cheaper tie to sell at a negative price; so it must be discarded. We conclude that the expensive ties sell for $4.00 and the cheaper ones for $2.50.

*Check:* At these prices there were $240/4 = 60$ expensive ties sold and $225/2.50 = 90$ of the cheap line. Notice that $60 + 90 = 150$.

**15**  Let $r$ be the rowing speed in still water. Then $r + 3$ is his speed downstream and $r - 3$ his speed upstream. The time spent rowing downstream is $10/(r + 3)$ and that upstream $10/(r - 3)$. Hence

$$\frac{10}{r + 3} + \frac{1}{2} + \frac{10}{r - 3} = 3.$$

After we simplify that equation, we have

$$r^2 - 8r - 9 = 0$$

$$(r - 9)(r + 1) = 0$$

$$r = 9 \text{ mph.}$$

The downstream speed is 12 mph and the upstream speed 6 mph.

Check: $\dfrac{10}{12} + \dfrac{1}{2} + \dfrac{10}{6} = 3$

19  Let $t$ be the time needed by Painter B, so that $t + 2$ is the time for Painter A. Thus, in an hour's time they paint $\dfrac{1}{t} + \dfrac{1}{t + 2}$ of the room, and this must equal $\dfrac{1}{12/5} = \dfrac{5}{12}$. Thus

$$\frac{1}{t} + \frac{1}{t + 2} = \frac{5}{12}$$

$$t + 2 + t = \frac{5}{12} t(t + 2)$$

$$24t + 24 = 5t^2 + 10t$$

$$5t^2 - 14t - 24 = 0$$

$$(t - 4)(5t + 6) = 0.$$

Therefore, it will take Painter B 4 hours and A 6 hours.

Check: $\dfrac{1}{4} + \dfrac{1}{6} = \dfrac{5}{12}$; so, working together, it takes $2\,\dfrac{2}{5}$ hours.

# (7)

## *Functions and relations*

### (7.1) (Graphing in two dimensions)

Up to this point we have been dealing with equations in one variable. Suppose, however, we are faced with an equation like $2x + 3y = 4$ in the two variables $x$ and $y$? A solution of such an equation must involve replacing both $x$ and $y$ with real numbers in such a way that the assertion of the equation is true. Thus when $x = 1/2$ and $y = 1$ we have $2(1/2) + 3(1) = 1 + 3 = 4$; so the pair of numbers $1/2$ and $1$ is a solution of the equation $2x + 3y = 4$. The pair $(1/2,1)$ is an example of an *ordered pair* of real numbers; that is, a pair $(a,b)$ of real numbers with $a$ denoting the first number in the pair and $b$ the second. We say that $(a,b) = (c,d)$ when both $a = c$ and $b = d$. Thus we want $(1,2) = (1,2)$ but $(1,2) \neq (2,1)$.

In order to visualize the solutions of an equation in two variables we must learn how to graph ordered pairs of real numbers. To do this we start with a horizontal and a vertical line intersecting at a point $O$, called the *origin*. The horizontal line is called the *horizontal axis*, or the *x-axis*; the vertical line is called the *vertical axis*, or the *y-axis*. The two lines are jointly referred to as the

*ordered pair*

*origin*

*horizontal axis, x-axis*

*vertical axis, y-axis*

*coordinate axes.* The system of coordinates we are about to introduce is called a *Cartesian,* or *rectangular coordinate, system,* and the plane determined by the coordinate axes is called the *Cartesian plane.*

The coordinate axes are then made into number lines, with the origin serving as the graph of 0 on both axes. The right half of the horizontal axis and the top half of the vertical axis are reserved for positive numbers. The two axes need not have the same scale. At this point we have described in words the situation illustrated by Figure 7.1. The coordinate axes divide the plane into four regions called *quadrants,* which are numbered as shown in Figure 7.2. (It should be noted that the quadrants do not include the coordinate axes.)

To draw the graph of the point $(a,b)$, you graph $a$ on the horizontal axis, $b$ on the vertical axis, construct perpendiculars to these points, and then see where the perpendiculars meet. Their point of intersection $P$ is defined to be the *graph* of $(a,b)$. In turn, the ordered pair $(a,b)$ represents the *coordinates* of the point $P$, with $a$ called its *abscissa* and $b$ its *ordinate* (see Figure 7.3). The graphs of the three points—$(1,3)$, $(4,-2)$, and $(-1,8)$—are displayed in Figure 7.4.

The *graph* of an equation in $x$ and $y$ is constructed by plotting all ordered pairs $(a,b)$ such that the equation is satisfied by $x = a$ and $y = b$. Thus $(1,5)$ represents coordinates for a point on the graph of $y = 2x + 3$ since $2(1) + 3 = 5$, but $(-3,5)$ does not since $2(-3) + 3 = -3 \neq 5$. To locate a point on the graph whose abscissa is $-3$, you just observe that the ordinate of such a point must satisfy the equation with $x = -3$. It follows that $(-3,-3)$ gives the desired point.

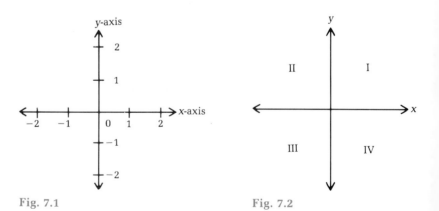

Fig. 7.1                    Fig. 7.2

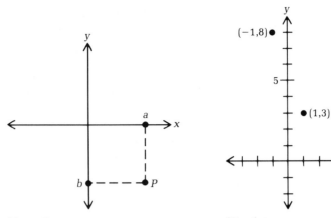

Fig. 7.3                    Fig. 7.4

To get a little practice in plotting the graphs of equations, let's consider equations of the form

$ax + by = c \quad (a \neq 0 \text{ or } b \neq 0).$

*linear equations*

It can be shown that the graph of such an equation is a straight line: for that reason, these equations are called *linear equations*. Recall now that a line is completely determined by any two points; so the graph of a linear equation may be constructed by locating any two of its points and then drawing a straight line through them. The graph of $2x + 3y = 6$ is constructed by observing that

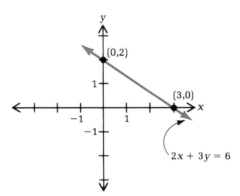

Fig. 7.5

203 (∘) Graphing in two dimensions

when $x = 0$, $y = 2$ and when $y = 0$, $x = 3$. Thus $(3,0)$ and $(0,2)$ are both on the graph, and we need only draw the line determined by these points (see Figure 7.5).

## Exercise set 7.1

*Draw the graph of each of the following sets of points.*

( ∘ )  **1** $\{(1,3), (3,1)\}$  **2** $\{(1,-5), (-6,2)\}$

**3** $\{(-3,4), (3,5), (-2,-1)\}$  **4** $\{(-5,-3), (7,-1), (0,4)\}$

**5** $\{(1,3), (1,5), (1,7)\}$  **6** $\{(-1,-1), (0,-1), (2,-1)\}$

*Observe that each of the following sets of points is finite because of the restrictions given on x or y. Draw the graph of each set.*

( ∘ )  **7** $\{(x,y) \mid y = x + 2, x = 0, 1, 2\}$

**8** $\{(x,y) \mid y = 3x - 4, x = 3, 5, 7\}$

**9** $\{(x,y) \mid 3x - 5y = 8, x = -1, 0, 1, 2\}$

**10** $\{(x,y) \mid 2x + 3y = 4, x = -3, -2, -1, 2\}$

**11** $\{(x,y) \mid y = 4x + 1, y = 0, 1, 2, 3\}$

**12** $\{(x,y) \mid y = 3x - 8, y = -1, 1, 2, 3, 4\}$

*Decide (without graphing) the quadrant in which the graph of each ordered pair of real numbers will lie.*

( ∘ )  **13** $(4,7)$  **14** $(-3,-6)$

**15** $(2,-8)$  **16** $(-3,1)$

**17** $(a^2 + 1, a^2 + 2)$  **18** $(a^2 + 1, -a^2 - 3)$

*Draw the graph of each of the following equations.*

( ∘ )  **19** $y = 3x + 5$  **20** $y = -2x + 1$

**21** $3x - 4y = 12$  **22** $2x + 3y = 6$

**23** $\dfrac{x}{2} + \dfrac{y}{3} = 1$  **24** $\dfrac{x}{5} - \dfrac{y}{3} = 1$

To plot the graph of a set like $\{(x,y) \mid y = x + 1, x \geq 1\}$, you simply use all ordered pairs $(x,y)$ that satisfy the equation with $x \geq 1$. The graph shown in Figure 7.6 is drawn by noting that when $x = 1$, $y = 2$ and when $x = 3$, $y = 4$. The condition $x \geq 1$ requires that the point $(1,2)$ be part of the graph. The use of a *closed* dot for $(1,2)$ indicates this. If we were to graph $\{(x,y) \mid y = x + 1, x > 1\}$, we would indicate

(1,2) with an *open* dot to show that it is not to be counted as part of the graph.

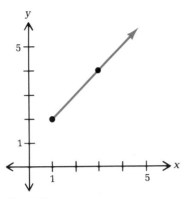

**Fig. 7.6**

*Use the foregoing procedure to draw the graph of each of the sets of points in problems 24 to 30.*

(∘)  25  $\{(x,y) \mid 5y + 3x = 10, x < 2\}$

26  $\{(x,y) \mid 4y - 2x = 4, x \le 3\}$

27  $\{(x,y) \mid x = 3y + 5, x > 6\}$

28  $\{(x,y) \mid 2x - y = 4, x \ge -2\}$

29  $\{(x,y) \mid 3x - 5y = 8, -1 \le x \le 6\}$

30  $\{(x,y) \mid 2y + x = 2, 0 < x \le 4\}$

(∘)

*Sample solutions for exercise set 7.1*

1

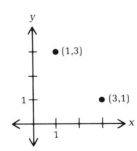

205 (∘) Graphing in two dimensions

**7**

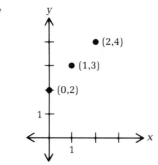

**13**  Points whose x- and y-coordinates are both positive must lie in quadrant I.

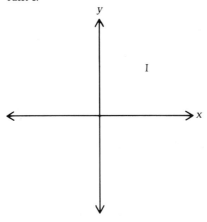

**19**  We must locate two points on the graph of $y = 3x + 5$. To do this we note that when $x = 0$, $y = 5$ and when $x = -2$, $y = -1$. Plotting the

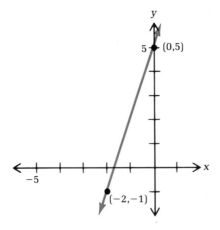

graphs of these two points and drawing a line through them produces the desired graph.

25  When $x = 2$, $5y + 6 = 10$, and $5y = 4$, so that $y = 4/5$. When $x = 0$, $y = 2$. Thus, we want the line determined by the points $(2,4/5)$ and $(0,2)$ with $(2,4/5)$ omitted as well as any points to its right.

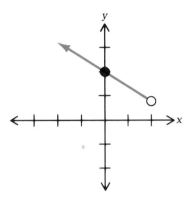

# (7.2) (Relations and functions)

*relations*

*domain*

*range*

Ordered pairs of real numbers were discussed in section 7.1, and it was mentioned that the solution set of an equation in two variables *is* a set of such ordered pairs. Sets of ordered pairs of real numbers are important enough mathematically to warrant a special name: they are called *relations*. The set of numbers that appear as first components of ordered pairs of a relation $R$ is called the *domain* of $R$, and the set of numbers that appear as second components is called the *range* of $R$. Thus if

$$R = \{(1,2), (3,2), (3,4)\}$$

then the domain of $R$ is $\{1,3\}$ and its range is $\{2,4\}$ as shown following.

Elements of domain
$$\{(1,2),\ (3,2),\ (3,4)\}$$
Elements of range

The graph of $R$ is displayed in Figure 7.7.

It is important that you realize a relation is an *arbitrary* set of ordered pairs of real numbers. It might be the set of *all* ordered

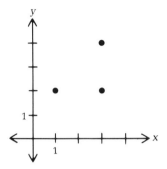

Fig. 7.7

pairs of real numbers, the empty set, a finite set of ordered pairs, or possibly some infinite set of ordered pairs of real numbers. Quite frequently we shall be dealing with relations that arise as the solution set of an equation in two variables. When this happens we shall say that the equation *defines* the relation $R$ if $R$ is its solution set. Thus the equation $x^2 + y^2 = 1$ defines the relation $\{(x,y)|x^2 + y^2 = 1\}$; similarly, $y = 3x + 5$ defines the relation $\{(x,y)|y = 3x + 5\}$.

If a relation is defined by an equation in $x$ and $y$ and if its domain is not otherwise specified, the understanding is that its domain shall consist of all values of $x$ for which the equation has a real solution. Thus the domain of $\{(x,y) \mid x^2 + y^2 = 1\}$ consists of the closed interval $[-1,1]$ since for $x^2 > 1$ the equation $x^2 + y^2 = 1$ has no real solution. In contrast to this, the relation $\{(x,y) \mid y = 3x + 5\}$ has as domain the set of all real numbers. On the other hand, the equation $y = 3x + 5$ $(x = 1, 2, 3)$ defines a relation whose domain is $\{1, 2, 3\}$ since this domain was specified by the restriction on $x$. Figure 7.8 shows the graphs of these last two relations. As a further example we point out that the domain of the relation defined by the equation $xy = 1$ is $\{x \mid x \in R, x \neq 0\}$. To see this we note that for $x \neq 0$, the equation $xy = 1$ can be solved by setting $y = 1/x$, and for $x = 0$ there is no solution.

Let's have a closer look at the relations $R = \{(x,y) \mid x^2 + y^2 = 1\}$ and $S = \{(x,y) \mid y = 3x + 5\}$. There is a fundamental difference between these two relations. Substituting $x = 0$ in the equation $x^2 + y^2 = 1$ produces

$$0^2 + y^2 = 1$$

$$y^2 = 1$$

$$y = \pm 1;$$

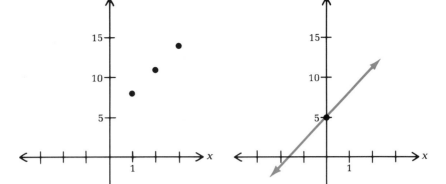

$\{(x,y) \mid y = 3x + 5, x = 1, 2, 3\}$

$\{(x,y) \mid y = 3x + 5\}$

Fig. 7.8

*function*

so the ordered pairs $(0,1)$ and $(0,-1)$ are both in $R$. Thus $R$ has a number in its domain that appears as the first component of more than one of its ordered pairs. On the other hand, each replacement value of $x$ in the equation $y = 3x + 5$ determines a unique value of $y$; so each number in the domain of $S$ appears as the first component of exactly one ordered pair in $S$. Thus when $x = 0$, $y = 3 \cdot 0 + 5 = 5$; so $(0,5)$ is the only ordered pair in $S$ that has $0$ as its first component. A relation is called a *function* if each number in its domain appears as the first component of exactly one of its ordered pairs. In other words:

$(\circ)$   *A function is a relation with the property that no two of its ordered pairs have the same first component.*

This is an extremely important concept that deserves your careful attention.

Let's recapitulate. The relation $R = \{(x,y) \mid x^2 + y^2 = 1\}$ is *not* a function because $(0,1)$ and $(0,-1)$ are both in $R$; in other words, because $0$ appears as the first component of two ordered pairs in $R$. However, the relation $S = \{(x,y) \mid y = 3x + 5\}$ *is* a function because each value of $x$ determines a unique value of $y$; thus when $x = 0$, $y = 5$, when $x = 1$, $y = 8$, and so on. As a further example, let us consider the relation $T$ defined by $x^2 + y^2 = 1$ $(y \geq 0)$. This relation *is* a function because of the restriction that $y$ be non-negative. If the equation is solved for $y$, the result is

$$y^2 = 1 - x^2$$

$$y = \sqrt{1 - x^2}$$

and thus we find that each value of $x$ determines at most one value of $y$; so we do have a function.

In general, to decide whether the relation defined by an equation in two variables is a function, you simply solve the equation for $y$. If $y$ is completely determined by $x$ you have a function; otherwise, you do not have a function. Thus if you solve for $y$ in the equation $x^2 + y^2 = 1$, you obtain $y = \pm \sqrt{1 - x^2}$; so this does not define a function. It was the restriction $y \geq 0$ that made $T$ into a function! To see whether a finite relation is a function you simply examine its ordered pairs. Thus $\{(1,3), (2,4), (1,5)\}$ is not a function since 1 appears as the first component of two ordered pairs, whereas $\{(1,3), (2,4), (3,6)\}$ is a function because no two of its ordered pairs have the same first component.

Relations and functions are often specified by capital or small letters: $R, S, T, f, g, h$, and so on. For each number $a$ in the domain of the function $f$ there is exactly one number $b$ such that $(a,b) \in f$. It is convenient to specify this number $b$ by writing $b = f(a)$ (read "$f$ of $a$"). For example, if $f = \{(1,3), (2,4), (3,4)\}$, we have $f(1) = 3$, $f(2) = 4$ and $f(3) = 4$. If the function $f$ is defined by the equation $y = 3x + 5$, one often combines the symbol $f$ with the variable $x$ to write $f(x) = 3x + 5$. When this is done, the understanding is that for each real number $a$ in the domain of $f$, $f(a)$ is the number obtained by replacing $x$ with $a$ in the equation. Thus

$$f(0) = 3 \cdot 0 + 5 = 5,$$

$$f(1) = 3 \cdot 1 + 5 = 8,$$

$$f(2) = 3 \cdot 2 + 5 = 11.$$

This same thing was done with polynomials when we represented a polynomial with a symbol $P(x)$ and defined $P(a)$ to be the number obtained by replacing $x$ with $a$ in the polynomial.

*one-one function*

Another important concept is that of a *one-one function*, which is a function $f$ that has the property that when $a_1 \neq a_2$, $f(a_1) \neq f(a_2)$. Thus $f = \{(1,3), (2,4), (3,5)\}$ is one-one because $f(1)$, $f(2)$ and $f(3)$ are all different, whereas $g = \{(1,3), (2,3), (3,5)\}$ is *not* one-one because $g(1) = g(2)$.

$(\circ)$ *A function is one-one if and only if no two of its ordered pairs have the same second component.*

Let's look at some examples defined by equations. The function defined by the equation $y = 3x + 5$ is one-one because each value of $y$ determines a unique value of $x$. In other words, if the equation is solved for $x$, the result is

$$3x = y - 5$$

$$x = \frac{1}{3}(y - 5)$$

and $x$ is determined uniquely by $y$. Thus when $y = 2$, $x = \frac{1}{3}(2 - 5)$

$= -1$, and the ordered pair $(-1,2)$ is the only ordered pair in the function that has 2 as its second component. On the other hand, the function $g$ defined by $x^2 + y^2 = 1$ $(y \geq 0)$ is *not* one-one. To see this, set $y = 0$ and notice that $x^2 + 0^2 = 1$, $x^2 = 1$, so $x = \pm 1$. Thus $(1,0)$ and $(-1,0)$ are both in $g$. In fact, this may also be seen by solving the equation $x^2 + y^2 = 1$ for $x$ and noting that

$$x^2 = 1 - y^2$$
$$x = \pm \sqrt{1 - y^2}$$

thereby showing that there can be two values of $x$ associated with a single value of $y$.

The graph of a relation can also be used to decide whether the relation is a function of $x$. To say that a relation $R$ is *not* a function is to say that for some number $a$ there are two ordered pairs $(a,b)$, $(a,c)$ in $R$ with $b \neq c$. This says that the vertical line $x = a$ intersects the graph of $R$ in more than one place.

$(\circ)$ *A relation $R$ is a function if and only if no vertical line intersects the graph of $R$ in more than one place.*

This principle is illustrated in Figures 7.9 and 7.10. The relation in Figure 7.9 is a function because a vertical line can intersect its graph in one point at most; the one in Figure 7.10 is not a function since a vertical line intersects its graph in more than one point.

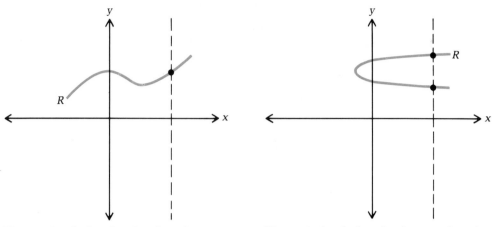

Fig. 7.9 A relation that is a function          Fig. 7.10 A relation that is not a function

## Exercise set 7.2

In problems 1 to 10, you are to (1) specify the domain and range of each relation; (2) decide whether or not each relation is a function; and (3) if the given relation is a function, determine whether or not it is one-one.

**( ∘ )**

1  $\{(2,5), (3,7), (5,2), (6,5)\}$

2  $\{(2,4), (3,-2), (4,1), (2,1)\}$

3  $\{(1,1), (1,2), (1,3)\}$

4  $\{(1,1), (2,1), (3,1\}$

5  $\{(1,1), (2,3), (3,4)\}$

6  $\{(0,5), (5,0)\}$

7  $\{(0,2), (1,3)\}$

8  $\{(0,0), (1,1), (2,2), (0,1)\}$

9  $\{(1,2)\}$

10  $\{(1,2), (2,2)\}$

Obtain the following numerical results.

**( ∘ )**

11  $f(1)$ for $f(x) = x^2 + 7x - 2$

12  $f(2)$ for $f(x) = \dfrac{x + 1}{x^2}$

13  $g(0)$ for $g(x) = \dfrac{1}{x^2 + 2}$

14  $h(-1)$ for $h(x) = \dfrac{x^3 - x}{1 + x^2}$

**( ∘ )**  15  $F(2) - F(1)$ for $F(x) = x^2 + 3x + 2$

**16** $G(1) - G(0)$ for $G(x) = \dfrac{1}{x+1}$

**17** $f(3) - f(-1)$ for $f(x) = x^2 + 2x$

**18** $f(2) - f(-2)$ for $f(x) = x^2 - x + 8$

Suppose now that you are given the function $f(x) = x^2 + 2x + 2$ and asked to form the expression $\dfrac{f(1+h) - f(1)}{h}$ $(h \neq 0)$. The following reasoning will show you that the answer reduces to $h + 4$.

$$f(1 + h) = (1 + h)^2 + 2(1 + h) + 2$$
$$= 1 + 2h + h^2 + 2 + 2h + 2$$
$$= h^2 + 4h + 5$$
$$f(1) \qquad = \qquad 5$$

$$f(1 + h) - f(1) = h^2 + 4h$$

Substituting $h^2 + 4h$ in $\dfrac{f(1+h) - f(1)}{h}$, we get $\dfrac{h^2 + 4h}{h}$, or $h + 4$.

In problems 19 to 24, find the value of $\dfrac{f(a + h) - f(a)}{h}$ $(h \neq 0)$ for the indicated value of a.

**(∘)** **19** $f(x) = 1 + x - 3x^2$; $\quad a = 2$

**20** $f(x) = 1 + x^2$; $\quad a = -1$

**21** $f(x) = \dfrac{1}{x}$; $\quad a = 1$

**22** $f(x) = \dfrac{1}{x+1}$; $\quad a = 2$

Now for some relations defined by equations. In each of the remaining problems (1) determine the domain of the relation defined by the equation; (2) decide whether the relation is a function; and (3) if the relation is a function, decide whether it is one-one.

**(∘)** **23** $x^2 - y^2 = 1$      **24** $4x^2 + 9y^2 = 1$

**25** $3x + 2y = 6$      **26** $3x^2 - y = 2$

**27** $xy = 1$      **28** $x(x + 3)y = 1$

**29** $|y| = |x|$      **30** $y^2 - x^2 = 0$

**31** $y = \dfrac{1}{x^2 + 1}$      **32** $x^3 + y^3 = 1$

**33** $\sqrt{x} + \sqrt{y} = 1$      **34** $y\sqrt{1 + x} = 1$

**1** This is a function, but it is not one-one since it contains the ordered pairs (2,5) and (6,5). Its domain is {2,3,5,6} and its range is {2,5,7}.

**11** $f(1) = 1^2 + 7(1) - 2 = 6$

**15** $F(2) = 2^2 + 3 \cdot 2 + 2 = 4 + 6 + 2 = 12$ and $F(1) = 1^2 + 3 \cdot 1 + 2 = 1 + 3 + 2 = 6$. Hence, $F(2) - F(1) = 12 - 6 = 6$.

**19**
$$f(2 + h) = 1 + (2 + h) - 3(2 + h)^2$$
$$= 1 + 2 + h - 12 - 12h - 3h^2$$
$$= -9 - 11h - 3h^2$$
$$f(2) = -9$$

$$f(2 + h) - f(2) = \quad -11h - 3h^2$$
$$\frac{f(2 + h) - f(2)}{h} = \quad -11 - 3h.$$

**23** Given the equation $x^2 - y^2 = 1$, we determine the domain of the associated relation by solving for $y$. Thus, $y^2 = x^2 - 1$, and $y = \pm \sqrt{x^2 - 1}$. Now $\sqrt{x^2 - 1}$ is real only when $x^2 - 1 \geq 0$; that is, when $(x + 1)(x - 1) \geq 0$, or when $x + 1$ and $x - 1$ have the same sign or are zero.

**a** In the first case, $(x + 1)(x - 1) = 0$ when $x = 1$ or $x = -1$.

**b** In the second case, $x + 1 > 0$ and $x - 1 > 0$. Then $x > -1$ and $x > 1$; so $x > 1$.

**c** In the third case, $x + 1 < 0$ and $x - 1 < 0$. Then $x < -1$ and $x < 1$; so $x < -1$.

It follows that $x \geq 1$ or $x \leq -1$; so the domain is $\{x \mid x \leq -1\} \cup \{x \mid x \geq 1\}$. For $x < -1$ or $x > 1$ the equation yields two distinct values of $y$; so the associated relation is not a function.

## (7.3) (The distance formula)

*distance*

Any two distinct points $P_1(x_1,y_1)$ and $P_2(x_2,y_2)$ determine a line segment in the plane—the line segment for which $P_1$ and $P_2$ are endpoints. The *distance* between $P_1$ and $P_2$ is defined to be the length of this line segment. Our purpose in this section is to develop a formula to compute this distance.

Let us begin with a concrete example. Suppose we are asked to find the distance between $P_1(1,2)$ and $P_2(3,5)$. We construct a line through $P_1$ parallel to the $x$-axis and a line through $P_2$ parallel to the *y*-axis; then we note that these two lines intersect at the point $P(3,2)$. This is illustrated in Figure 7.11. The triangle $PP_1P_2$ is a right triangle, and the theorem of Pythagoras tells us that the distance $d$ between $P_1$ and $P_2$ is given by $d^2 = a^2 + b^2$ when $a$ is the length of the base of the triangle and $b$ its altitude. Now $a = 3 - 1 = 2$ and $b = 5 - 2 = 3$; so $d^2 = 4 + 9 = 13$, and $d = \sqrt{13}$.

More generally, for $P_1(x_1,y_1)$ and $P_2(x_2,y_2)$ we may draw lines parallel to the *x*-axis through $P_1$ and parallel to the *y*-axis through $P_2$. These two lines intersect in a point $P(x_2,y_1)$. The case where $x_1 < x_2$ and $y_1 < y_2$ is illustrated in Figure 7.12, but the formula we will develop works in all cases. Triangle $PP_1P_2$ is a right triangle whose base is $x_2 - x_1$ and whose altitude is $y_2 - y_1$. By the theorem of Pythagoras, the distance $d$ between $P_1$ and $P_2$ is given by $d^2 = (x_2 - x_1)^2 + (y_2 - y_1)^2$; so

(∘)  **DF.1**  $d = \sqrt{(x_2 - x_1)^2 + (y_2 - y_1)^2}$

*distance formula*

This is known as the *distance formula,* and in all cases it gives the distance between the points $P_1$ and $P_2$. Thus the distance between $(-3,5)$ and $(8,-1)$ is

$$d = \sqrt{[8 - (-3)]^2 + (-1 - 5)^2} = \sqrt{(11)^2 + (-6)^2}$$
$$= \sqrt{121 + 36} = \sqrt{157}$$

and that between $(5,4)$ and $(5,1)$ is

$$d = \sqrt{(5 - 5)^2 + (1 - 4)^2}$$
$$= \sqrt{0^2 + (-3)^2} = \sqrt{9} = 3.$$

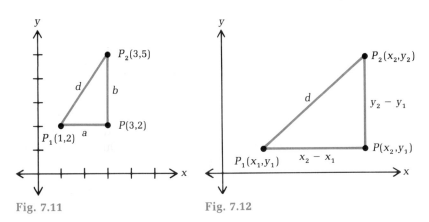

Fig. 7.11          Fig. 7.12

215 (∘) The distance formula

## Exercise set 7.3

*Find the distance between each of the given pairs of points.*

$(\circ)$    **1**   (3,1) and (6,5)          **2**   (3,1) and $(-1,4)$

       **3**   $(-2,3)$ and $(5,-1)$         **4**   $(3,-7)$ and $(2,-4)$

       **5**   $(-1,-5)$ and $(-6,-2)$       **6**   $(-3,1)$ and $(-5,-2)$

       **7**   (1,3) and $(1 + \sqrt{2}, 3 - \sqrt{7})$     **8**   $(-2 - \sqrt{3}, 4)$ and $(-2,4 + \sqrt{33})$

       **9**   (0,0) and (5,7)         **10**   $(a,b)$ and (0,0)

The theorem of Pythagoras states that in a right triangle, the square of the length of the hypotenuse equals the sum of the squares of the lengths of the other two sides. The converse of this theorem is also true. It states that if the sides of a triangle have lengths $a$, $b$, $c$ and if $c^2 = a^2 + b^2$, then the triangle is, in fact, a right triangle.

*In problems 11 to 16, you will be given the vertices of a triangle. Use the converse of the theorem of Pythagoras to decide whether or not each triangle is a right triangle, and sketch each triangle on the Cartesian plane.*

$(\circ)$    **11**   $(0,-5)$, (2,3), (6,2)       **12**   (1,2), (4,2), (1,6)

       **13**   (3,1), (5,3), $(-1,4)$        **14**   $(5,-2)$, $(-1,0)$, (1,4)

       **15**   (3,1), (5,3), $(-1,5)$        **16**   $(5,-2)$, $(-1,0)$, (1,6)

Suppose now that you are asked to find an equation whose graph is the set of points equidistant from the points $(-1,2)$ and (4,1). To say that the point $(x,y)$ is equidistant from the given two points is to say that $\sqrt{(x + 1)^2 + (y - 2)^2} = \sqrt{(x - 4)^2 + (y - 1)^2}$. Solving for $y$ in terms of $x$, we square both members to obtain $(x + 1)^2 + (y - 2)^2 = (x - 4)^2 + (y - 1)^2$. We now simplify this as follows:

$$x^2 + 2x + 1 + y^2 - 4y + 4 = x^2 - 8x + 16 + y^2 - 2y + 1$$

$$2x + 1 \quad\quad - 4y + 4 = \quad - 8x + 16 \quad\quad - 2y + 1$$

$$2x - 4y + 5 = \quad - 8x - 2y + 17$$

$$- 2y \quad = \quad - 10x + 12$$

$$y \quad = \quad 5x - 6.$$

*Use this technique to find an equation whose graph is the set of points equidistant from each of the following pairs of points.*

$(\circ)$    **17**   $(-3,6)$ and (1,5)        **18**   (3,4) and $(-2,2)$

       **19**   (4,6) and (0,0)         **20**   $(-1,3)$ and (0,0)

1 Using the distance formula, we obtain

$$d = \sqrt{(6-3)^2 + (5-1)^2} = \sqrt{3^2 + 4^2}$$

$$= \sqrt{9 + 16} = \sqrt{25} = 5.$$

11 The lengths of the three sides of the triangle are computed using the distance formula. Since we are after the squares of the lengths, we might just as well compute them, for then we avoid having to write radical signs. We have

$$a^2 = (2-0)^2 + [3-(-5)]^2 = 2^2 + 8^2 = 4 + 64 = 68;$$

$$b^2 = (6-0)^2 + [2-(-5)]^2 = 6^2 + 7^2 = 36 + 49 = 85;$$

$$c^2 = (6-2)^2 + (2-3)^2 = 4^2 + (-1)^2 = 16 + 1 = 17.$$

Since $a^2 + c^2 = 68 + 17 = 85 = b^2$, we deduce that the triangle must indeed be a right triangle. To sketch its graph, we set up Cartesian coordinates, and it looks like this.

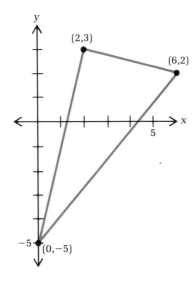

17 We want to find all points $(x,y)$ such that $(x+3)^2 + (y-6)^2 = (x-1)^2 + (y-5)^2$. Solving this equation for $y$, we obtain

$$x^2 + 6x + 9 + y^2 - 12y + 36 = x^2 - 2x + 1 + y^2 - 10y + 25$$

$$6x + 9 - 12y + 36 = -2x + 1 - 10y + 25$$

$$6x - 12y + 45 = -2x - 10y + 26$$

$$-2y = -8x - 19$$

$$y = 4x + \frac{19}{2}.$$

# (7.4) (Equations for straight lines)

In Section 7.1 you learned how to plot the graph of an equation in the form

$$ax + by + c = 0 \qquad (a,b,c \in R; a \neq 0 \text{ or } b \neq 0).$$

Such an equation is called a *linear equation* since its graph is a straight line. Now we are going to reverse the process. That is, given a straight line in the Cartesian plane, we are going to find a linear equation whose graph is that line. However, it should be noted that multiplying an equation by a nonzero constant does not change its graph; so there is no unique equation whose graph is a given straight line. In fact, there are infinitely many of them!

*slope*

The notion of the *slope* of a line segment turns out to be useful in our discussion. In order to have some insight into this concept, we ask you to picture a ladder leaning against a tall building. (Figure 7.13 can help you.) Suppose the base of the ladder is fixed at a distance of 4 feet from the building but that the ladder is of variable length so that its inclination can be changed. The inclination of the ladder can be measured by considering the height $h$ of the top of the ladder from the ground. We define the *slope*

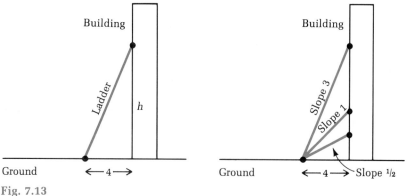

Fig. 7.13

of the ladder to be the number $h/4$; that is,

$$\text{Slope of ladder} = \frac{\text{distance of top of ladder from ground}}{\text{distance of base of ladder from building}}.$$

If the top of the ladder is 4 feet from the ground, the ladder has slope 1; if it is 12 feet from the ground the slope is 3; if it is 2 feet from the ground, the slope is $1/2$, and so on. As the slope increases, the ladder is inclined more steeply; as the slope decreases toward 0, the ladder gets closer to being horizontal.

In general, the slope of the line segment determined by the points $P_1(x_1,y_1)$ and $P_2(x_2,y_2)$ $(x_1 \neq x_2)$ is defined by the rule

$$m = \frac{y_2 - y_1}{x_2 - x_1}.$$

*positive slope*
*zero slope*
*negative slope*

*Positive slope* corresponds to a line segment that is rising as it is traversed from left to right. *Zero slope* corresponds to a horizontal line segment. *Negative slope* corresponds to a line segment that is falling as it is traversed from left to right. (See Figure 7.14.)

To define the *slope* of a nonvertical line $L$, one chooses any two distinct points $P_1(x_1,y_1)$ and $P_2(x_2,y_2)$ on $L$ and defines the slope of $L$ to be that of the line segment $\overline{P_1P_2}$. It is an easy exercise in geometry to show that the slope of $L$ does not depend on the particular points $P_1$ and $P_2$ that are chosen. In case $P_1$ and $P_2$ are picked so that $x_2 - x_1 = 1$, the slope has an interesting geometric interpretation, since then

$$m = \frac{y_2 - y_1}{x_2 - x_1} = \frac{y_2 - y_1}{1} = y_2 - y_1$$

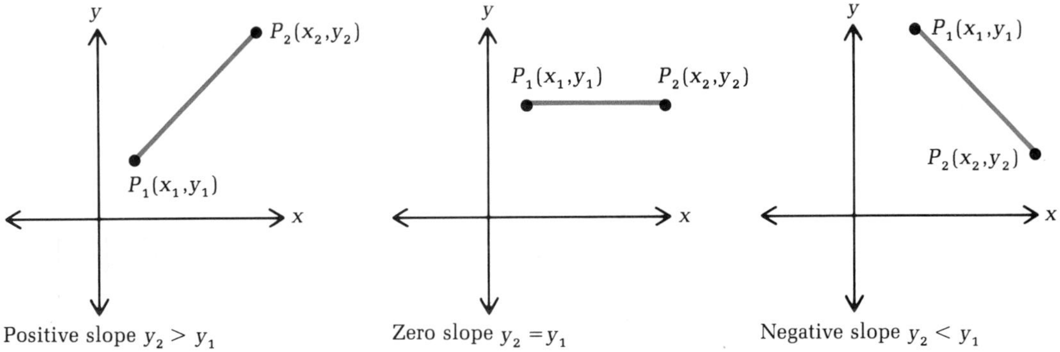

Positive slope $y_2 > y_1$      Zero slope $y_2 = y_1$      Negative slope $y_2 < y_1$

Fig. 7.14

so that $y_2 = y_1 + m$. Imagine a particle moving from left to right along $L$. As it moves one unit in the positive $x$-direction, it will move $m$ units vertically. This shows again that a line with positive slope is inclined *upward* (from left to right), whereas one with negative slope is inclined *downward*. This observation may also be used to sketch the graph of a line that has a certain slope and passes through some specified point. Suppose you are asked to sketch the graph of a line with slope $m = -3$ passing through $(2,4)$. To locate a second point on the line, you just observe that when $x = 2 + 1 = 3$, $y = 4 + m = 4 - 3 = 1$ so that the point $(3,1)$ is also on the line. This particular graph is shown in Figure 7.15. We shall not define the slope of a vertical line.

*intercept*
*y-intercept*
*x-intercept*

A second important notion is that of *intercept*. A nonvertical line is said to have intercept, or *y-intercept*, $b$ if it crosses the $y$-axis at $(0,b)$; similarly, the *x-intercept* of a nonhorizontal line is $a$ if the line crosses the $x$-axis at $(a,0)$.

Now it's time to relate all this information to equations for straight lines. The linear equation

$$Ax + By + C = 0 \qquad (A, B, C \in R; A \neq 0 \text{ or } B \neq 0)$$

*standard form*

is said to be in *standard form*.

Another useful form for a linear equation is

$$y = mx + b \qquad (m, b \in R);$$

*slope-intercept form*

this is called *slope-intercept form*, and the reason for the name is quite simple. When $x = 0$, $y = b$, so that the $y$-intercept of the graph of this equation is $b$. Since $(1, b + m)$ is also on its graph, we see that its slope is

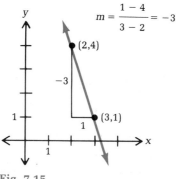

$$m = \frac{1 - 4}{3 - 2} = -3$$

(2,4)

−3

1

(3,1)

1

**Fig. 7.15**

220 ($\circ$) Functions and relations

$$\frac{(m+b) - b}{1 - 0} = m.$$

Thus, the graph of the equation $y = mx + b$ is a straight line with slope $m$ and intercept $b$. To determine the slope and intercept of the graph of $2x - y = 6$, we solve for $y$, obtaining $y = 2x - 6$. From this we see that the slope is $2$ and the intercept is $-6$.

A third, and extremely useful, form for linear equations is called the *point-slope form*. Suppose we want the equation of a line with slope $m$ that passes through the point $(x_1, y_1)$. We claim that the answer is

*point-slope form*

$$y - y_1 = m(x - x_1).$$

To see this, note first that we have indeed written down a linear equation; furthermore, if $x = x_1$, $y = y_1$ the equation is satisfied, so that its graph is a straight line passing through $(x_1, y_1)$. Putting this equation in slope-intercept form, we obtain

$$y = mx + (y_1 - mx_1)$$

and from this we see that its slope is $m$. Since there can be only one line with slope $m$ passing through the point $(x_1, y_1)$ we have found its equation. To illustrate this concretely, we shall find an equation for the line with slope $4/5$ passing through the point $(-3, 2)$. In point-slope form we obtain $y - 2 = (4/5)(x + 3)$. If we were asked for an equivalent equation in standard form we would write $5y - 10 = 4x + 12$, $4x - 5y + 22 = 0$.

At this point we have introduced three forms for linear equations:

$( \circ )$    **LEQ.1** $Ax + By + C = 0$     *(A or B $\neq$ 0)*      *(Standard form)*

$( \circ )$    **LEQ.2** $y = mx + b$                     *(Slope-intercept form)*

$( \circ )$    **LEQ.3** $y - y_1 = m(x - x_1)$        *(Point-slope form)*

The second equation represents a line with slope $m$ and intercept $b$, whereas the third equation yields a line with slope $m$ passing through the point $(x_1, y_1)$. It is also worth mentioning that the graph of $x = a$ is a vertical line with $x$-intercept $a$, and that of $y = b$ is a horizontal line with $y$-intercept $b$.

The notion of slope can be used to determine when two lines are parallel or perpendicular. It can be shown that two nonvertical lines are parallel if and only if they have the same slope. Similarly,

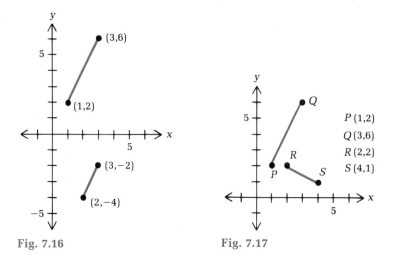

Fig. 7.16    Fig. 7.17

two such lines are perpendicular if and only if the product of their slopes is −1. The line segments shown in Figure 7.16 are parallel in that they both have slope 2. In Figure 7.17, the slope of $\overline{PQ}$ is $\dfrac{6 - 2}{6 - 1} = 2$, while that of $\overline{RS}$ is $\dfrac{1 - 2}{4 - 2} = -\dfrac{1}{2}$. Since $2\left(-\dfrac{1}{2}\right) = -1$, the segments are perpendicular.

As a further example, let $L$ be the line determined by the points $(1,2)$ and $(3,8)$. Suppose you are asked to find equations for the

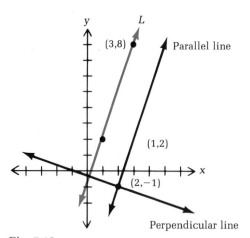

Fig. 7.18

222 (∘) Functions and relations

lines parallel and perpendicular to $L$, passing through the point $(2,-1)$. This situation is illustrated in Figure 7.18. The first step is to observe that the slope of $L$ is $\dfrac{8-2}{3-1} = \dfrac{6}{2} = 3$. Using the point-slope form for a straight line, we see that an equation for the parallel line is $y + 1 = 3(x - 2)$. The slope of the perpendicular line is the negative reciprocal of the slope of $L$; namely, it is $-\dfrac{1}{3}$.

Thus its equation can be written in the form $y + 1 = -\dfrac{1}{3}(x - 2)$.

## Exercise set 7.4

*In problems 1 to 8, find the slope of the line containing the given two points.*

( ○ )  **1**  $(3,-2)$ and $(-1,5)$        **2**  $(4,1)$ and $(5,3)$

**3**  $(-2,1)$ and $(2,9)$        **4**  $(-1,3)$ and $(0,-4)$

**5**  $(0,1)$ and $(3,-1)$        **6**  $\left(\dfrac{5}{2},1\right)$ and $(0,3)$

**7**  $\left(\dfrac{4}{3},2\right)$ and $\left(\dfrac{2}{3},1\right)$        **8**  $(3,2)$ and $(5,2)$

*In 9 to 14, put each equation into slope-intercept form. Specify the slope and the y-intercept of each line.*

( ○ )  **9**  $3x - 4y = 5$        **10**  $7x - y = 3$

**11**  $\dfrac{3}{2}x + \dfrac{4}{3}y - 2 = 0$        **12**  $\dfrac{7}{5}y - \dfrac{5}{4}x + 6 = 0$

**13**  $y - 1 = -2(x + 3)$        **14**  $y + \dfrac{2}{3} = -\dfrac{3}{5}\left(x + \dfrac{5}{3}\right)$

*In 15 to 20, put each equation in standard form.*

( ○ )  **15**  $y = -\dfrac{5}{7}x + 1$        **16**  $y = 3x + 4$

**17**  $y - 3 = -2(x + 1)$        **18**  $y + 7 = 2(x + 1)$

**19**  $x = 4$        **20**  $y = 4$

*Write an equation (in standard form) for each of the following lines.*

(∘)  **21**  slope −1, *y*-intercept 2

**22**  slope 5, *y*-intercept 4

**23**  slope 230, *y*-intercept 3

**24**  slope $-\dfrac{7}{5}$, *y*-intercept $\dfrac{5}{7}$

**25**  slope 1, *y*-intercept −8

**26**  slope −1, *y*-intercept −10

(∘)  **27**  slope 2, passing through (1,5)

**28**  slope 3, passing through (−1,2)

**29**  slope $\dfrac{3}{2}$, passing through (2,0)

**30**  slope $-\dfrac{7}{6}$, passing through (1,6)

To find an equation for the line determined by a given pair of points, you first find its slope and then use the point-slope form of a linear equation. We illustrate this by finding an equation for the line determined by $\left(\dfrac{5}{2},3\right)$ and $\left(-\dfrac{3}{2},1\right)$. The slope is given by

$$m = \frac{1-3}{-\dfrac{3}{2}-\dfrac{5}{2}} = \frac{-2}{-4} = \frac{1}{2}.$$

Substituting into the point-slope form, we get $y-3 = \dfrac{1}{2}\left(x-\dfrac{5}{2}\right)$, which, in standard form, becomes $4y - 2x - 7 = 0$. Note that if the point $\left(-\dfrac{3}{2},1\right)$ had been used in place of $\left(\dfrac{5}{2},3\right)$, an equivalent equation would have resulted. Try it!

*In the next few problems, find an equation (in standard form) for the line determined by each pair of points.*

(∘)  **31**  (2,3) and (3,2)        **32**  (2,3) and (3,−2)

**33**  (−1,5) and (7,−6)        **34**  (−1,3) and (−5,−1)

**35**  (0,6) and (7,−1)        **36**  (0,3) and (−2,0)

*The following problems involve the use of slope to determine when two lines are parallel or perpendicular.*

( ○ ) **37** Show that the line passing through the points (1,3) and (−3,5) is parallel to the one passing through (−8,1) and $\left(9,-\dfrac{15}{2}\right)$. Draw the graph of both lines.

**38** Show that the line passing through the points (−3,−5) and (−2,8) is parallel to the one passing through (0,6) and (2,32). Draw the graph of both lines.

**39** Show that the line passing through the points (1,3) and (−3,5) is perpendicular to the one passing through the points (−8,1) and (9,35). Draw the graph of both lines.

**40** Show that the line passing through the points (−3,−5) and (−2,8) is perpendicular to the one passing through the points (0,6) and $\left(2,\dfrac{76}{13}\right)$. Draw the graph of both lines.

**41** Find an equation for the line passing through (1,−1) and parallel to $3x - 5y + 7 = 0$.

**42** Find an equation for the line passing through (5,−4) and parallel to $4x + \dfrac{1}{2}y + 1 = 0$.

( ○ ) **43** Find an equation for the line passing through (1,−1) and perpendicular to $3x - 5y + 7 = 0$.

**44** Find an equation for the line passing through (5,−4) and perpendicular to $4x + \dfrac{1}{2}y + 1 = 0$.

**45** Write equations for the lines parallel and perpendicular to $x = 3$, that pass through the point (5,4).

**46** Write equations for the lines parallel and perpendicular to $y = -7$ that pass through the point (−3,1).

**47** Let $L$ be the line determined by the points (2,5) and (5,−2). Write equations for the lines parallel and perpendicular to $L$ that pass through the point (1,1). Draw the graphs of all three lines on the same coordinate axes.

**48** Let $L$ be the line determined by the points $\left(\dfrac{3}{2},1\right)$ and $\left(\dfrac{7}{4},2\right)$. Write equations for the lines parallel and perpendicular to $L$ that pass through (4,0). Graph all three lines on the same coordinate axes.

Consider the triangle whose vertices are (1,2), (3,5), and (4,0). The slopes of its sides are given by $m_1 = \dfrac{5-2}{3-1} = \dfrac{3}{2}$, $m_2 = \dfrac{0-5}{4-3}$

$= -5$, and $m_3 = \dfrac{0 - 2}{4 - 1} = -\dfrac{2}{3}$. Since $m_1 m_3 = -1$, we see that the triangle in question is, in fact, a right triangle.

*Use the notion of slope to solve each of the following problems.*

**49**  Let $(-4,1)$, $(1,6)$, and $(-1,-2)$ be the vertices of a triangle. Show that the triangle is a right triangle and draw its graph.

**50**  Let $(1,3)$, $(-3,5)$, and $(0,1)$ be vertices of a triangle. Show that the triangle is a right triangle and draw its graph.

$(\circ)$  **51**  Let $(-2,5)$, $(0,7)$, $(6,1)$, and $(8,3)$ be vertices of a quadrilateral $Q$. Show that $Q$ is a parallelogram and draw its graph.

**52**  Let $(-6,-1)$, $(-2,2)$, $(-1,-2)$, and $(3,1)$ be vertices of a quadrilateral $Q$. Show that $Q$ is a parallelogram and draw its graph.

**53**  Let $(-2,5)$, $(0,3)$, $(0,7)$, and $(2,5)$ be vertices of a quadrilateral $Q$. Show that $Q$ is a rectangle and draw its graph.

**54**  Let $(-6,-1)$, $(-2,2)$, $(-3,-5)$, and $(1,-2)$ be vertices of a quadrilateral $Q$. Show that $Q$ is a rectangle and draw its graph.

**55**  Let $P(-2,2)$ and $Q(3,3)$ be points in the plane. Determine all possible values of $k$ so that if $R = R(4,k)$, the triangle $PQR$ is a right triangle with $\overline{PR}$ or $\overline{QR}$ the hypotenuse.

**56**  Let $P(4,1)$ and $Q(-2,3)$ be points in the plane. Determine all possible values of $k$ so that if $R = R(1,k)$, the triangle $PQR$ is a right triangle with $\overline{PR}$ or $\overline{QR}$ the hypotenuse.

$(\circ)$

*Sample solutions for exercise set 7.4*

**1**  Substituting into the formula for slope we get $m = \dfrac{5 - (-2)}{-1 - 3} = \dfrac{5 + 2}{-4} = -\dfrac{7}{4}$.

**9**  To put the equation $3x - 4y = 5$ into slope-intercept form, we must solve for $y$. We therefore write

$$-4y = -3x + 5$$

$$y = \frac{3}{4}x + \left(-\frac{5}{4}\right).$$

Since this last equation is in slope-intercept form, we see that the slope is $3/4$, and the $y$-intercept $-5/4$.

**15**  We are given the equation $y = -\dfrac{5}{7}x + 1$. To put this in standard form, we write

$$7y = -5x + 7$$

$$5x + 7y - 7 = 0.$$

You should realize that there is no single correct answer for this one. The equation $10x + 14y - 14 = 0$ is equally correct, as is $\dfrac{5}{7}x + y - 1 = 0$. In fact, there are infinitely many correct answers. When possible, it is preferable to seek an equation whose coefficients are integers.

**21**  In slope-intercept form, the desired equation is $y = -x + 2$. In standard form this becomes $x + y - 2 = 0$.

**27**  In point-slope form, we have $y - 5 = 2(x - 1)$. This becomes

$$y - 5 = 2x - 2$$

$$-2x + y - 3 = 0.$$

**31**  The slope of the line is given by $m = \dfrac{2 - 3}{3 - 2} = -1$. Hence, its equation is

$$y - 3 = -1(x - 2)$$

$$y - 3 = -x + 2$$

$$x + y - 5 = 0.$$

**37**  We need only show that the two lines have the same slope. The slope of the first line is given by

$$m_1 = \frac{5 - 3}{-3 - 1} = \frac{2}{-4} = -\frac{1}{2}$$

and that of the second line by

$$m_2 = \frac{-\dfrac{15}{2} - 1}{9 - (-8)} = \frac{-\dfrac{17}{2}}{17} = \frac{-17}{34} = -\frac{1}{2}.$$

The graph looks like this.

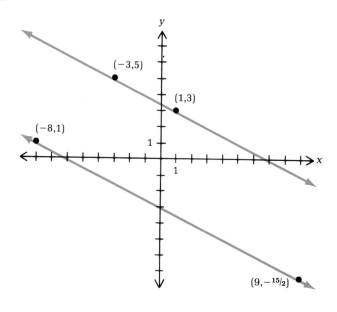

**43** We begin by obtaining the slope of the line $3x - 5y + 7 = 0$. To do this, we solve for $y$ and get

$$-5y = -3x - 7$$

$$y = \frac{3}{5}x + \frac{7}{5}.$$

The slope of $3x - 5y + 7 = 0$ is $\frac{3}{5}$, and the slope of any perpendicular line is $-\frac{5}{3}$. Thus, an equation for the line we are after is $y + 1 = -\frac{5}{3}(x - 1)$. In standard form this becomes $5x + 3y - 2 = 0$.

**51** The time spent drawing a graph is seldom wasted since you will have something concrete to think about when you go on to the algebra. This is what the graph looks like.

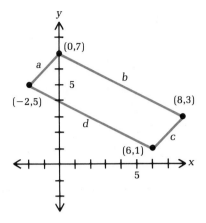

The graph certainly *looks* like a parallelogram. In order to prove that it really is, we shall compute the slopes of the four sides of $Q$, with the sides labeled as in the graph.

**Side a** $\dfrac{7-5}{0+2} = \dfrac{2}{2} = 1$

**Side b** $\dfrac{3-7}{8-0} = \dfrac{-4}{8} = -\dfrac{1}{2}$

**Side c** $\dfrac{3-1}{8-6} = \dfrac{2}{2} = 1$

**Side d** $\dfrac{1-5}{6+2} = \dfrac{-4}{8} = -\dfrac{1}{2}$

This shows that sides $a$ and $c$ are parallel, as are sides $b$ and $d$; so $Q$ is indeed a parallelogram.

## (7.5) (Graphing quadratic functions)

Having spent some time discussing linear functions and their graphs, we now turn to quadratic functions; these are functions of the form

$$f(x) = ax^2 + bx + c \qquad (a \neq 0).$$

We begin our discussion with the concrete example $f(x) = x^2 - 9$.

The graph of $f$ consists of those points in the plane whose coordinates are in the form $x$ and $f(x)$; in other words, the graph of $f$ is the solution set (in the real number system) of the equation $y = x^2 - 9$. If we observe that $f(0) = 0^2 - 9 = -9$, we see that $(0,-9)$ is on the graph of $f$. Similarly, $f(1) = 1^2 - 9 = -8$ shows $(1,-8)$ to be on the graph. Continuing in this manner we have the following points on the graph of $f$: $(-4,7)$, $(-3,0)$, $(-2,-5)$, $(-1,-8)$, $(0,-9)$, $(1,-8)$, $(2,-5)$, $(3,0)$, $(4,7)$. Figure 7.19 displays the graph of these points.

Our basic instinct is that the graph of $f$ ought to be a nice, smooth curve. To convince ourselves that this is correct, we plot the following additional points: $\left(-\dfrac{7}{2}, \dfrac{13}{4}\right)$, $\left(-\dfrac{5}{2}, -\dfrac{11}{4}\right)$, $\left(-\dfrac{3}{2}, -\dfrac{27}{4}\right)$, $\left(-\dfrac{1}{2}, -\dfrac{35}{4}\right)$, $\left(\dfrac{1}{2}, -\dfrac{35}{4}\right)$, $\left(\dfrac{3}{2}, -\dfrac{27}{4}\right)$, $\left(\dfrac{5}{2}, -\dfrac{11}{4}\right)$, $\left(\dfrac{7}{2}, \dfrac{13}{4}\right)$. Figure 7.20 shows the effect of adding these points to the graph of $f$, and in Figure 7.21 we have connected the points with a smooth curve. This kind of curve is called a *parabola*, and it will be discussed in this, as well as the next, section.

*parabola*

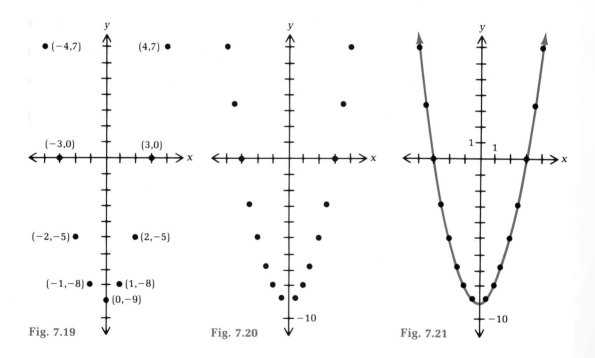

Fig. 7.19    Fig. 7.20    Fig. 7.21

230 (∘) Functions and relations

There are a few things you should notice about the graph of $f(x) = x^2 - 9$. First of all, the point $(0,-9)$ is the lowest point on the graph. For obvious reasons, we shall call such a point the *turning point* of the parabola. Secondly, with respect to the $y$-axis, the graph is symmetric in the following sense: If the plane were rotated about the $y$-axis so that the positive $x$-axis became the negative $x$-axis (in other words, through an angle of 180°), the parabola would be unchanged. A rotation of this kind is called a *reflection* about the $y$-axis, and it is pictured in Figure 7.22. When you look at Figure 7.22, you should observe that points on the $y$-axis remain fixed, and a point that has coordinates $(a,b)$ with $a \neq 0$ gets moved to where $(-a,b)$ used to be. The $y$-axis is called the *axis of symmetry* of the graph of $f$. Incidentally, Figure 7.21 shows that the domain of $f$ is the set of all real numbers, while its range is $\{y \mid y \geq -9\}$.

Though we have not proved it, it is certainly true that any quadratic equation $y = ax^2 + bx + c$ ($a \neq 0$) has a parabola for its graph. These parabolas need not be symmetric about the $y$-axis, and they may even be turned upside down.

To illustrate this, we look at the graph of $f(x) = -2x^2 + 6x - 5$. Routine computation shows the points $(-1,-13)$, $(0,-5)$, $(1,-1)$, $(2,-1)$, $(3,-5)$, $(4,-13)$ to be on the graph, and the obvious symmetry leads us to plot $\left(\dfrac{3}{2}, -\dfrac{1}{2}\right)$, too, thus producing the graph indicated in Figure 7.23. Notice that the function $f(x) = -2x^2 + 6x - 5$ has for its domain all real numbers and for its range $\{y \mid y \leq -\dfrac{1}{2}\}$. Its graph is a parabola opening downward, with turning point $\left(\dfrac{3}{2}, -\dfrac{1}{2}\right)$

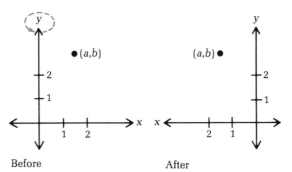

Before      After

Fig. 7.22 **Reflection about the y-axis**

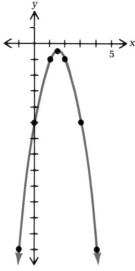

**Fig. 7.23**

and with the line $x = \dfrac{3}{2}$ as its axis of symmetry.

The graph of a relation of the form $x = ay^2 + by + c \ (a \neq 0)$ is a parabola turned on its side. We illustrate this by drawing the

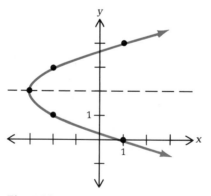

**Fig. 7.24**

graph of $x = y^2 - 4y + 1$. Points on its graph are located by assigning a value to $y$ and then solving for $x$. Thus, when $y = 0$, $x = 1$; when $y = 1$, $x = 1^2 - 4 \cdot 1 + 1 = -2$. Continuing in this manner we obtain the following points: $(1,0)$, $(-2,1)$, $(-3,2)$, $(-2,3)$, $(1,4)$. The graph is shown in Figure 7.24. Notice that this graph is a parabola opening to the right. Its turning point is $(-3,2)$, and it has the line $y = 2$ as its axis of symmetry.

## Exercise set 7.5

*In the first four problems, assume that $x \in \{-3, -2, -1, 0, 1, 2, 3\}$, and draw the graph of the indicated equation.*

$(\circ)$   **1**   $y = x^2 - 5x$                     **2**   $y = -x^2 + 3x$

  **3**   $y = -2x^2 + 3x + 1$          **4**   $y = \dfrac{1}{2}x^2 + 2x + 2$

*We now remove the restriction on the values of $x$, and ask you to draw the graph of each of the following equations.*

$(\circ)$   **5**   $y = x^2 - 1$                      **6**   $y = x^2 - 2$

  **7**   $y = x^2 + 4$                      **8**   $y = x^2 + 2$

  **9**   $y = -x^2$                        **10**   $y = -x^2 + 4$

  **11**   $y = -x^2 - 1$                   **12**   $y = -x^2 + 2$

  **13**   $y = 4x^2 - 1$                    **14**   $y = 4x^2 - 2$

  **15**   $y = -3x^2 + 6$                   **16**   $y = -3x^2 - 3$

  **17**   $y = x^2 - 4x$                    **18**   $y = x^2 + 6x$

  **19**   $f(x) = x^2 + 2x + 3$             **20**   $f(x) = x^2 - 2x + 1$

$(\circ)$   **21**   $f(x) = -x^2 + 8x - 10$        **22**   $f(x) = -x^2 + 6x - 2$

  **23**   $y = 2x^2 + 4x - 3$               **24**   $y = -3x^2 + 12x - 10$

  **25**   $g(x) = x^2 + x + 1$              **26**   $h(x) = x^2 + 3x - 2$

  **27**   $y = -2x^2 + x + 3$               **28**   $y = -3x^2 + 3x + 1$

*Now, let us draw the graph of a few relations of the type $x = ay^2 + by + c$   $(a \neq 0)$.*

  **29**   $x = 3y^2$                        **30**   $x = -2y^2$

$(\circ)$   **31**   $x = -y^2 - 5$                 **32**   $x = y^2 + 1$

  **33**   $x = y^2 + 4y - 2$                **34**   $x = y^2 - 6y + 2$

Suppose we draw the graph of $y = \dfrac{1}{2}x^2$, $y = x^2$ and $y = 2x^2$ on the same coordinate axes. We arrive at the following graph.

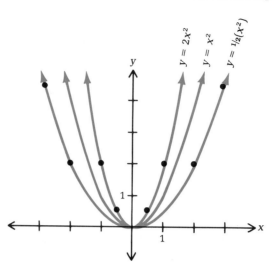

Problems 35 and 36 will give you some practice in graphing what happens as we allow $a$ to vary in an equation of the form $y = ax^2 + c$.

**35** Plot $y = -\dfrac{1}{3}x^2$, $y = -x^2$, and $y = -3x^2$ on the same coordinate axes.

**36** Plot $y = \dfrac{1}{4}x^2 - 1$, $y = x^2 - 1$, and $y = 4x^2 - 1$ on the same coordinate axes.

The next two problems should give you an idea of what happens if we allow $c$ to vary in an equation of the type $y = ax^2 + bx + c$.

( ∘ ) **37** Plot $y = x^2$, $y = x^2 - 2$ and $y = x^2 + 3$ on the same coordinate axes.

**38** Plot $y = x^2 - 2x$, $y = x^2 - 2x + 1$, and $y = x^2 - 2x - 2$ on the same coordinate axes.

1

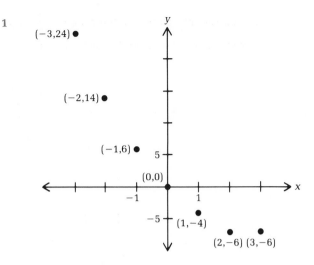

5    After establishing that the following points, (0,−1), (1,0), (2,3), (−1,0), and (−2,3), are on it, our graph looks like this.

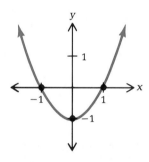

**21**  Plotting the points $(0,-10)$, $(2,2)$, $(3,5)$, $(4,6)$, $(5,5)$, $(6,2)$, and $(8,-10)$, we get the following graph.

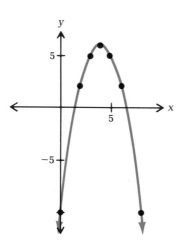

**31**  To plot $x = -y^2 - 5$, we replace $y$ by $-2$, $-1$, $0$, $1$, and $2$ in order to see that the points $(-9,-2)$, $(-6,-1)$, $(-5,0)$, $(-6,1)$, and $(-9,2)$ all lie on the graph. The graph looks like this.

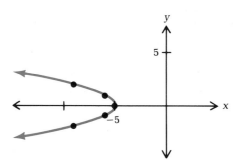

**37** Notice how the graph of $y = x^2 + c$ slides up and down as $c$ is changed.

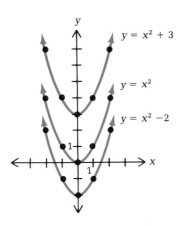

# (7.6) (Sketching parabolas)

In this section we shall examine the graph of the quadratic equation

$$y = ax^2 + bx + c \qquad (a \neq 0)$$

more closely, with a view toward learning some shortcuts in the graphing process. The graph of such an equation is a parabola that either opens upward or downward (see Figure 7.25). It is help-ful to know which! By examining $y$ as $x$ gets very large, it is clear that the parabola opens upward if $a > 0$ and downward if $a < 0$.

*turning point*

*axis of symmetry*

Let's recall some terminology from Section 7.5. A parabola that opens upward has a lowest point and one that opens downward, a highest point. Such a point is called the *turning point* of the parabola. The vertical line that passes through the turning point is the *axis of symmetry* of the parabola. This means that if the Cartesian plane is rotated 180° using this line as an axis, the parabola will be unchanged. It would certainly be nice to display the turning point of a parabola on its graph. In Section 7.5, we accomplished this simply by plotting enough points so that the turning point could be located. It turns out that a systematic technique can be developed that involves nothing more than completing the square.

To illustrate the process, let us find the turning point of the graph of $y = -x^2 + 3x - 2$. This graph clearly opens downward;

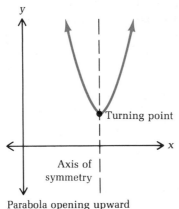

Parabola opening upward            Parabola opening downward

**Fig. 7.25**

so its turning point will be the highest point on the graph. To locate it, we write $-x^2 + 3x - 2 = -[x^2 - 3x + 2]$. To complete the square we add and subtract $(3/2)^2 = 9/4$ to the expression in the brackets (note that this does not change its value), and we obtain

$$-x^2 + 3x - 2 = -\left[ x^2 - 3x + \frac{9}{4} - \frac{9}{4} + 2 \right]$$

$$= -\left[ \left( x - \frac{3}{2} \right)^2 - \frac{1}{4} \right]$$

$$= -\left( x - \frac{3}{2} \right)^2 + \frac{1}{4}.$$

The function $f(x) = -x^2 + 3x - 2 = -\left( x - \dfrac{3}{2} \right)^2 + \dfrac{1}{4}$ will be as

large as possible when $x - \dfrac{3}{2} = 0$. Hence, the turning point of the

graph of $y = -x^2 + 3x - 2$ is at $\left( \dfrac{3}{2}, \dfrac{1}{4} \right)$.

One may also locate the turning point of the graph of the general quadratic equation $y = ax^2 + bx + c \quad (a \neq 0)$ by a similar process. The first step is to write $ax^2 + bx + c = a\left[ x^2 + \dfrac{b}{a}x + \dfrac{c}{a} \right]$. To complete the square, the expression $b^2/4a^2$ must be added to and subtracted from the expression in brackets, so that

$$ax^2 + bx + c = a\left[ x^2 + \frac{b}{a}x + \frac{b^2}{4a^2} - \frac{b^2}{4a^2} + \frac{c}{a} \right]$$

$$= a\left[ \left( x + \frac{b}{2a} \right)^2 + \left( \frac{c}{a} - \frac{b^2}{4a^2} \right) \right]$$

$$= a\left( x + \frac{b}{2a} \right)^2 + \left( c - \frac{b^2}{4a} \right).$$

For $a > 0$, the expression $a\left( x + \dfrac{b}{2a} \right)^2 \geq 0$; so the function $f(x)$ $= ax^2 + bx + c$ will have its lowest value when $x + \dfrac{b}{2a} = 0$; in other words, when $x = -\dfrac{b}{2a}$. Similar reasoning shows that when $a < 0$, $f(x)$ has its greatest value when $x = -\dfrac{b}{2a}$. Thus the abscissa of the turning point is in all cases $x = -\dfrac{b}{2a}$. To locate its ordinate, one just observes that $f\left( -\dfrac{b}{2a} \right) = c - \dfrac{b^2}{4a}$.

*intercepts*

*Intercepts* of parabolas are defined exactly as they were for straight lines. The *y-intercept* of the graph of the equation $y = ax^2 + bx + c$ is clearly $c$ since it crosses the $y$-axis at $(0,c)$. The *x-intercepts* represent the points (if any) that the parabola crosses the $x$-axis; they are located by finding the real roots of the quadratic equation $ax^2 + bx + c = 0$. Here, is a step-by-step procedure for sketching the graph of the quadratic function $f(x) = ax^2 + bx + c$ $(a \neq 0)$.

**Step 1:** If $a > 0$ note that the graph is a parabola opening upward, and if $a < 0$ it opens downward.

**Step 2:** Locate the turning point by noting that its abscissa is $-\dfrac{b}{2a}$ and its ordinate $f\left( -\dfrac{b}{2a} \right)$. Use a broken line to draw the axis of symmetry $x = -\dfrac{b}{2a}$.

**Step 3:** Plot one or two points on each side of the turning point. It may be convenient to make use of the axis of symmetry here.

**Step 4:** If appropriate, plot the intercepts.

**Step 5:** Sketch the graph.

We will illustrate the foregoing procedure with the graph of the function $f(x) = 3x^2 - 2x + 1$. Here, $a = 3$, $b = -2$, $c = 1$; so the graph is a parabola opening upward. The turning point is obtained by setting $x = -\dfrac{b}{2a} = -\dfrac{-2}{2 \cdot 3} = \dfrac{1}{3}$ and noting that $f\!\left(\dfrac{1}{3}\right) = 3\left(\dfrac{1}{3}\right)^2 - 2\left(\dfrac{1}{3}\right) + 1 = \dfrac{3}{9} - \dfrac{2}{3} + 1 = \dfrac{2}{3}$. The turning point is, therefore, $\left(\dfrac{1}{3}, \dfrac{2}{3}\right)$, and the axis of symmetry is the vertical line $x = \dfrac{1}{3}$. Plotting a point or two on both sides of the turning point, we observe that $(-1,6)$, $(0,1)$, and $(1,2)$ are on the graph. Thus we already have enough information to sketch the graph (see Figure

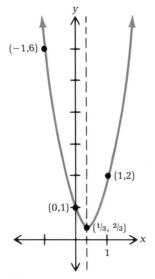

Fig. 7.26

7.26). Notice that we have plotted the $y$-intercept, and there are no $x$-intercepts for this one.

Obvious modifications of the five-step procedure should enable you to graph equations of the form $x = ay^2 + by + c$ ($a \neq 0$). Just bear in mind that if $a > 0$ the graph is a parabola opening to the right, and if $a < 0$ it opens to the left, as shown in Figure 7.27. As an example, let's look at $x = 2y^2 - 3y + 1$. To begin with, notice that the graph is a parabola opening to the right. To find the coordinates of its turning point, we set $y = -\dfrac{-3}{2 \cdot 2} = \dfrac{3}{4}$ and note that $2\left(\dfrac{3}{4}\right)^2 - 3\left(\dfrac{3}{4}\right) + 1 = \dfrac{9}{8} - \dfrac{9}{4} + 1 = -\dfrac{9}{8} + 1 = -\dfrac{1}{8}$; so the turning point is $\left(-\dfrac{1}{8}, \dfrac{3}{4}\right)$, and the axis of symmetry is the horizontal line $y = \dfrac{3}{4}$. The $x$-intercept is evidently 1, and the $y$-intercepts are obtained by solving the equation

$$2y^2 - 3y + 1 = 0$$

$$(2y - 1)(y - 1) = 0$$

$$y = \frac{1}{2} \text{ or } y = 1.$$

We now know that the $y$-intercepts are $\dfrac{1}{2}$ and 1.

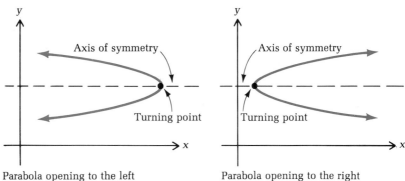

Parabola opening to the left

Parabola opening to the right

Fig. 7.27

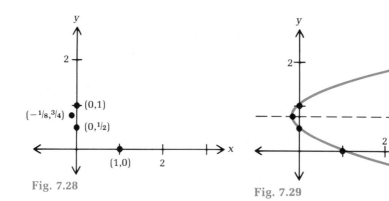

Fig. 7.28

Fig. 7.29

Thus far we have coordinates for the points $\left(-\frac{1}{8}, \frac{3}{4}\right)$, $(1,0)$, $\left(0, \frac{1}{2}\right)$, and $(0,1)$ on the graph. These points are plotted in Figure 7.28. To obtain a better sketch, let us determine the point on the graph whose ordinate is 2. When $y = 2$, $x = 2 \cdot 4 - 3 \cdot 2 + 1 = 9 - 6 = 3$; so the point has coordinates $(3,2)$. The graph is shown in Figure 7.29.

## Exercise set 7.6

In problems 1 to 10, use the methods of this section to find the y-intercepts, the x-intercepts (if any), and the coordinates of the turning point of the graph of each equation.

**1** $y = x^2 - 5x + 4$      **2** $y = x^2 + 6x + 8$

**3** $y = -x^2 - 9x$      **4** $y = -x^2 + 3x$

**5** $f(x) = 6 - x - x^2$      **6** $g(x) = x^2 - 4x + 5$

**7** $g(x) = 2x^2 - x + 2$      **8** $h(x) = 3 - 4x - 4x^2$

**9** $y = 3x^2 + \frac{7}{2}x + 1$      **10** $y = 9 + \frac{2}{5}x - 3x^2$

In problems 11 to 20, you are to sketch each graph by using the methods of this section. In other words, you are to plot the turning point, the intercepts, the axis of symmetry, and (if necessary) one or two points on each side of the turning point. This information is then to be used to sketch the graphs of the equations stated in problems 1 to 10. Naturally, you may use the computations you have already made for these problems.

*Use the techniques of this section to sketch the graph of each of the equations in problems 21 to 24.*

$(\circ)$    **21**    $x = -3y^2 + 5y + 2$          **22**    $x = y^2 - 3y - 4$

      **23**    $x = 2y^2 + 3y - 4$          **24**    $x = -4y^2 - 3y$

The problem of finding coordinates for the turning point of the graph of $y = ax^2 + bx + c$   ($a \neq 0$) is equivalent to finding the greatest or least value of the function $f(x) = ax^2 + bx + c$. This useful idea comes up in a variety of situations.

**Problem:** A man has 40 feet of fencing. He wishes to fence in a rectangular area in the middle of his yard to use as a garden. What dimensions should his garden be in order for its area to be as large as possible?

**Solution:** Let $x$ be the length of one side of the garden. Then $x$ is also the length of the opposite side. If we are not to run out of fencing, the perimeter of the garden had best be 40 feet. Hence, each remaining side must have length $\frac{1}{2}(40 - 2x) = 20 - x$. Thus, the garden looks like this.

$$20 - x$$

$x$                                          $x$

$$20 - x$$

Since each side must have positive length, we clearly have $x > 0$ and $20 - x > 0$, so that $0 < x < 20$. The area is therefore given by the function $A(x) = x(20 - x) = -x^2 + 20x$ with domain $(0,20)$. The largest value of $A(x)$ occurs when $x = -\dfrac{20}{2(-1)} = 10$, which shows that the garden should be square with each side 10 feet long.

*Use this technique for the remaining problems.*

$(\circ)$    **25**    Two numbers add up to 22. What are the numbers if their product is to be as large as possible?

      **26**    Find two numbers whose sum is 17 and whose product is as large as possible.

**27** Find two numbers whose difference is 9 and whose product is as small as possible.

**28** Find two numbers whose difference is 22 and whose product is as small as possible.

**29** A man installing a rectangular swimming pool wants its perimeter to be 120 feet and its area as large as possible. What should the dimensions of the pool be?

**30** A man is building a rectangular barn. He wants the floor space to be as large as possible but wants the perimeter of the barn to be 160 feet. What dimensions should the barn have?

( ∘ ) **31** A man wishes to fence in a rectangular garden. He is going to put the garden next to his house; so he will only have to fence in three sides. What dimensions should the garden have if he has 30 feet of fencing and wants the area to be as large as possible? *Hint:* The following picture may help.

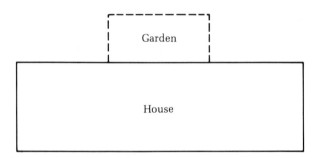

**32** A man has just bought a large dog. He wishes to fence in a rectangular area of his yard as a play area for the dog. He is going to use 5 feet of one side of his house as part of one side of the area, because this will include a door that will give him access to the play area. The remainder of the area will, of course, be fenced in. The situation is this:

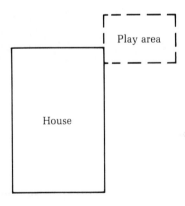

If he has 43 feet of fencing, what should the dimensions of the play area be in order to produce the most room for the dog?

**33** A man is installing a swimming pool. He wants its perimeter to be 200 feet, with the area as large as possible. The pool will be shaped in the form of a rectangle surmounted at each end by a semicircle, like this.

What should the dimensions of the pool be? *Hint:* The formula for the area of a circle is given by $A = \pi r^2$ and that for its circumference by $C = 2\pi r$ where $r$ is its radius.

**34** Do problem 33 with the understanding that the pool will be in the shape of a rectangle surmounted by a semicircle at just one end, like this.

A projectile shot upward from the ground at velocity $v$ feet per second obeys the equation $s = vt - 16t^2$ where $t$ is the time measured in seconds from the time of the shot and $s$ measures in feet the distance of the projectile from the ground.

*Use the foregoing information in problems 35 and 36.*

( ∘ ) **35** A projectile is shot upward from the ground at a velocity of 80 feet per second. How high will it go, and how long will it take before it hits the ground?

**36** How fast would you have to shoot a projectile in order for it to reach a maximum height of 144 feet?

( ∘ )

*Sample solutions*
*for exercise set 7.6*

**1** We are given the equation $y = x^2 - 5x + 4$. The $y$-intercept of its graph is clearly 4. Its $x$-intercepts are simply the real roots (if any) of the quadratic equation $x^2 - 5x + 4 = 0$. These can easily be determined by the factorization $(x - 4)(x - 1) = 0$; thus, they are seen to be 4 and 1. To obtain the coordinates of the turning point we set $x = -\dfrac{-5}{2} = \dfrac{5}{2}$ and note that $y$

$= \left(\dfrac{5}{2}\right)^2 - 5\left(\dfrac{5}{2}\right) + 4 = \dfrac{25}{4} - \dfrac{25}{2} + 4 = -\dfrac{25}{4} + 4 = -\dfrac{9}{4}$. In summary,

then, the answer is that the *y-intercept* is 4, the *x-intercepts* are 1 and

4, and the *turning point* is $\left(\dfrac{5}{2}, -\dfrac{9}{4}\right)$.

**11**  The intercepts and turning point asked for in problem 1 allow us to make this sketch of the graph.

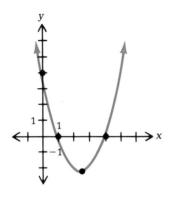

**21**  The equation we are to plot is $x = -3y^2 + 5y + 2$. The *x-intercept* of its graph is at 2. To obtain the *y-intercepts* we solve the equation

$3y^2 - 5y - 2 = 0$

$(3y + 1)(y - 2) = 0$

$$y = -\dfrac{1}{3} \quad \text{or} \quad y = 2.$$

Thus, the *y-intercepts* are at $-\dfrac{1}{3}$ and 2. Next we find coordinates for

the turning point. When $y = -\dfrac{5}{2(-3)} = \dfrac{5}{6}$

$x = -3\left(\dfrac{5}{6}\right)^2 + 5\left(\dfrac{5}{6}\right) + 2 = -3\left(\dfrac{25}{36}\right) + \dfrac{25}{6} + 2$

$= -\dfrac{25}{12} + \dfrac{25}{6} + 2 = \dfrac{25}{12} + 2 = \dfrac{49}{12}.$

The turning point has coordinates $\left(\dfrac{49}{12}, \dfrac{5}{6}\right)$. This is the graph we get when we plot these points.

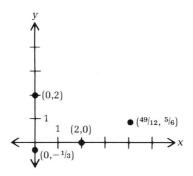

To obtain a more accurate graph of the parabola, we note finally that when $y = -1$, $x = -3(1) + 5(-1) + 2 = -3 - 5 + 2 = -6$. The parabola opens to the left, as shown below.

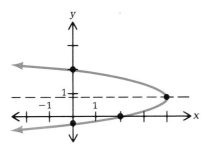

**25** Let $x$ be one of the numbers. The other one must then be $22 - x$. We want $x(22 - x) = 22x - x^2$ to be as large as possible. This event occurs at the turning point of the graph of $y = -x^2 + 22x$. Thus, we want $x = -\dfrac{22}{2(-1)} = 11$. It follows that both numbers should be 11.

**31**  Here is our picture.

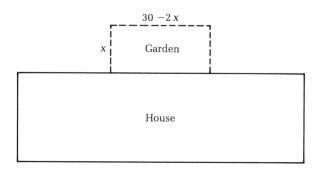

Let $x$ be the length of each side perpendicular to the house. The side parallel to the house must then have length $30 - 2x$ if we are to use up 30 feet of fencing. Since both $x$ and $30 - 2x$ must be positive, we have $0 < x < 15$; so the area of the garden is given by $A(x) = x(30 - 2x) = -2x^2 + 30x \quad (0 < x < 15)$. The area will be as large as possible when $x = -\dfrac{30}{2(-2)} = \dfrac{30}{4} = \dfrac{15}{2}$. Thus we want the side perpendicular to the house to be $7\dfrac{1}{2}$ feet long and the side parallel to the house 15 feet long.

**35**  The function that gives the height in feet of the projectile after $t$ seconds is $s(t) = 80t - 16t^2$. It will reach its maximum height when $t = -\dfrac{80}{2(-16)} = \dfrac{80}{32} = \dfrac{5}{2}$ seconds, and this maximum height will be

$$s = 80\left(\frac{5}{2}\right) - 16\left(\frac{5}{2}\right)^2 = 200 - 16\left(\frac{25}{4}\right) = 200 - 100$$

$$= 100 \text{ feet.}$$

To say that the projectile is on the ground is to say that $s = 0$. Solving the equation $80t - 16t^2 = 0$ we obtain

$$16t(5 - t) = 0$$

$$t = 0 \text{ or } t = 5.$$

The projectile will therefore return to earth after 5 seconds. Incidentally, this also shows that the domain of the function $s(t)$ is $[0,5]$ and its range is $[0,100]$.

## (7.7) (Circles, ellipses, and hyperbolas)

When you worked with equations of the form $y = ax^2 + bx + c$ and $x = ay^2 + by + c$ you were beginning your study of equations of the second degree. In the present section you will see some other interesting kinds of second-degree equations.

We begin by sketching the graph of the equation $x^2 + y^2 = 4$. To obtain the domain of the relation defined by this equation we solve for $y$, obtaining

$$y^2 = 4 - x^2$$

$$y = \pm\sqrt{4 - x^2}.$$

The condition for $y$ to be a real number is that $4 - x^2 \geq 0$, or that $-2 \leq x \leq 2$. Thus the domain of the relation defined by $x^2 + y^2 = 4$ is the closed interval $[-2, 2]$. When $x = 2$, $y = \pm\sqrt{4 - 4} = 0$; so the point $(2,0)$ is on the graph. Similar reasoning shows that $(1, \sqrt{3})$, $(1, -\sqrt{3})$, $(0,2)$, $(0,-2)$, and $(-2,0)$ *all* lie on this graph. These points are plotted in Figure 7.30. Connecting the points with a smooth curve (and, if necessary, plotting additional points), we see that the graph of the equation $x^2 + y^2 = 4$ is a *circle* with radius 2 and center the origin (see Figure 7.31). That this is true is also shown by the fact that the point $(x,y)$ is 2 units from the origin if and only if $\sqrt{(x - 0)^2 + (y - 0)^2} = \sqrt{x^2 + y^2} = 2$, which is equivalent to saying that $x^2 + y^2 = 4$. The same type of reasoning will show that the graph of

*circle*

**QE.3** $x^2 + y^2 = r^2 \qquad (r > 0)$

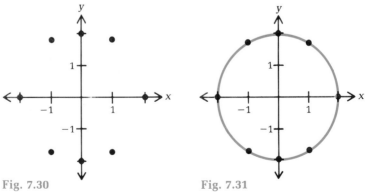

Fig. 7.30    Fig. 7.31

is a circle with radius $r$ and center the origin.

The equation $9x^2 + 4y^2 = 36$ is an example of the second type of equation we shall discuss. Solving for $y$, we obtain

$$4y^2 = 36 - 9x^2 = 9(4 - x^2)$$

$$y^2 = \frac{9}{4}(4 - x^2)$$

$$y = \pm\frac{3}{2}\sqrt{4 - x^2}.$$

It follows that the relation defined by our equation has domain $[-2,2]$. Substituting $x = 0, \pm\frac{1}{2}, \pm1, \pm\frac{3}{2}, \pm2$ in the original equation and solving for $y$, we obtain the points $(0,\pm3)$, $\left(\pm\frac{1}{2},\pm\frac{3}{4}\sqrt{15}\right)$, $\left(\pm1,\pm\frac{3}{2}\sqrt{3}\right)$, $\left(\pm\frac{3}{2},\pm\frac{3}{4}\sqrt{7}\right)$, and $(\pm2,0)$. These points are plotted in Figure 7.32 and connected with a smooth curve in Figure 7.33. The resulting oval-shaped figure is called an *ellipse*. It can be thought of as a circle that has been "stretched" along one of the coordinate axes.

*ellipse*

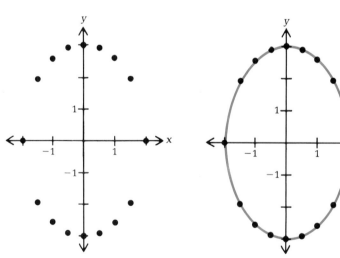

Fig. 7.32                    Fig. 7.33

More generally, if $a$, $b$, and $c$ are positive numbers with $a \neq b$, the graph of the equation

**QE.4** $a^2 x^2 + b^2 y^2 = c^2$

is an ellipse. Such a graph can be sketched quite easily from your general knowledge of the shape of an ellipse once you have the location of its intercepts. When $x = 0$, $b^2 y^2 = c^2$; so the $y$-intercepts are $\pm \sqrt{\dfrac{c}{b}}$; similarly, the $x$-intercepts are $\pm \sqrt{\dfrac{c}{a}}$.

To illustrate, let us consider the equation $x^2 + 4y^2 = 3$. When $x = 0$, $y = \pm\dfrac{1}{2}\sqrt{3}$, and when $y = 0$, $x = \pm\sqrt{3}$. Using just these points together with our knowledge that the graph is an ellipse, we sketch the graph as shown in Figure 7.34.

This brings us to the third and final type of equation we wish to discuss. We illustrate this type by sketching the graph of the equation $x^2 - 4y^2 = 4$. When $x = 0$, $-4y^2 = 4$, which is impossible in the real number system. Thus there are no $y$-intercepts. On the other hand, when $y = 0$, $x = \pm 2$; so the graph has $x$-intercepts $\pm 2$. Solving for $y$, we obtain $y = \pm\dfrac{1}{2}\sqrt{x^2 - 4}$; so $x^2 - 4 \geq 0$, and $x \geq 2$ or $x \leq -2$. It follows that the domain of the relation defined by our equation is $\{x \mid x \leq -2\} \cup \{x \mid x \geq 2\}$. Substituting $x = \pm 2$, $\pm 3$, $\pm 4$, $\pm 5$, $\pm 6$ and solving for $y$ we obtain $(\pm 2, 0)$, $\left(\pm 3, \pm\dfrac{1}{2}\sqrt{5}\right)$,

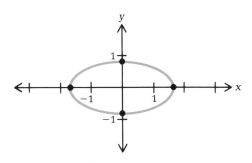

Fig. 7.34

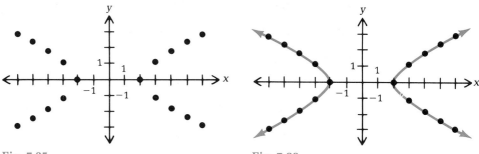

Fig. 7.35                                  Fig. 7.36

$(\pm 4, \pm \sqrt{3})$, $\left(\pm 5, \pm \dfrac{1}{2}\sqrt{21}\right)$, and $(\pm 6, \pm \sqrt{8})$ as points on the graph.

These points are plotted in Figure 7.35. It is important that you connect them correctly with *two* smooth curves, as illustrated in Figure 7.36. The resulting graph is called a *hyperbola*. When plotting the graph of a hyperbola, it is helpful first to draw a pair of straight lines called *asymptotes*. The graph of $x^2 - 4y^2 = 4$, together with its asymptotes, is shown in Figure 7.37. The asymptotes are the graph of the equation $x^2 - 4y^2 = 0$; that is, the straight lines $x = 2y$ and $x = -2y$. They are helpful because of the way that the hyperbola "approaches" its asymptotes. As the magnitude of $x$ gets larger and larger, the hyperbola gets arbitrarily close to its asymptotes without ever touching them.

In general the graph of

**QE.5**  $a^2 x^2 - b^2 y^2 = c^2$    $(a, b, c > 0)$

*hyperbola*

*asymptotes*

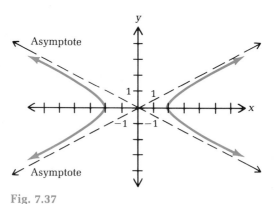

Fig. 7.37

is a hyperbola opening to the right and left. Its $x$-intercepts are obtained by setting $y = 0$ and noting that $a^2 x^2 = c^2$, so that $x = \pm \dfrac{c}{a}$.

Similarly the graph of

**QE.6** $b^2 y^2 - a^2 x^2 = c^2$ $\qquad (a, b, c > 0)$

is a hyperbola opening up and down (you can see one in Figure 7.39). Its $y$-intercepts are at $\pm \dfrac{c}{b}$. In either case, the asymptotes are the straight lines $a^2 x^2 - b^2 y^2 = 0$, or more simply

$$y = \pm \frac{a}{b} x.$$

As a final illustration, we consider the graph of $3y^2 - x^2 = 9$. This will clearly be a hyperbola opening up and down. When $x = 0$, $3y^2 = 9$ and $y = \pm\sqrt{3}$. The asymptotes are the straight lines $3y^2 - x^2 = 0$, which yields $x = \pm\sqrt{3}y$. In Figure 7.38 this much is sketched in. Finally, setting $y = \pm 2$ and $\pm 3$ and solving for $x$ we obtain $(\pm\sqrt{3}, \pm 2)$, $(\pm\sqrt{21}, \pm 3)$, which are the points used in Figure 7.39 to sketch the graph.

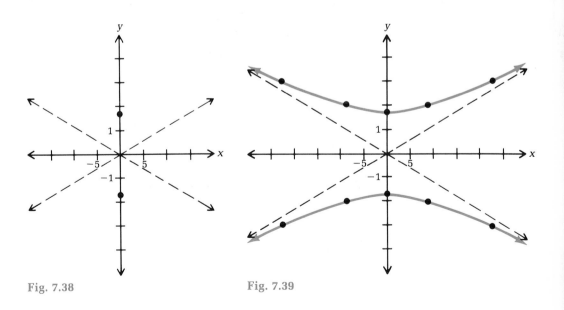

Fig. 7.38         Fig. 7.39

253 (∘) Circles, ellipses, and hyperbolas

One of the more interesting facts about parabolas, circles, ellipses, and hyperbolas is that they can all be obtained by "slicing" a cone with a plane; for this reason, they are sometimes called *conic sections*. Figure 7.40 shows how they are derived from the cone.

*conic sections*

| Curve | Conic section | Equation | Graph |
|---|---|---|---|
| Parabola |  | $y = ax^2 + bx + c$ <br> or <br> $x = ay^2 + by + c$ <br> $(a \neq 0)$ | 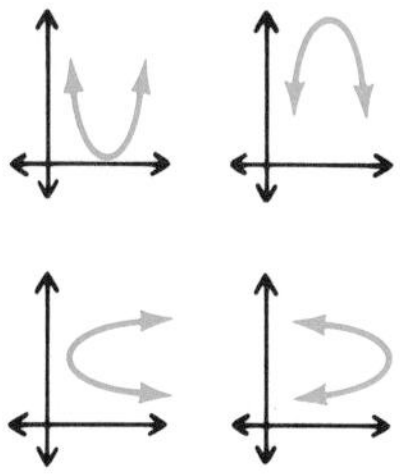 |
| Circle | 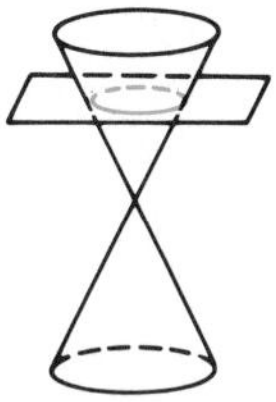 | $x^2 + y^2 = r^2$ <br> $(r > 0)$ | 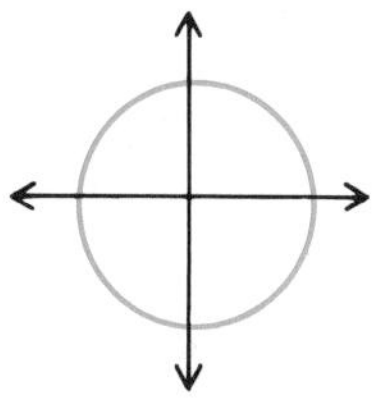 |
| Ellipse |  | $a^2 x^2 + b^2 y^2 = c^2$ <br> $(a, b, c > 0; a \neq b)$ | 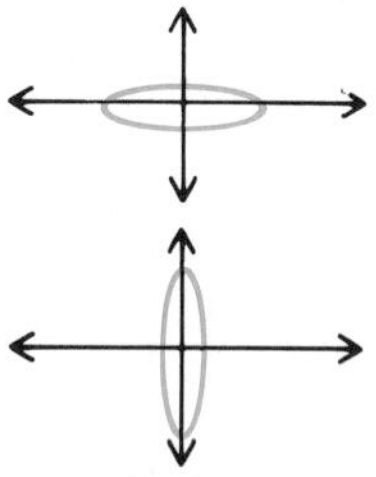 |
| Hyperbola |  | $a^2 x^2 - b^2 y^2 =$ <br> or <br> $b^2 y^2 - a^2 x^2 = c^2$ <br> $(a, b, c > 0)$ | 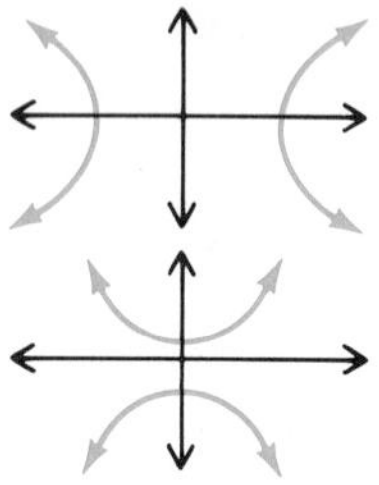 |

Fig. 7.40

# Exercise set 7.7

*Look at each equation and decide without graphing whether the curve is a parabola, a circle, an ellipse, or a hyperbola. Then find the intercepts and the domain of the associated relation.*

$(\circ)$  
1  $x^2 + y^2 = 5$        2  $x^2 + y^2 = 6$

3  $3x^2 + 3y^2 = 4$      4  $4x^2 + 4y^2 = 6$

5  $3x^2 + y^2 = 9$       6  $x^2 + 3y^2 = 9$

$(\circ)$  
7  $-2x^2 - y^2 = -4$     8  $-x^2 - 6y^2 = -6$

9  $\dfrac{x^2}{6} + \dfrac{y^2}{9} = 1$       10  $\dfrac{x^2}{4} + \dfrac{y^2}{3} = 1$

$(\circ)$  
11  $2x^2 - 5y^2 = 3$      12  $4x^2 - y^2 = 8$

13  $5y^2 - 2x^2 = 3$      14  $y^2 - 4x^2 = 8$

15  $\dfrac{x^2}{9} - \dfrac{y^2}{16} = 1$      16  $\dfrac{x^2}{25} - \dfrac{y^2}{9} = 1$

17  $y = 3x^2$          18  $x = -2y^2$

19  $3x^2 - 5y^2 = -5$     20  $5y^2 - 4x^2 = -4$

*The remaining problems ask you to do some graphing. For problems 21 to 40, sketch the graphs of the equations in problems 1 to 20. Naturally, you should make use of the work you did when you solved these problems before.*

$(\circ)$  
41  Sketch the graph of the equation $x^2 + 4y^2 = 0$; more generally, sketch the graph of $ax^2 + by^2 = 0$ $(a, b > 0)$.

42  Try to sketch the graph of $x^2 + 4y^2 = -4$. More generally, describe the graph of $ax^2 + by^2 = -c$ $(a, b, c > 0)$.

$(\circ)$

*Sample solutions for exercise set 7.7*

1  This is a circle whose radius is $\sqrt{5}$. Its $x$-intercepts are $\pm\sqrt{5}$, as are its $y$-intercepts. Thus it follows that the domain of the associated relation is $\{x \mid -\sqrt{5} \le x \le \sqrt{5}\}$. This can also be obtained by solving for $y$ and noting that $y = \pm\sqrt{5 - x^2}$, so that $5 - x^2 \ge 0$, $-\sqrt{5} \le x \le \sqrt{5}$.

7  The graph of this curve is an ellipse since it is of the form $a^2 x^2 + b^2 y^2 = c^2$ with $a \ne b$. When $x = 0$, $-y^2 = -4$; so $y = \pm 2$. When $y = 0$, $-2x^2 = -4$, $x^2 = 2$, and $x = \pm\sqrt{2}$. Thus the $x$-intercepts are $\pm\sqrt{2}$ and the $y$-intercepts $\pm 2$. To find the domain, note that $y^2 = 4 - 2x^2$, so that $y = \pm\sqrt{4 - 2x^2}$. Thus $4 - 2x^2 \ge 0$ when $-\sqrt{2} \le x \le \sqrt{2}$. This shows that the domain of the associated relation is $\{x \mid -\sqrt{2} \le x \le \sqrt{2}\}$.

11  This one is a hyperbola opening to the right and left. Setting $y =$

0, you find the x-intercepts are at $\pm\sqrt{\dfrac{3}{2}}$. Solving for $y$ will produce

$5y^2 = 2x^2 - 3$; so $y = \pm\sqrt{\dfrac{2x^2 - 3}{5}}$. It follows that $2x^2 - 3 \geq 0$, $x^2 \geq \dfrac{3}{2}$

and, finally, that the associated relation has as its domain $\left\{x \mid x \geq \sqrt{\dfrac{3}{2}}\right\}$

$\cup \left\{x \mid x \leq -\sqrt{\dfrac{3}{2}}\right\}$.

**21** In problem 1 you noted that the graph is a circle with radius $\sqrt{5}$. This is all the information you need to sketch its graph.

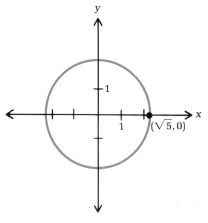

**27** From problem 7 we know that the points $(\pm\sqrt{2},0)$ and $(0,\pm2)$ are on the graph of $-2x^2 - y^2 = -4$. We may now use that information to sketch the graph since we know it is an ellipse.

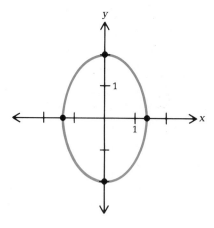

**31** From problem 11 we know the graph is a hyperbola opening to the right and left with $x$-intercepts at $\pm\sqrt{\dfrac{3}{2}}$. The equations for the asymptotes are obtained from $2x^2 - 5y^2 = 0$ and are $y = \pm\sqrt{\dfrac{2}{5}}\,x$. When $x = \pm 2$, $8 - 5y^2 = 3$, $5y^2 = 5$, and $y = \pm 1$. Thus $(\pm 2, \pm 1)$ lie on the graph of the equation. Similarly, $(\pm 3, \pm\sqrt{3})$ are also on the graph. When we use this information, this is the graph we get.

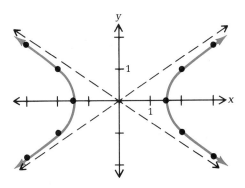

**41** Solving for $y$, we obtain $4y^2 = -x^2$, $y^2 = -\dfrac{1}{4}x^2$, or $y = \pm\dfrac{1}{2}\sqrt{-x^2}$. But $\sqrt{-x^2}$ is a real number only when $x = 0$. It follows that the graph of $x^2 + 4y^2 = 0$ is the single point $(0,0)$. Similar reasoning shows the graph of $ax^2 + by^2 = 0$ $(a,\, b > 0)$ to be the single point $(0,0)$.

## (7.8) (Direct and inverse variation)

When two variables $x$ and $y$ are related by an equation of the form

**V.1** $y = kx$   ($k$ a positive constant)

*direct variation*
*constant of variation*
*constant of proportionality*

we say that $y$ *varies directly* as $x$ or that $y$ is *directly proportional* to $x$. The constant $k$ is called the *constant of variation* or the *constant of proportionality.* Similarly, we say that $y$ varies directly as the *square* of $x$ or that $y$ is directly proportional to the *square* of $x$ when

**V.2** $y = kx^2$   ($k$ a positive constant).

In general, when we say that $y$ varies directly as the n*th power* of $x$ or that $y$ is directly proportional to the n*th power* of $x$ $(n > 0)$, we mean that

**V.3** $y = kx^n$  ($k$ a positive constant).

Functions defined by equations of this form are a useful tool in physics, chemistry, geometry, and other subjects that attempt to talk about the world in a precise way. For example, in geometry, the formula $A = \pi r^2$ for the area of a circle tells us that the area varies directly as the square of its radius, the constant of variation being $\pi$. Similarly, the formula $I = 0.08D$ for the income $I$ from an investment of $D$ dollars for one year at 8 percent simple interest states that the income from such an investment is directly proportional to the amount of the investment, with 0.08 the constant of proportionality. As a third illustration, we mention that Kepler's third law states that the distance of a planet from the sun varies directly as the two-thirds power of the time it takes for the planet to make a complete revolution around the sun.

*joint variation*

It can happen that $y$ varies directly with respect to more than one variable. We then speak of *joint variation*. Thus, to say that *y varies jointly* as $x$ and the square of $w$ means that for some positive constant $k$ $y = kxw^2$. The formula $V = \pi r^2 h$ for the volume $V$ of a right circular cylinder with radius $r$ and height $h$ tells us that the volume varies jointly with the height and the square of the radius of its base.

An equation of the form

**V.4** $y = \dfrac{k}{x}$  ($k$ a positive constant)

*inverse variation*

is an example of *inverse variation*. To say that *y varies inversely* as the $n$th power of $x$, or that $y$ is *inversely proportional* to the $n$th power of $x$ $(n > 0)$ is to say that for some positive constant $k$, $y = k/x^n$. To illustrate this we consider an automobile traveling 80 miles at a speed of $v$ miles per hour in a time of $t$ hours. The equation $80 = vt$ when put in the form $v = \dfrac{80}{t}$ shows that the speed varies inversely to the time of the trip.

It is also possible for $y$ to vary directly with one variable and inversely with another. Thus, to say that $y$ varies directly as the square of $w$ and inversely as the fourth power of $z$ is to say that $y = kw^2/z^4$ ($k$ a positive constant). A concrete example of this relationship is provided by *intensity of illumination,* in that the intensity $I$ of a screen illuminated by a light source varies directly as the power $P$ of the source and inversely as the square of its distance $s$ from the source. This says that $I = kP/s^2$ with $k$ a positive constant.

In problems involving direct and inverse variation, the constant of variation must either be given to you (as in $A = \pi r^2$) or you must be given enough data to compute its value. For example, let us suppose that $y$ varies directly as the square of $x$, and that $y = 3$ when $x = 2$. Then $y = kx^2$, and $3 = k(2^2)$. From this we see that $k = 3/4$; so the equation relating $x$ and $y$ is $y = (3/4)x^2$. Here is a worked-out problem to illustrate this idea.

Problem: The time $T$ it takes to fill a tank varies inversely as the square of the diameter $d$ of the pipe used to fill the tank. If it takes 10 hours to fill the tank with a 1-inch pipe, how long will it take with a 2-inch pipe?

Solution: We know that $T = k/d^2$ ($k$ a positive constant) and that $10 = k/1^2$. Hence $k = 10$, $T = 10/d^2$, and when $d = 2$, $T = 10/2^2 = 10/4 = 5/2$. It would take 2-1/2 hours with a 2-inch pipe.

## Exercise set 7.8

*Translate each of the following variational relationships into an equation.*

(∘) 1 The distance $D$ traveled by an automobile with constant velocity varies directly as the time $t$.

2 The area $A$ of a square varies directly as the square of the length $b$ of its base.

3 The circumference $C$ of a circle varies directly as the length $r$ of its radius.

4 When an object is dropped from a tower, the distance $d$ it falls is directly proportional to the square of the time $t$.

5 Your salary $S$ is directly proportional to the number of hours $H$ that you work.

6 The pressure $P$ of an enclosed gas at constant temperature varies inversely as the volume $V$.

7 The intensity $H$ of an electric field varies inversely as the square of the distance $d$ from a charge.

8 Under constant acceleration the velocity $v$ of a car varies directly as the time $t$.

**9** The volume $V$ of a right circular cone varies jointly as its height $h$ and the square of the radius $r$ of its base.

**10** The area $A$ of a triangle varies jointly as its base $b$ and its height $h$.

*In problems 11 to 16, find the constant k of variation and then write down the equation that expresses the stated relationship.*

(∘) **11** The quantity $y$ varies directly as the cube of $x$, and $y = 1$ when $x = 2$.

**12** The quantity $y$ varies directly as the square of $r$ and $y = 5$ when $r = 2$.

**13** The quantity $y$ varies inversely as $x$ and $y = 2$ when $x = 6$.

**14** The quantity $y$ varies inversely as $w$ and $y = 2$ when $w = \dfrac{1}{2}$.

**15** $W$ varies jointly as $d$ and the square of $t$, and $W = 1$ when $d = 2$ and $t = 3$.

**16** $I$ varies directly as the square of $R$ and inversely as the cube of $E$. When $E = 2$ and $R = \dfrac{1}{2}$, $I = 3$.

*Now for a few more challenging problems.*

(∘) **17** If $y$ varies directly as $x$ and inversely as the square of $t$, and if $y = \dfrac{3}{2}$ when $x = 3$ and $t = 2$, find the value of $y$ when $x$ and $t$ are doubled.

**18** If $d$ varies jointly as $s$ and the cube of $t$ and if $d = 54$ when $s = 2$ and $t = 3$, find $d$ when $s = 1$ and $t = 2$.

**19** In problem 1, assume that the car travels 3 miles in 9 minutes. How far will it travel in 2 hours?

**20** In problem 4, assume that the object drops 64 feet in 2 seconds. How far will it drop in 5 seconds?

**21** The weight $W$ of a body is inversely proportional to the square of its distance $x$ from the center of the earth. If an object weighs 100 pounds on the surface of the earth, how much will it weigh at a height 100 miles above the earth's surface? *Note:* Use 4,000 miles for the radius of the earth.

**22** The *period* of a pendulum (the time for a complete oscillation) varies directly with the square root of its length. If a 12-foot pendulum has a period of 10 seconds, find the period of a 3-foot pendulum.

If $y$ is directly proportional to $x$, then the graph of the equation $y = kx$ is a straight line passing through the origin. But when $y$ is inversely proportional to $x$, the situation changes dramatically. The graph of $y = \dfrac{k}{x}$ turns out to be a hyperbola that has the coordinate axes as asymptotes. This should give you enough of a hint for the last two problems.

( ∘ )    23   Sketch the graph of the equation $y = \dfrac{1}{x}$.

      24   Sketch the graph of the equation $y = \dfrac{4}{x}$.

( ∘ )

*Sample solutions
for exercise set 7.8*

1   By definition of direct variation, $D = kt$ where $k$ is a positive constant.

11   We know that $y = kx^3$ and $y = 1$ when $x = 2$. Hence $1 = k(2^3) = 8k$, and $k = \dfrac{1}{8}$. The desired equation is $y = \dfrac{x^3}{8}$.

17   The equation described in the problem is $y = \dfrac{kx}{t^2}$. You know that $\dfrac{3}{2} = \dfrac{k \cdot 3}{2^2} = \dfrac{3k}{4}$, so that $6k = 12$ and $k = 2$. Thus $y = \dfrac{2x}{t^2}$, and when $x = 6$ and $t = 4$, we have $y = \dfrac{2 \cdot 6}{4^2} = \dfrac{12}{16} = \dfrac{3}{4}$.

23   We are to sketch the graph of $y = \dfrac{1}{x}$. The domain of the relation defined by $y = \dfrac{1}{x}$ is $\{x \mid x \in R,\, x \neq 0\}$. The following points are all on the graph: $\left(\dfrac{1}{4},4\right)$, $\left(\dfrac{1}{2},2\right)$, $(1,1)$, $\left(2,\dfrac{1}{2}\right)$, $\left(3,\dfrac{1}{3}\right)$, $\left(-\dfrac{1}{4},-4\right)$, $\left(-\dfrac{1}{2},-2\right)$, $(-1,-1)$, $\left(-2,-\dfrac{1}{2}\right)$, $\left(-3,-\dfrac{1}{3}\right)$. The next step is to plot some of these points in the plane.

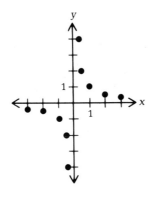

Finally, we use our knowledge that the graph is a hyperbola whose asymptotes are the coordinate axes to connect the points with two smooth curves. The result looks like this.

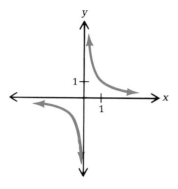

# (8)

# *Exponential and logarithmic functions*

## (8.1) (Scientific notation)

*scientific notation*

People often have to work with extremely large or extremely small numbers. In order to manipulate such numbers, it is useful to express them in a form called *scientific notation*. To illustrate the idea, let us consider the following numbers:

$$25.4 = (2.54)10^1 \qquad\qquad 0.254 = (2.54)10^{-1}$$

$$254 = (2.54)10^2 \qquad\qquad 0.0254 = (2.54)10^{-2}$$

$$2{,}540 = (2.54)10^3 \qquad\qquad 0.00254 = (2.54)10^{-3}$$

You should observe that multiplication of a number by $10^n$ ($n \in N$) moves its decimal point $n$ places to the *right*, and multiplication by $10^{-n}$ moves it $n$ places to the *left*. Each of the numbers above has been expressed in scientific notation, which means that it has been written in the form

$$a \cdot 10^n \qquad (1 \le a < 10; \; n \in I).$$

To write a number in scientific notation, the factor $a$ is obtained by placing a decimal point after the first nonzero digit that occurs as one reads the number from left to right. The integer $n$ that appears in the second factor produces the power of 10 needed to shift the decimal point from its location in $a$ to its location in the given number. Let's illustrate this concretely. To put 41,300,000 in scientific notation, we observe that we must write it in the form $41,300,000 = (4.13)10^n$ for suitable $n \in I$. To determine $n$ we write down our number and look at where the decimal point would be in scientific notation:

4 1 3 0 0 0 0 0.
↑
Scientific notation decimal point

We then count the number of places the decimal point must be shifted to go from scientific notation to regular notation:

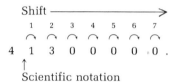

Since the decimal must be shifted seven places to the right, we have that $41,300,000 = (4.13)10^7$.

Similarly, to express 0.000203 in scientific notation we write

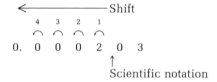

and observe that the decimal point must be shifted four units to the left. Hence $0.000203 = (2.03)10^{-4}$.

A number in scientific notation is expressed in regular notation by using the fact that multiplication by $10^n$ shifts the decimal point an appropriate number of places to the right or left. Thus $(3.2)10^5 = 320,000$ (decimal point shifted 5 places to right), and $(3.2)10^{-3} = 0.0032$ (decimal point shifted 3 places to left).

We now pause to see how very large and very small numbers can arise. For example, the sun is 93,000,000 miles from the earth. In scientific notation this is $(9.3)10^7$. The closest star, Alpha Centauri, is approximately $(2.6)10^{13}$ miles from Earth. But you don't have to talk about astronomy to get into numbers of this magnitude. For example, the total amount of personal consumer expenditures in the United States in a recent year was approximately $(4.138)10^{11}$ dollars. If you want some small numbers to think about, consider the fact that a red blood cell measures about $(2.9)10^{-4}$ inches in diameter, and the diameter of some viruses is $10^{-6}$ inches.

Scientific notation often helps when you are faced with a computation that involves large or small numbers like the ones we just mentioned. For example, to evaluate $96,800,000/0.0000242$, we put the numbers in scientific notation and note that

$$\frac{(9.68)10^7}{(2.42)10^{-5}} = \frac{9.68}{2.42} \cdot \frac{10^7}{10^{-5}} = 4 \cdot 10^{12} = 4,000,000,000,000.$$

## Exercise set 8.1

*Express in regular notation.*

$(\circ)$   1   $(3.8)10^3$         2   $(7.6)10^4$

3   $(2.39)10^{-2}$         4   $(3.01)10^{-4}$

5   $(5)10^{12}$         6   $(7.5)10^{10}$

7   $(3.1)10^{-1}$         8   $(2.41)10^{-2}$

9   $(12.23)10^3$         10   $(138.0)10^{-2}$

*Express in scientific notation.*

$(\circ)$   11   314         12   439

13   25,000,000         14   93,400,000

15   0.0031         16   0.59

$(\circ)$   17   0.000004         18   0.00000000000801

19   14.001         20   3.00001

To show how scientific notation can be used as a computational aid, let's express $(2 \cdot 10^{-3})/(5 \cdot 10^{-7})$ as a number in scientific notation. We have

$$\frac{2}{5} \cdot \frac{10^{-3}}{10^{-7}} = (0.4)(10^{-3+7}) = (4 \cdot 10^{-1})10^4 = 4 \cdot 10^3.$$

*Express each of the following as a number in scientific notation.*

$(\circ)$   21   $\dfrac{(4 \cdot 10^3)(3 \cdot 10^6)^2}{(2 \cdot 10^8)(6 \cdot 10^{-4})}$         22   $\dfrac{(5 \cdot 10^{-4})(8 \cdot 10^{-2})}{(2 \cdot 10^3)(2 \cdot 10^{-4})^3}$

$$23 \quad \frac{(340{,}000)(0.0016)(120)}{(0.0006)(3{,}200)(0.000017)} \qquad 24 \quad \frac{(18{,}000)(0.12)(0.24)}{(0.0024)(0.036)(20)}$$

*The purpose of the remaining problems is to familiarize you with the use of scientific notation. Express each answer in scientific notation. Problems 25 to 28 involve a unit of measurement called a light year, which is the name for the distance traveled by a beam of light in one year's time. A light year is approximately $6 \cdot 10^{12}$ miles.*

$(\circ)$  25  The star Sirius (the brightest visible star) is 8.7 light years from Earth. What is this distance in miles?

26  The star Vega is 26 light years from the earth. Express this distance in miles.

27  A standard unit of astronomical distance is the *parsec*, which is 3.26 light years. How many miles are there in a parsec?

28  Two new galaxies were discovered recently by the Italian astronomer, Paolo Maffei, at an estimated distance of 3 million light years from the earth. Compute their distance in miles.

29  A recent US gross national product was $974.1 billion. Express this figure in scientific notation.

30  A typical annual national defense expenditure for the United States is $73,581 billion. Express this figure in scientific notation.

*In the remaining problems, use the fact that a micron is about 1/25,000 of an inch, and a millimicron is 1/1,000 of a micron.*

$(\circ)$  31  The virus that causes foot-and-mouth disease in animals is about 20 millimicrons in diameter. Express its diameter in inches.

32  *Silicosis* is an incurable chronic disease of the lungs caused by inhaling fine particles of silicon. The particles that do the most damage are between 1 and 3 microns in diameter. How would you express the size of the particles in inches?

$(\circ)$

*Sample solutions for exercise set 8.1*

1  Multiplication by $10^3$ shifts the decimal point three places to the right; so $(3.8)\,10^3 = 3{,}800$.

11  $314 = (3.14)10^2$

17  $0.000004 = 4 \cdot 10^{-6}$

$$21 \quad \frac{(4 \cdot 10^3)(3 \cdot 10^6)^2}{(2 \cdot 10^8)(6 \cdot 10^{-4})} = \frac{4 \cdot 10^3 \cdot 9 \cdot 10^{12}}{12 \cdot 10^8 \cdot 10^{-4}} = \frac{36 \cdot 10^{15}}{12 \cdot 10^4} = 3 \cdot 10^{11}$$

25  If one light year is $6 \cdot 10^{12}$ miles, then 8.7 light years is

$$(8.7)(6 \cdot 10^{12}) = (52.2)10^{12} = [(5.22)10^1]10^{12}$$

$$= (5.22)10^{13} \text{ miles}.$$

31  Let's begin by computing the length of a millimicron in inches. Using the fact that 1 micron is $1/25{,}000$ of an inch, we have that a millimicron is

$$\frac{1}{25{,}000} \cdot \frac{1}{1{,}000} = \frac{1}{25} \cdot \frac{1}{1{,}000} \cdot \frac{1}{1{,}000} = (0.04)(10^{-3})(10^{-3})$$

$$= (4 \cdot 10^{-2})(10^{-6}) = 4 \cdot 10^{-8} \text{ inches}.$$

Hence 20 millimicrons is $(20)(4 \cdot 10^{-8}) = (80)10^{-8} = 8 \cdot 10^{-7}$ inches.

# (8.2) (Inverses of functions)

Given a relation $S$, it is often useful to define a new relation $S^{-1}$ (read "$S$ inverse" or "the inverse of $S$") by the following rule.

( ∘ )  $S^{-1} = \{(b,a) \mid (a,b) \in S\}$

Thus, the ordered pairs of $S^{-1}$ are obtained by interchanging the components of the ordered pairs of $S$. If $S = \{(1,3), (2,-2), (3,4)\}$, then $S^{-1} = \{(3,1), (-2,2), (4,3)\}$. Note that the domain of $S$ equals the range of $S^{-1}$, and the range of $S$ equals the domain of $S^{-1}$. We shall be especially interested in forming inverses of relations that are defined by equations. Before doing so, however, it will be instructive to have a geometric interpretation of the process.

The graph of $S$ is shown in part ($a$) of Figure 8.1. The graph of $S^{-1}$ can be obtained by simply interchanging the roles of $x$ and $y$ in the plane. The horizontal axis is now called the $y$-axis and the vertical axis the $x$-axis. Furthermore, ordered pairs must be written in the reverse order to ensure that the first component gives the $x$-coordinate and the second component the $y$-coordinate of each point. But now [see Figure 8.1, part ($b$)] the orientation is all wrong. In order to get the Cartesian plane to look the way we are used to seeing it, we first *reflect* it about the vertical axis (in other words, we rotate it through an angle of 180° using the vertical axis as the axis of rotation) and then *rotate* it clockwise through an angle of 90° about the origin, as illustrated in Figure 8.1, ($c$) and ($d$). At this point we have provided a geometric construction for the graph of $S^{-1}$.

Now, let's consider the relation $F$ defined by the equation $y = 2x + 3$. Then $F = \{(x,y) \mid y = 2x + 3\}$ and $F^{-1} = \{(y,x) \mid y = 2x + 3\}$. Since it is customary to write $x$ for the first component of an ordered pair and $y$ for the second component, we interchange the roles of $x$ and $y$ to obtain $F^{-1} = \{(x,y) \mid x = 2y + 3\}$. Thus if $F$ is defined by the equation $y = 2x + 3$, $F^{-1}$ is defined by the equation $x = 2y + 3$ that we obtained by interchanging the roles of $x$ and $y$. In Figure 8.2 we see how the graph of $F^{-1}$ is obtained from that of $F$, and in Figure 8.3 both graphs are displayed on the same coordinate system.

As our next illustration we consider the relation $G$ defined by the equation $y = x^2 - 4$. The inverse relation is defined by the equation $x = y^2 - 4$, and its graph is obtained from that of $G$, as shown in Figure 8.4. Finally, we look at the relation $T = \{(x,y) \mid x^2$

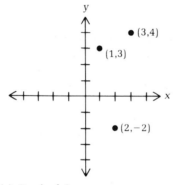

**(a)** Graph of $S$.

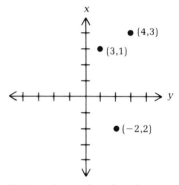

**(b)** Interchange the roles of $x$ and $y$ in the plane.

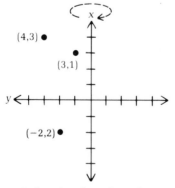

**(c)** Reflect the plane about the vertical axis.

Fig. 8.1

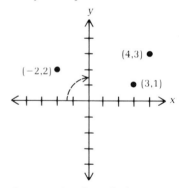

**(d)** Rotate the plane clockwise through an angle of 90° about the origin to obtain the graph of $S^{-1}$.

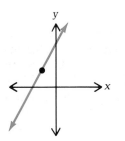

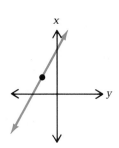

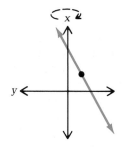

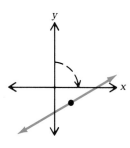

**(a)** Graph of $F$.

Fig. 8.2

**(b)** Interchange $x$ and $y$.

**(c)** Reflect about the vertical axis.

**(d)** Rotate 90° to produce graph of $F^{-1}$.

$+ 9y^2 = 9\}$. Its graph is the ellipse shown in Figure 8.5, and the graph of $T^{-1}$ is the graph of $9x^2 + y^2 = 9$; it is displayed in Figure 8.6.

Let's have another look at the relations $F = \{(x,y) \mid y = 2x + 3\}$ and $G = \{(x,y) \mid y = x^2 - 4\}$. These relations are both clearly functions, but when you examine part $(d)$ of Figures 8.2 and 8.4, you can see that $F^{-1}$ is a function but $G^{-1}$ is not. Thus, the inverse relation to a function may or may not be a function itself. To see why $G^{-1}$ is not a function we observe that the vertical line $x = 0$ intersects the graph of $G$ at the points $(0,2)$ and $(0,-2)$. This occurs because $G(2) = G(-2) = 0$, which says that $G$ is not a one-one function.

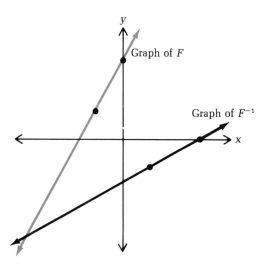

Fig. 8.3

269 (∘) Inverses of functions

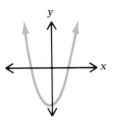 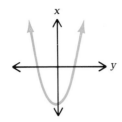 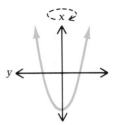 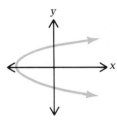

**(a)** Graph of $G$

**(b)** Interchange $x$ and $y$

**(c)** Reflect about vertical axis

**(d)** Rotate 90°; graph of $G^{-1}$

Fig. 8.4

( ○ )   *In fact, for any function f, the inverse relation $f^{-1}$ is a function if and only if f is one-one.*

The proof of this statement is left for the exercises.

As we have already mentioned, if the function $F$ is defined by the equation $y = 2x + 3$, then $F^{-1}$ is defined by $x = 2y + 3$. Solving for $y$, we obtain $2y = x - 3$; so $y = (1/2)(x - 3)$. Thus if $F(x) = 2x + 3$, then $F^{-1}(x) = (1/2)(x - 3)$.

( ○ )   *In general, if an equation defines a one-one function f, then the equation formed by interchanging x and y defines the inverse function $f^{-1}$. Furthermore, the domain of f equals the range of $f^{-1}$, and the range of f equals the domain of $f^{-1}$.*

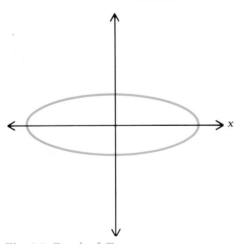

Fig. 8.5  Graph of $T$

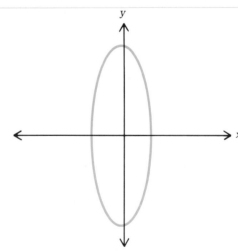

Fig. 8.6  Graph of $T^{-1}$

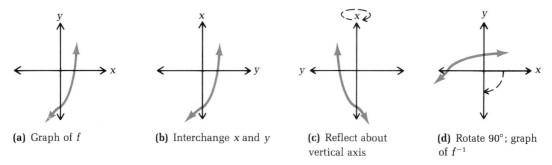

**(a)** Graph of $f$    **(b)** Interchange $x$ and $y$    **(c)** Reflect about vertical axis    **(d)** Rotate 90°; graph of $f^{-1}$

**Fig. 8.7**

Suppose $f(x) = x^3 - 4$. Distinct values of $x$ yield distinct values of $f(x)$; so $f$ is a one-one function. Now $f$ is defined by the equation $y = x^3 - 4$, and $f^{-1}$ is defined by $x = y^3 - 4$. Solving this equation for $y$ produces $y^3 = x + 4$, or $y = \sqrt[3]{x + 4}$. This shows that $f^{-1}(x) = \sqrt[3]{x + 4}$. The graph of $f^{-1}$ is constructed from that of $f$ in Figure 8.7.

If the equation $y = x^2 + 1$ is solved for $x$, the result is $x^2 = y - 1$, $x = \pm\sqrt{y - 1}$. This shows that more than one value of $x$ can be associated with the same value of $y$; so the function $f(x) = x^2 + 1$ is not one-one. Of course this can also be seen by observing that $f(-1) = 2 = f(1)$. On the other hand, consider the function $A(x) = \pi x^2$ that expresses the area of a circle with radius $x$. The domain of this function is $R^+$, and for that reason it is a one-one function. Its inverse is the function that gives the radius of a circle that has area $x$. Now $A$ is defined by the equation $y = \pi x^2$, and $A^{-1}$ by $x = \pi y^2$. Noting that $y > 0$ we see that $y^2 = x/\pi$ and $y = \sqrt{x/\pi}$. Thus $A^{-1}(x) = \sqrt{x/\pi}$; so a circle that has area $x$ will have radius $\sqrt{x/\pi}$. The graphs of $A$ and $A^{-1}$ are displayed in Figure 8.8.

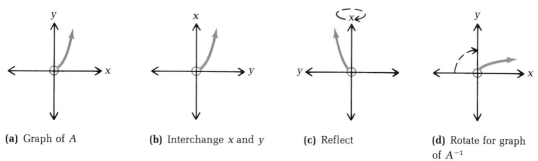

**(a)** Graph of $A$    **(b)** Interchange $x$ and $y$    **(c)** Reflect    **(d)** Rotate for graph of $A^{-1}$

**Fig. 8.8**

271 (∘) Inverses of functions

# Exercise set 8.2

In problems 1 to 10, sketch the graph of the given relation. Then use the methods of the text to construct the graph of the inverse relation.

$( \circ )$    1   $S = \{(x,y) \mid 2x + 3y = 4\}$        2   $S = \{(x,y) \mid x + 4y = 1\}$

3   $S = \{(-1,4),\ (1,2),\ (1,3)\}$        4   $S = \{(-2,1),\ (1,1),\ (1,4)\}$

5   $S = \{(x,y) \mid y = 1 - x^2\}$        6   $S = \{(x,y) \mid y = 6 - 2x^2\}$

7   $S = \{(x,y) \mid x^2 + 4y^2 = 4\}$        8   $S = \{(x,y) \mid 3x^2 + y^2 = 9\}$

9   $S = \{(x,y) \mid x^2 - 4y^2 = 4\}$        10   $S = \{(x,y) \mid 3x^2 - y^2 = 9\}$

Each of the following equations defines a one-one function f. Express $f^{-1}(x)$ in terms of x.

$( \circ )$    11   $y = 3x + 2$        12   $y = 5x - 3$

13   $y = \dfrac{1}{4}x + \dfrac{3}{2}$        14   $y = -\dfrac{2}{3}x + 4$

15   $4x + 7y = 10$        16   $3x - 5y = 8$

17   $y = \dfrac{1}{x+1}\ (x \neq -1)$        18   $y = \dfrac{1}{x-2}\ (x \neq 2)$

In problems 19 to 24, sketch the graph of each function, then sketch the graph of the inverse relation. Determine whether the given function is one-one. If it is one-one, express the inverse function in terms of x and specify its domain.

$( \circ )$    19   $f(x) = x^2 \quad (x \leq 0)$        20   $g(x) = x^2 - 1 \quad (x < 0)$

21   $h(x) = \sqrt{x + 1}$        22   $f(x) = \sqrt{x - 3}$

23   $f(x) = 4 - x^2$        24   $h(x) = (x - 2)^2$

Decide whether each of the following functions is one-one, and determine the inverse function of those that are one-one.

$( \circ )$    25   $f(x) = x^3 + 1$        26   $g(x) = x^5 - 3$

27   $f(x) = x^4 + 2$        28   $h(x) = x^4 + 8$

29   $g(x) = 2x^2 - 9 \quad (x \geq 0)$        30   $g(x) = x^4 - 16 \quad (x > 0)$

31   $f(x) = -\sqrt{x + 1}$        32   $g(x) = -\sqrt{x - 3}$

Many one-one functions arise from practical problems in which both the function and its inverse have geometric or physical interpretations. A geometric example was provided at the end of the section. As a further example we ask you to consider an object that is dropped from a height of 100 feet. Its height above the ground at the end of $x$ seconds is given by the function $h(x) = 100 - 16x^2$. The object will reach the ground when $0 = 100 - 16x^2$. This says that $4x^2 = 25$, or that $x = \dfrac{5}{2}$. Since $h(x)$ cannot be negative, the domain of this function is the closed interval $\left[0, \dfrac{5}{2}\right]$ and its range is $[0,100]$. The function is defined by the equation $y = 100 - 16x^2$, and its inverse is defined by $x = 100 - 16y^2$. Solving this equation for $y$, we have $16y^2 = 100 - x$, $4y = \sqrt{100 - x}$; so $y = \dfrac{1}{4}\sqrt{100 - x}$. Thus, the inverse function $h^{-1}(x) = \dfrac{1}{4}\sqrt{100 - x}$ has domain $[0,100]$, range $\left[0, \dfrac{5}{2}\right]$, and it expresses the time at which the object has height $x$ feet.

*Each of problems 33 to 42 involves a one-one function that has some physical or geometric interpretation. Determine the inverse function and give an appropriate interpretation of what it represents. Unless otherwise specified, give the domain and range of the inverse function.*

( ○ ) **33** An automobile is traveling at a speed of 60 mph. The distance it travels at the end of $x$ hours is given by the function $d(x) = 60x$.

**34** An automobile is traveling at a speed of 40 mph on a 400-mile trip. The function that expresses the distance to the destination at the end of $x$ hours is $d(x) = 400 - 40x$.

**35** If an object is dropped from a height of 200 feet, the distance it falls after $x$ seconds is given by $d(x) = 16x^2$.

**36** If an object is dropped from a height of 800 feet, its height at the end of $x$ seconds is given by $h(x) = 800 - 16x^2$.

( ○ ) **37** The circumference $C$ of a circle of radius $x$ is given by the function $C(x) = 2\pi x$.

**38** The area $A$ of a triangle with altitude 6 and base $x$ is $A(x) = 3x$.

**39** The volume $V$ of a sphere with radius $x$ is $V(x) = \dfrac{4}{3}\pi x^3$.

**40** The surface area $A$ of a sphere of radius $x$ is $A(x) = 4\pi x^2$.

*In problems 41 and 42 you need not determine the domain and range of the inverse function.*

**41** The function that expresses $x°$ Fahrenheit as a Centigrade temperature is $C(x) = \dfrac{5}{9}(x - 32)$.

**42** The weight $W$ (in pounds) of a certain object at a distance of $x$ miles from the surface of the earth is given by $W(x) = \dfrac{(1.6)10^9}{(x + 4{,}000)^2}$.

We stated earlier that for any function $f$, the inverse relation $f^{-1}$ is a function if and only if the function $f$ is one-one. The remaining two exercises are devoted to the proof of this statement.

$(\circ)$  **43**  Show that if $f$ is a one-one function, then $f^{-1}$ is a function.

**44**  Let $f$ be a function. Show that if $f^{-1}$ is a function, then $f$ must be one-one.

$(\circ)$

*Sample solutions
for exercise set 8.2*  **1**  The graph of the function and its inverse looks like this.

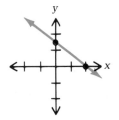

**(a)** Graph of $S$

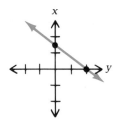

**(b)** Interchange $x$ and $y$

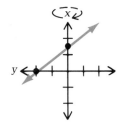

**(c)** Reflect

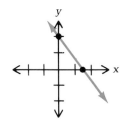

**(d)** Rotate for graph of $S^{-1}$

**11**  The function $f^{-1}$ is defined by the equation $x = 3y + 2$. Solving for $y$, we obtain $3y = x - 2$; so $y = \dfrac{1}{3}(x - 2)$. Hence $f^{-1}(x) = \dfrac{1}{3}(x - 2)$.

**19**  We have the condition $x \le 0$, which ensures that $f$ is a one-one function. It is defined by the equation $y = x^2$, and its inverse by $x = y^2$. The condition $x \le 0$ on $f$ translates to $y \le 0$ on $f^{-1}$. Hence $y = -\sqrt{x}$, and the inverse function is given by $f^{-1}(x) = -\sqrt{x}$. The domain of $f^{-1}$ is $\{x \mid x \ge 0\}$.

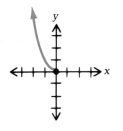

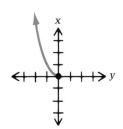

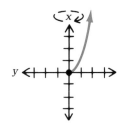

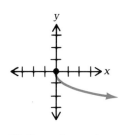

**(a)** Graph of $f$    **(b)** Interchange $x$ and $y$    **(c)** Reflect    **(d)** Rotate for graph of $f^{-1}$

**25**   Distinct values of $x$ yield distinct values of $x^3$; so the given function is one-one. The inverse function is defined by the equation $x = y^3 + 1$. Solving this for $y$ produces $y^3 = x - 1$; so $y = \sqrt[3]{x-1}$. Thus $f^{-1}(x) = \sqrt[3]{x-1}$.

**33**   Because it makes no sense to speak of negative distance or negative time in this problem, the domain and range of $d$ are each $\{x \mid x \geq 0\}$. It follows that $d^{-1}$ has domain and range equal to $\{x \mid x \geq 0\}$. Since $d$ is defined by $y = 60x$, $d^{-1}$ comes from $x = 60y$; so $d^{-1}(x) = \dfrac{x}{60}$. The function $d^{-1}$ expresses the length of time it would take for the automobile to travel a distance of $x$ miles.

**37**   The function $C(x)$ is defined by the equation $y = 2\pi x$; so $C^{-1}$ is defined by $x = 2\pi y$. Thus $C^{-1}(x) = \dfrac{x}{2\pi}$; it gives the radius of a circle with circumference $x$, and its domain and range are both clearly $R^+$.

**43**   Let $f$ be a one-one function, and form $f^{-1} = \{(a,b) \mid (b,a) \in f\}$. To show that $f^{-1}$ is a function we must show that if $(a,b)$, $(a,c) \in f^{-1}$, then $b = c$. But if $(a,b)$, $(a,c) \in f^{-1}$, then $(b,a) \in f$ and $(c,a) \in f$. Since $f$ is one-one, $b = c$.

# (8.3) (Exponential functions)

In Chapter 2 we defined $b^x$ for $b \in R$ and $x$ a natural number. In Chapter 5 this definition was extended to define $b^x$ for $b \neq 0$ and $x$ an *arbitrary* integer. It was then extended to the case where $x \in Q$, but here we had to restrict $b$ to $R^+$ in order to be certain that $b^x$ was a real number. In the present section we shall indicate how to define $b^x$ for $b > 0$ and $x$ an irrational number. A thorough treatment of this topic involves ideas that are not covered in this book, but we will at least give you some notion of what is involved.

To illustrate the definition of $b^x$ for $b > 0$, $x \in P$ we shall define $3^{\sqrt{2}}$. The procedure is based on the following facts about the real number system, which we state without proof.

( ◦ ) *If $w$ and $x \in Q$ and $w < x$, then $3^w < 3^x$.*

( ◦ ) *The number $\sqrt{2}$ has a nonterminating decimal expansion $\sqrt{2}$ = 1.414214 ....*

What this means is that

$1 < \sqrt{2} < 2,$

$1.4 < \sqrt{2} < 1.5,$

$1.41 < \sqrt{2} < 1.42,$

$1.414 < \sqrt{2} < 1.415$, and so on.

The further we go out in the decimal expansion of $\sqrt{2}$, the closer we get to $\sqrt{2}$. The point is that each of the numbers 1, 1.4, 1.41, 1.414,... is rational so we may compute

$3^{1.4} \approx 4.655$

$3^{1.41} \approx 4.707$

$3^{1.414} \approx 4.727$

$3^{1.4142} \approx 4.728.$

The symbol $\approx$ is used in place of an equals sign to emphasize the fact that these are only approximations. By going far enough out in the decimal expansion of $\sqrt{2}$ we obtain rational numbers $r_1$, $r_2, \ldots, r_n, \ldots$ that come as close as we desire to $\sqrt{2}$. The corresponding numbers $3^{r_1}, 3^{r_2}, \ldots, 3^{r_n}, \ldots$ come arbitrarily close to a number that we define to be $3^{\sqrt{2}}$. Since $1.4142 < \sqrt{2} < 1.4143$ we already know that $3^{1.4142} < 3^{\sqrt{2}} < 3^{1.4143}$. Using the facts that $3^{1.4142} \approx 4.728$ and $3^{1.4143} \approx 4.729$, we see that $3^{\sqrt{2}}$ is between 4.728 and 4.729. Thus, out to three decimal places its decimal expansion is 4.728. By going out a sufficient number of places in the decimal expansion of $\sqrt{2}$ we may compute as many decimal places as we want in the expansion of $3^{\sqrt{2}}$. This is the sort of thing that a computer can do very nicely, but before the chapter is over you will have a technique of your own for doing it.

The point to all this is that for $b > 0$ and $x \in R$, a real number $b^x$ can be defined in such a way that the laws of exponents are valid. To refresh your memory we repeat them here. Let $a$ and $b \in R^+$ and $w, x \in R$.

$(\circ)$    E.1   $b^w \cdot b^x = b^{w+x}$

$(\circ)$    E.2   $(b^x)^w = b^{xw}$

$(\circ)$    E.3   $(ab)^x = a^x b^x$

$(\circ)$    E.4   $\dfrac{b^x}{b^w} = b^{x-w}$

Thus

$5^{\sqrt{3}} \cdot 5^{\pi} = 5^{\sqrt{3}+\pi}$    (by E.1),

$(8^{\sqrt{2}})^{\sqrt{2}} = 8^2 = 64$    (by E.2),

$5^{\pi} \cdot 4^{\pi} = 20^{\pi}$    (by E.3),

$\dfrac{2^{1+\sqrt{7}}}{2^{3+\sqrt{7}}} = 2^{(1+\sqrt{7})-(3+\sqrt{7})} = 2^{-2} = \dfrac{1}{4}$    (by E.4).

We now direct our attention to a fixed positive number $b$. For each real number $x$, the formation of $b^x$ associates with $x$ a unique second real number. This tells us that the relation $f = \{(x, b^x) \mid x \in R\}$ is a *function*. The function $f(x) = b^x$ is called the *exponential function with base $b$*. Since $b^x$ is defined for all real numbers $x$, the domain of this function is $R$. For $b \neq 1$, its range turns out to be $R^+$. To obtain some geometric intuition about exponential functions we present the graphs of $y = 3^x$ and $y = \left(\dfrac{1}{3}\right)^x$ in Figure 8.9. The graph of $f(x) = 3^x$ is obtained by observing that $f(-2) = \dfrac{1}{9}, f(-1) = \dfrac{1}{3}, f(0) = 1, f(2) = 9$. The points $\left(-2, \dfrac{1}{9}\right)$, $\left(-1, \dfrac{1}{3}\right)$, $(0,1)$, $(1,3)$, $(2,9)$ are plotted and then connected with a smooth curve as in Figure 8.9 ($a$). Similarly, in Figure 8.9 ($b$) the

*exponential function with base b*

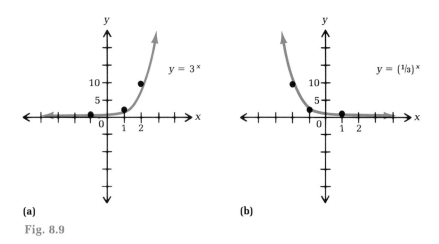

**(a)**

**(b)**

Fig. 8.9

points $(-2,9)$, $(-1,3)$, $(0,1)$, $\left(1,\dfrac{1}{3}\right)$, $\left(2,\dfrac{1}{9}\right)$ are plotted and connected with a smooth curve to produce the graph of $f(x) = \left(\dfrac{1}{3}\right)^x$. The graph of $y = 3^x$ is typical of $y = b^x$ for $b > 1$ and that of $y = \left(\dfrac{1}{3}\right)^x$ typifies the case in which $0 < b < 1$. Inspection of these graphs should at least make plausible the assertion that if $x < w$

$b^x < b^w$     (in case $b > 1$) and

$b^x > b^w$     (when $0 < b < 1$).

It is immediately apparent that if $x \ne w$, $b^x \ne b^w$; so exponential functions are, in fact, one-one. Now let us have some examples that illustrate how exponential functions arise.

First, we consider a bacteria culture. Under ideal conditions the number of bacteria present at the end of $x$ days is given by the function $N(x) = N_0 b^x$ where $N_0$ is the number initially present and $b$ is a constant that is determined by the type of bacteria, the temperature, and so on. These functions are called *exponential growth functions*; they are exponential functions with base $b$, $b > 1$. Suppose you start with a culture containing 10,000 bacteria, and that after three days you count 2,160,000. To determine the value of $b$, notice that $N(x) = 10,000b^x$ and that $N(3) = 2,160,000$; so that

*exponential growth functions*

$$2,160,000 = 10,000b^3$$

$$b^3 = \frac{2,160,000}{10,000} = 216$$

$$b = \sqrt[3]{216} = 6.$$

The exponential growth function for this culture is therefore $N(x)$ $= 10,000(6^x)$. At the end of one day there were $N(1) = 10,000(6^1)$ $= 60,000$ bacteria present; at the end of four days there should be $N(4) = 10,000(6^4) = (10,000)(1,296) = (10^4)(1.296)(10^3) = (1.296)10^7$ bacteria.

Our second example involves the decay of a radioactive substance. With the passage of time some of the atoms of a radioactive element break up and turn into atoms of other elements. The function that gives the amount $A(x)$ of a radioactive element remaining at time $x$ looks like $A(x) = A_0 b^x$ where $A_0$ is the amount initially present and $b$ is a constant that depends on the type of element. Functions such as these are called *exponential decay functions*; they are exponential functions with base $b$, $0 < b < 1$. The *half-life* of a radioactive substance is the time it takes for half its atoms to decay; in other words, it is the time $x$ for which $A(x) = \frac{1}{2}A_0$. The most common isotope of radium has a half-life of approximately 1,600 years. This fact may be used to compute the constant $b$ that appears in its decay function. We have that

*exponential decay functions*

$$A(x) = A_0 b^x$$

$$A(1,600) = \frac{1}{2}A_0 = A_0 b^{1,600}.$$

Hence

$$b^{1,600} = \frac{1}{2}$$

$$b = \left(\frac{1}{2}\right)^{1/1600};$$

so $A(x) = A_0\left[\left(\frac{1}{2}\right)^{1/1600}\right]^x$. If you started with 0.05 grams of radi-

um, then after 800 years there would remain $A(800) = 0.05\left(\frac{1}{2}\right)^{800/1600}$

$$= 0.05 \left( \frac{1}{2} \right)^{1/2} \approx (0.05)(0.7) = 0.035 \text{ grams.}$$

## Exercise set 8.3

*Use the laws of exponents to simplify each answer, and if possible express each result as an integer or a fraction in lowest terms.*

( ∘ )   1   $(10^{\sqrt{3}})^{\sqrt{3}}$                              2   $(2^{\sqrt{2}})^{\sqrt{8}}$

      3   $(7^{1/\pi})^{\pi}$                              4   $(2^{\pi})^{1/\pi}$

      5   $3^{1+\sqrt{2}} \cdot 3^{1-\sqrt{2}}$                    6   $2^{3+2\pi} \cdot 2^{1-2\pi}$

      7   $\dfrac{10^{3-\sqrt{5}}}{10^{4-\sqrt{5}}}$                    8   $\dfrac{10^{3-5\pi}}{10^{2-5\pi}}$

( ∘ )   9   $(3^{3+\sqrt{2}} \cdot 3^{-3+\sqrt{2}})^{\sqrt{2}}$          10   $(2^{1+\sqrt{2}})^{1-\sqrt{2}}$

*Fill in the missing component of each ordered pair so that the ordered pair will be on the graph of the corresponding equation.*

( ∘ )   11   $y = 2^{x}$        $(-1,?), (0,?), (2,?)$

     12   $y = 4^{x}$        $(-2,?), (1,?), \left( \dfrac{1}{2},? \right)$

     13   $y = 5^{x}$        $(-3,?), (-1,?), (2,?)$

     14   $y = \left( \dfrac{1}{2} \right)^{x}$        $(-2,?), (2,?), (3,?)$

     15   $y = 2^{-x}$        $(-3,?), (-2,?), (4,?)$

     16   $y = 3^{-x}$        $(-3,?), (-2,?), (4,?)$

( ∘ )   17   $y = 10^{x}$        $(?,1), \left( ?,\dfrac{1}{10} \right), (?,1{,}000)$

     18   $y = 10^{x}$        $\left( ?,\dfrac{1}{100} \right), (?,10), (?,100)$

     19   $y = 2^{x}$        $\left( ?,\dfrac{1}{4} \right), \left( ?,\dfrac{1}{8} \right), (?,32)$

     20   $y = 3^{x}$        $(?,81), \left( ?,\dfrac{1}{3} \right), (?,1)$

*Plot the graph of each of the following functions for* $-2 \le x \le 2$.

( ○ ) 21   $f(x) = 10^x$        22   $f(x) = 4^x$

     23   $f(x) = 2^x$        24   $f(x) = 5^x$

     25   $f(x) = \left(\dfrac{1}{4}\right)^x$        26   $f(x) = \left(\dfrac{1}{2}\right)^x$

     27   $g(x) = 4^{-x}$        28   $g(x) = 2^{-x}$

( ○ ) 29   Compare the graphs of problems 25 and 27. Can you conclude anything about the functions $f(x) = \left(\dfrac{1}{4}\right)^x$ and $g(x) = 4^{-x}$?

30   Compare the graphs of problems 26 and 28. Can you conclude anything about the functions $f(x) = \left(\dfrac{1}{2}\right)^x$ and $g(x) = 2^{-x}$? Extend your observation to the functions $f(x) = \left(\dfrac{1}{b}\right)^x$ and $g(x) = b^{-x}$.

*The remaining problems involve physical situations in which exponential functions arise.*

( ○ ) 31   A bacteria culture contains 5,000 bacteria. At the end of 6 hours (1/4 day) the count is 15,000. Determine the growth function that gives the size of the culture at the end of $x$ days. What will the size be at the end of 1 day?

32   A bacteria colony has an initial size of 4,000. At the end of 2 hours the count is 5,760. Determine the growth function that gives the size of the culture at the end of $x$ hours. What will the size be at the end of 3 hours?

33   A certain radioactive element (radium 221) has a half-life of 30 seconds. If you start with 1 milligram of this element, how much would remain at the end of 60 seconds? How much would remain at the end of 90 seconds?

34   A certain radioactive element (radon 219) has a half-life of 4 seconds. If you started with $1/250$ of a milligram of this element, how much would remain at the end of 8 seconds? How much would remain at the end of 10 seconds?

( ○ ) 35   A substance whose temperature is 170° F. is placed in a room whose temperature is 70° F. After 30 minutes, the temperature of the substance is 100° F. What will its temperature be after one hour? *Hint:* The function that expresses the temperature $T(x)$ at the end of $x$ hours is $T(x) = 70 + T_0(b^x)$ where $T_0$ and $b$ are constants to be determined from the conditions of the problem.

**36** An object whose temperature is 50° F. is placed in an oven that maintains a temperature of 350° F. After 10 minutes the temperature of the object is 80° F. What will its temperature be at the end of 30 minutes? *Hint:* The function that expresses the temperature $T(x)$ at the end of $x$ minutes is $T(x) = 350 - T_0(b^x)$ where $T_0$ and $b$ are constants to be determined from given conditions.

**37** The function that gives the atmospheric pressure $p(x)$, in inches of mercury, at an altitude of $x$ miles above sea level is given approximately by $p(x) = 30.0(0.8)^x$. What is the atmospheric pressure at sea level? What is the pressure at 2 miles above sea level?

**38** When a wound is healing, an approximation for the area $A(x)$ remaining unhealed after $x$ days is given by $A(x) = A_0 b^x$. If one-fourth of the wound is healed after ten days, how much of it will be healed at the end of twenty days? Here $A_0$ represents the original area of the wound, and $b$ is a constant that must be determined from the given data.

(∘)

*Sample solutions*
*for exercise set 8.3*

**1** $(10^{\sqrt{3}})^{\sqrt{3}} = 10^{\sqrt{3}\sqrt{3}} = 10^3 = 1{,}000$

**9** $(3^{3+\sqrt{2}} \cdot 3^{-3+\sqrt{2}})^{\sqrt{2}} = (3^{3+\sqrt{2}-3+\sqrt{2}})^{\sqrt{2}}$

$$= (3^{2\sqrt{2}})^{\sqrt{2}} = 3^{2\sqrt{2}\sqrt{2}}$$

$$= 3^4 = 81$$

**11** We observe that $2^{-1} = \dfrac{1}{2}$, $2^0 = 1$, and $2^2 = 4$; so the ordered pairs should read $\left(-1, \dfrac{1}{2}\right)$, $(0,1)$, $(2,4)$.

**17** Noting that $10^0 = 1$, $10^{-1} = \dfrac{1}{10}$, and $10^3 = 1{,}000$, we have the ordered pairs $(0,1)$, $\left(-1, \dfrac{1}{10}\right)$, $(3,1{,}000)$.

**21** Since integer powers of 10 are especially easy to compute, we plot the points on the graph of $y = 10^x$ corresponding to $x = -2, -1, 0, 1, 2$ and connect them with a smooth curve. These points are $\left(-2, \dfrac{1}{100}\right)$, $\left(-1, \dfrac{1}{10}\right)$, $(0,1)$, $(1,10)$, and $(2,100)$. The graph looks like this.

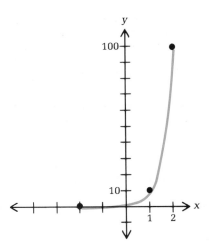

29  You should note that the two graphs are the same and conclude that $f = g$. This can also be seen from the fact that $4^{-x} = \dfrac{1}{4^x} = \left(\dfrac{1}{4}\right)^x$.

31  We know that the growth function is of the form $N(x) = N_0 b^x$, and that $5{,}000 = N(0) = N_0 b^0 = N_0$; so $N(x) = 5{,}000b^x$. We also have

$$15{,}000 = N\left(\dfrac{1}{4}\right) = 5{,}000b^{1/4},$$

$$b^{1/4} = \dfrac{15{,}000}{5{,}000} = 3$$

$$b = 3^4 = 81.$$

This shows that the growth function is $N(x) = 5{,}000(81^x)$. At the end of one day there will be $N(1) = 5{,}000(81) = 405{,}000 = (4.05)10^5$ bacteria present.

35  We are given that $T(x) = 70 + T_0(b^x)$. When $x = 0$, $T = 170$, so that $T(0) = 70 + T_0(b^0) = 170$. Hence, $70 + T_0 = 170$ and $T_0 = 100$. We also know that $T(1/2) = 100$; so

$$100 = 70 + 100b^{1/2}$$

$$30 = 100b^{1/2}$$

$$0.3 = b^{1/2}$$

$$b = (0.3)^2 = 0.09.$$

Thus $T(x) = 70 + 100(0.09)^x$, and $T(1) = 70 + 100(0.09) = 70 + 9 = 79$. At the end of one hour the temperature will be $79°$ F.

## (8.4) (Logarithmic functions)

*logarithmic function with base b*

In the last section we noted that the exponential function with base $b$ $(b > 0,\ b \neq 1)$ is a one-one function. The inverse of this function is called the *logarithmic function with base b*. The inverse to the exponential function

$$f = \{(x,y) \mid y = b^x\}$$

is obtained by interchanging the roles of $x$ and $y$ to write

$$f^{-1} = \{(x,y) \mid x = b^y\}.$$

If we agree to use the symbol $y = \log_b x$ (read "logarithm with base $b$ of $x$") to indicate the fact that $b^y = x$, we may write $f^{-1}$ in the form

$$f^{-1} = \{(x,y) \mid y = \log_b x\}.$$

The exponential function $f(x) = b^x$ has $R$ for its domain and $R^+$ for its range. It follows that the domain of $f^{-1}(x) = \log_b x$ is $R^+$, and its range is $R$.

It is important to be aware that the equations $b^y = x$ and $y = \log_b x$ are equivalent. You should be prepared to switch back and forth between the exponential and logarithmic forms of such equations. To illustrate, the exponential equation $2^3 = 8$ becomes $\log_2 8 = 3$ in logarithmic form, whereas $9^{3/2} = 27$ becomes $\log_9 27 = \dfrac{3}{2}$, and

$5^{-1} = \dfrac{1}{5}$ becomes $\log_5 \dfrac{1}{5} = -1$. Going in the other direction, $\log_{12} 1$

$= 0$ becomes $12^0 = 1$, $\log_{36} 6 = \dfrac{1}{2}$ becomes $(36)^{1/2} = 6$, and $\log_{2/3} \dfrac{9}{4}$

$= -2$ becomes $\left(\dfrac{2}{3}\right)^{-2} = \dfrac{9}{4}$.

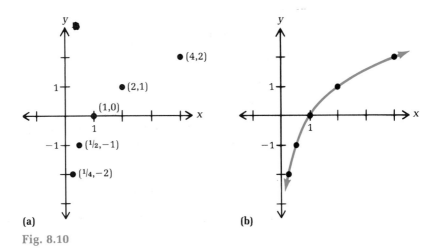

**(a)**                                    **(b)**

Fig. 8.10

To obtain some geometric feeling for logarithmic functions, let's sketch the graph of $y = \log_2 x$. In order to do this, we must obtain the value of $\log_2 x$ for enough values of $x$ to get an idea of what the graph looks like. We do this by considering the equivalent exponential equation $x = 2^y$. Using the fact that $\dfrac{1}{4} = 2^{-2}, \dfrac{1}{2} = 2^{-1}, 1 = 2^0, 2 = 2^1, 4 = 2^2$ we obtain the following values for $\log_2 x$:

| $x$ | $\dfrac{1}{4}$ | $\dfrac{1}{2}$ | 1 | 2 | 4 |
|---|---|---|---|---|---|
| $\log_2 x$ | $-2$ | $-1$ | 0 | 1 | 2 |

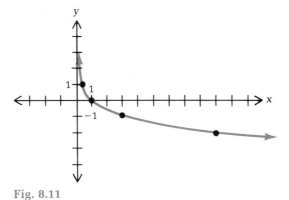

Fig. 8.11

285 (∘) Logarithmic functions

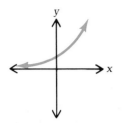

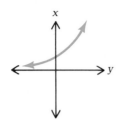

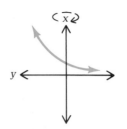

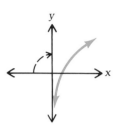

**(a)** Graph of $y = b^x$
$(b > 1)$

**(b)** Interchange $x$ and $y$

**(c)** Reflect

**(d)** Rotate for graph
of $y = \log_b x$
$(b > 1)$

Fig. 8.12

These points are plotted in Figure 8.10($a$), and in Figure 8.10($b$) they are connected with a smooth curve to produce a sketch of the graph of $y = \log_2 x$. This example typifies the graph of $y = \log_b x$ for $b > 1$.

To illustrate the case where $0 < b < 1$, we look at the graph of $y = \log_{1/3} x$. Again we consider the equivalent exponential equation $x = \left(\dfrac{1}{3}\right)^y$ in order to obtain values for $\log_{1/3} x$. Noting that $\dfrac{1}{9}$

$$= \left(\frac{1}{3}\right)^2, \frac{1}{3} = \left(\frac{1}{3}\right)^1, 1 = \left(\frac{1}{3}\right)^0, 3 = \left(\frac{1}{3}\right)^{-1}, 9 = \left(\frac{1}{3}\right)^{-2}, \text{ we}$$

have

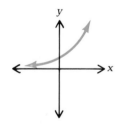

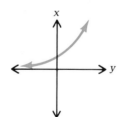

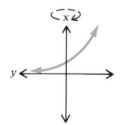

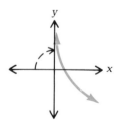

**(a)** Graph of $y = b^x$
$(0 < b < 1)$

**(b)** Interchange $x$ and $y$

**(c)** Reflect

**(d)** Rotate for graph
of $y = \log_b x$
$(0 < b < 1)$

Fig. 8.13

| $x$ | $\dfrac{1}{9}$ | $\dfrac{1}{3}$ | 1 | 3 | 9 |
|---|---|---|---|---|---|
| $\log_{1/3} x$ | 2 | 1 | 0 | $-1$ | $-2$ |

The graph of $y = \log_{1/3} x$ is shown in Figure 8.11.

Of course the methods of Section 8.2 can also be applied to construct the graph of $f^{-1}(x) = \log_b x$ from that of $f(x) = b^x$. Figure 8.12 illustrates this for the case where $b > 1$, and Figure 8.13 covers $0 < b < 1$.

## Exercise set 8.4

Express each of the following exponential equations in logarithmic form.

( ○ )  **1**  $2^5 = 32$          **2**  $3^2 = 9$

**3**  $3^0 = 1$          **4**  $8^0 = 1$

**5**  $4^{-2} = \dfrac{1}{16}$          **6**  $5^{-1} = \dfrac{1}{5}$

**7**  $4^{1/2} = 2$          **8**  $64^{1/2} = 8$

**9**  $\left(\dfrac{2}{3}\right)^{-2} = \dfrac{9}{4}$          **10**  $\left(\dfrac{4}{9}\right)^{3/2} = \dfrac{8}{27}$

Express each of the following logarithmic equations in exponential form.

( ○ )  **11**  $\log_4 16 = 2$          **12**  $\log_3 27 = 3$

**13**  $\log_4 2 = \dfrac{1}{2}$          **14**  $\log_9 27 = \dfrac{3}{2}$

**15**  $\log_{1/6} 36 = -2$          **16**  $\log_{9/10} \dfrac{10}{9} = -1$

**17**  $\log_{10} 1{,}000 = 3$          **18**  $\log_{10} 0.01 = -2$

**19**  $\log_{1/8} 16 = -\dfrac{4}{3}$          **20**  $\log_{1/27} 9 = -\dfrac{2}{3}$

Find the numerical value of each of the following.

( ○ )  **21**  $\log_7 49$          **22**  $\log_5 125$

**23**  $\log_9 9$          **24**  $\log_7 7$

**25** $\log_{1/2} 16$          **26** $\log_{1/3} 9$

**27** $\log_{12} \dfrac{1}{144}$          **28** $\log_{100} 10$

**29** $\log_b b^2$          **30** $\log_b \sqrt[3]{b^2}$

*Fill in the missing components in order to make each ordered pair a solution of the corresponding equation.*

$(\circ)$   **31** $y = \log_4 x$     $(64,?), (1,?), \left(\dfrac{1}{4},?\right)$

**32** $y = \log_{25} x$     $(5,?), \left(\dfrac{1}{5},?\right), (25,?)$

**33** $y = \log_3 x$     $(81,?), \left(\dfrac{1}{9},?\right), (3,?)$

**34** $y = \log_6 x$     $\left(\dfrac{1}{36},?\right), (216,?), (\sqrt{6},?)$

**35** $y = \log_{1/4} x$     $(2,?), (8,?), \left(\dfrac{1}{16},?\right)$

**36** $y = \log_{1/16} x$     $(2,?), (8,?), \left(\dfrac{1}{16},?\right)$

**37** $y = \log_7 x$     $(?,2), (?,-1), (?,0)$

**38** $y = \log_{100} x$     $\left(?,\dfrac{1}{2}\right), (?,-1), \left(?,\dfrac{3}{2}\right)$

*Solve for the indicated variable (b, x, or y).*

**39** $y = \log_5 25$          **40** $y = \log_4 \dfrac{1}{8}$

$(\circ)$   **41** $2 = \log_4 x$          **42** $3 = \log_9 x$

**43** $-3 = \log_3 x$          **44** $-2 = \log_7 x$

$(\circ)$   **45** $-2 = \log_b 49$          **46** $-3 = \log_b 1{,}000$

*Plot the graph of each of the following.*

**47** $y = \log_4 x$          **48** $y = \log_9 x$

**49** $y = \log_{1/2} x$          **50** $y = \log_{1/4} x$

1   In logarithmic form this becomes $\log_2 32 = 5$.

11   In exponential form we have $4^2 = 16$.

21   The question before us is, "Seven raised to what power equals 49?" Since the answer is 2, we have $\log_7 49 = 2$.

31   Using the fact that $4^3 = 64$, $4^0 = 1$, and $4^{-1} = 1/4$, we may fill in the missing components as follows: $(64,3)$, $(1,0)$, $\left(\dfrac{1}{4}, -1\right)$.

41   Converting the equation $2 = \log_4 x$ to exponential form produces $4^2 = x$; thus $x = 16$.

45   Upon conversion to exponential form we obtain $b^{-2} = 49$, $b^2 = 1/49$, and finally $b = 1/7$. The positive square root was chosen because the logarithmic function with base $b$ is the inverse of the corresponding exponential function with base $b$ ($b > 0$, $b \neq 1$).

## (8.5) (Properties of logarithms)

We have just seen that for $b > 0$ ($b \neq 1$) the equations $y = \log_b x$ and $x = b^y$ are equivalent in that they have the same solution sets. Thus for any positive number $x$, $\log_b x$ is the unique number $y$ such that $b^y = x$. Replacing $y$ with $b^x$ in this last equation produces the important fact that

(∘)   LO.1   $b^{\log_b x} = x$     $(x \in R,\, x > 0)$.

Thus $3^{\log_3 8} = 8$, $7^{\log_7 4} = 4$, and $10^{0.3010} = 2$ since $\log_{10} 2 = 0.3010$. Note that 0.3010 is only a decimal approximation for the irrational number $\log_{10} 2$. Though it is customary to use an "equals" sign in this context we really mean that $\log_{10} 2 \approx 0.3010$. If we replace $x$ with $b^y$ in the equation $y = \log_b x$, we obtain the symmetric result

(∘)   LO.2   $y = \log_b b^y$     $(y \in R)$.

To illustrate this we point out that $\log_7 7^{-5} = -5$, $\log_4 4^3 = 3$, and $\log_{10} 10^{\sqrt{2}} = \sqrt{2}$. You will use both equations LO.1 and LO.2 often throughout the chapter.

It is time to establish others of the basic properties of logarithms. Each of the following laws comes from the fact that logarithms are defined in terms of certain exponents.

($\circ$)　　LO.3 $\log_b(x_1 x_2) = \log_b x_1 + \log_b x_2$

For example, that $\log_3(9 \cdot 27) = \log_3(243) = \log_3 3^5 = 5$ and $\log_3 9 + \log_3 27 = \log_3 3^2 + \log_3 3^3 = 2 + 3 = 5$ illustrates the fact that $\log_3(9 \cdot 27) = \log_3 9 + \log_3 27$.

($\circ$)　　LO.4 $\log_b \dfrac{x_1}{x_2} = \log_b x_1 - \log_b x_2$

That $\log_2 \dfrac{32}{4} = \log_2 8 = 3$ and $\log_2 32 - \log_2 4 = 5 - 2 = 3$ shows that $\log_2 \dfrac{32}{4} = \log_2 32 - \log_2 4$.

($\circ$)　　LO.5 $\log_b x_1{}^r = r(\log_b x_1)$

This law is illustrated by the fact that $\log_4 2^6 = \log_4 64 = 3$ and $6(\log_4 2) = 6(1/2) = 3$. In the foregoing statements, $x_1$ and $x_2 > 0$, $b > 0$, $b \neq 1$, and $r \in R$.

Law LO.3 can be established by observing that if

$r_1 = \log_b x_1, \ r_2 = \log_b x_2$

then

$b^{r_1} = x_1$ and $b^{r_2} = x_2$.

By the laws of exponents,

$x_1 x_2 = b^{r_1} b^{r_2} = b^{r_1 + r_2}$.

In logarithmic form the equation $b^{r_1 + r_2} = x_1 x_2$ asserts that

$\log_b(x_1 x_2) = r_1 + r_2$.

Hence $\log_b(x_1 x_2) = \log_b x_1 + \log_b x_2$ as desired. Laws LO.4 and LO.5 are established in a similar manner.

To illustrate these laws, let's compute $\log_{10} 4$, $\log_{10} 5$ and $\log_{10} 6$ from the fact that $\log_{10} 2 = 0.3010$ and $\log_{10} 3 = 0.4771$. Using LO.3,

$$\log_{10} 6 = \log_{10}(2 \cdot 3) = 10g_{10}\, 2 + \log_{10} 3$$

$$= 0.3010 + 0.4771$$

$$= 0.7781.$$

To compute $\log_{10} 5$ we observe that $5 = 10/2$; so by LO.4,

$$\log_{10} 5 = \log_{10} 10 - \log_{10} 2$$

$$= 1 - 0.3010$$

$$= 0.6090.$$

Finally, we may obtain $\log_{10} 4$ by writing

$$\log_{10} 4 = \log_{10} 2^2 = 2(\log_{10} 2) = 2(0.3010)$$

$$= 0.6020.$$

## Exercise set 8.5

*Express each of the following as the sum or difference of logarithms of simpler expressions.*

$(\circ)$   1   $\log_3 \dfrac{xy^2}{z^3}$

2   $\log_{1/2} \dfrac{x(x+1)}{(x+2)^2}$

3   $\log_4 \dfrac{4(x^2 + x^3)}{\sqrt{x-2}}$

4   $\log_7 \dfrac{49w\sqrt{w+1}}{(w+2)^3}$

5   $\log_{16} \dfrac{4}{\sqrt{y-1}}$

6   $\log_3 \sqrt[3]{\dfrac{9(x+1)}{x^2}}$

The laws of logarithms may sometimes be used to write expressions involving logarithms as the logarithm of a single quantity. For example, we may use the fact that $2 = \log_6 36$ to write

$$2 \log_6 x - 3 \log_6 y + 2 = \log_6 x^2 - \log_6 y^3 + \log_6 36$$

$$= \log_6 \frac{36x^2}{y^3}.$$

*Express each of the following as the logarithm of a single quantity.*

$(\circ)$   7   $2 \log_3 x + 4 \log_3 y + 2$

9   $3 \log_5 (x+5) - \log_5 x - 1$

10   $3[\log_9 y - 4 \log_9 (y+1) + 1]$

8   $\dfrac{1}{2} \log_4 x + \dfrac{1}{4} \log_4 y + 1$

A *logarithmic equation* is simply an equation that involves logarithms. The solution involves applying the laws of logarithms to simplify the equation, and then rephrasing the equation in a form that does not involve logarithms. This is best shown by an example. To solve the equation

$$\log_{10} (x + 3) + \log_{10} (x - 6) = 1$$

we write the equivalent equation

$$\log_{10} (x + 3)(x - 6) = \log_{10} 10.$$

Using the fact that $f(x) = \log_{10} x$ is a one-one function we deduce that

$$(x + 3)(x - 6) = 10$$

$$x^2 - 3x - 18 = 10$$

$$x^2 - 3x - 28 = 0$$

$$(x + 4)(x - 7) = 0$$

$$x = -4 \text{ or } x = 7.$$

The solution $x = -4$ must be rejected because $\log_{10} (x + 3)$ is not defined for $x = -4$. On the other hand, $x = 7$ is a valid solution since $\log_{10} (7 + 3) + \log_{10} (7 - 6) = 1 + 0 = 1$.

*Solve each of the following logarithmic equations.*

**11** $\log_6 (x + 1) = 2$

**12** $\log_3 (x - 3) = 3$

$(\circ)$ **13** $\log_2 x + \log_2 (x + 2) = 3$

**14** $\log_5 x + \log_5 (x + 4) = 1$

**15** $\log_9 (x + 4) - \log_9 (x - 2) = \dfrac{1}{2}$

**16** $\log_2 (2x + 1) - \log_2 (x - 3) = 3$

*Use the fact that $\log_{10} 2 = 0.3010$ and $\log_{10} 7 = 0.8451$ to find $\log_{10}$ of each of the following numbers.*

$(\circ)$ **17** 8      **18** 16      **19** 14      **20** $\dfrac{7}{2}$      **21** $\dfrac{2}{7}$

$(\circ)$ **22** 28      **23** 400      **24** 14,000      **25** 25      **26** $\dfrac{1}{50}$

The number $5^{2\log_5 3}$ may be written as an integer by observing that $5^{2\log_5 3} = 5^{\log_5 3^2} = 5^{\log_5 9} = 9$.

*Express each of the following as an integer.*

$(\circ)$   27   $3^{2\log_3 5}$                         28   $4^{3\log_4 2}$

29   $10^{1+\log_{10} 3}$                  30   $10^{2+\log_{10} 5}$

The remaining problems can all be done by using the fact that $\log_b x$ is the unique real number $y$ such that $b^y = x$. For example, to show that

$$\log_2 x = 4\log_{16} x$$

we observe that

$$2^{4\log_{16} x} = (2^4)^{\log_{16} x} = 16^{\log_{16} x} = x;$$

so by the very definition of logarithms, $\log_2 x = 4\log_{16} x$.

$(\circ)$   31   Show that $\log_4 x = 2\log_{16} x$.

32   Show that $\log_3 x = 3\log_{27} x$.

33   For $b > 0$ (and $\neq 1$) show that $\log_b x = 2\log_{b^2} x$.

34   For $b > 0$ (and $\neq 1$) show that $\log_b x = 3\log_{b^3} x$.

$(\circ)$   35   Show that $\log_2 x = \log_6 x (1 + \log_2 3)$.

36   Show that for $a$, $b > 0$ and $a$, $b$, $ab \neq 1$ $\log_b x = (\log_{ab} x)(1 + \log_b a)$.

$(\circ)$

*Sample solutions for exercise set 8.5*

1   The idea is to use the laws of logarithms to simplify the expressions as much as possible. Thus we have

$$\log_3 \frac{xy^2}{z^3} = \log_3 xy^2 - \log_3 z^3 = \log_3 x + \log_3 y^2 - \log_3 z^3$$

$$= \log_3 x + 2\log_3 y - 3\log_3 z.$$

7   In order to use the laws of logarithms we must write each term as the logarithm of some expression. Thus we have

$$2\log_3 x + 4\log_3 y + 2 = \log_3 x^2 + \log_3 y^4 + \log_3 9$$

$$= \log_3 9x^2 y^4.$$

13   Here we are going to use the fact that if $\log_2 a = \log_2 b$, then $a = b$. Using the fact that $\log_2 8 = 3$, we may put the equation in the form

$$\log_2 x + \log_2 (x + 2) = \log_2 8$$

$$\log_2 x(x + 2) = \log_2 8.$$

It follows from this that $x(x + 2) = 8$, $x^2 + 2x - 8 = 0$. Factoring will produce $(x + 4)(x - 2) = 0$; so $x = -4$ or $x = 2$. The solution $x = -4$ must be rejected since $\log_2 x$ is not defined for this value of $x$. The solution set is $\{2\}$. When we check by substituting 2, we have $\log_2 2 + \log_2 (2 + 2) = \log_2 2 + \log_2 4 = 1 + 2 = 3$.

17    $\log_{10} 8 = \log_{10} 2^3 = 3(\log_{10} 2) = 3(0.3010) = 0.9030$

23    $\log_{10} 400 = \log_{10} (4 \cdot 100) = \log_{10} 4 + \log_{10} 100$

$$= 2(\log_{10} 2) + \log_{10} 10^2$$

$$= 2(0.3010) + 2$$

$$= 2.6020$$

27    $3^{2 \log_3 5} = 3^{\log_3 25} = 25$

31    We need only observe that $4^{2 \log_{16} x} = (4^2)^{\log_{16} x} = 16^{\log_{16} x} = x$.

35    We observe that

$$2^{\log_6 x + \log_6 x \log_2 3} = 2^{\log_6 x} \cdot 2^{\log_2 3 \log_6 x}$$

$$= 2^{\log_6 x} \cdot 3^{\log_6 x}$$

$$= 6^{\log_6 x} = x.$$

# (8.6) (Common logarithms)

*common logarithms*

Logarithms were originally developed to aid in performing numerical computations. The most important type of logarithm for this purpose is the logarithm with base 10. These are sometimes called *common logarithms* and will be denoted by the symbol $\log x$ as well as $\log_{10} x$. Thus to say that $y = \log x$ is equivalent to saying that $10^y = x$. In particular,

$\log 1{,}000 = 3$ because $10^3 = 1{,}000$,

$\log 100 = 2$ because $10^2 = 100$,

$\log 10 = 1$ because $10^1 = 10$,

$\log 1 = 0$ because $10^0 = 1$,

$\log 0.1 = -1$ because $10^{-1} = 0.1$,

$\log 0.01 = -2$ because $10^{-2} = 0.01$, and so on.

Indeed, for any integer $n$, $\log 10^n = n$. It is for this reason that common logarithms are used in computational problems.

There is a table of common logarithms in the backmatter of this book, and in Figure 8.14 we reproduce the part of the table that lists log $x$ for $3.0 \le x < 3.8$. The left column contains the first two digits of the numeral for $x$, and the column headings refer to the third digit. Thus to find log 3.14 you look at the intersection of the row labeled 3.1 and the column headed by 4, and you get the information that log $3.14 = 0.4969$. Similarly, log $3.49 = 0.5428$, log $3 = 0.4771$, and log $3.7 = 0.5682$.

A glance at the table of logarithms in the backmatter shows that it gives values of log $x$ for $1 \le x \le 9.99$. To obtain the value of log $x$ for $x$ outside this range, one simply writes $x$ in scientific notation as $x = a \cdot 10^n$ ($1 \le a < 10$; $n \in I$) and uses L0.3 (Section 8.5) to write

$$\log x = \log a + \log 10^n$$

$$= \log a + n.$$

Since $1 \le a < 10$, we can use the table of logarithms to find log $a$, and we are done. Thus

$$\log 34 = \log [(3.4)10^1] = \log 3.4 + 1$$

$$= 0.5315 + 1 = 1.5315,$$

$$\log 340 = \log [(3.4)10^2] = \log 3.4 + 2$$

$$= 0.5315 + 2 = 2.5315,$$

$$\log 34{,}000 = \log [(3.4)10^4] = \log 3.4 + 4$$

$$= 0.5315 + 4 = 4.5315, \text{ and so on.}$$

It is customary to write a logarithm as the sum of an integer and a decimal between 0.0000 and 0.9999. The integer is called

| $x$ | 0 | 1 | 2 | 3 | 4 | 5 | 6 | 7 | 8 | 9 |
|---|---|---|---|---|---|---|---|---|---|---|
| **3.0** | .4771 | .4786 | .4800 | .4814 | .4829 | .4843 | .4857 | .4871 | .4886 | .4900 |
| **3.1** | .4914 | .4928 | .4942 | .4955 | .4969 | .4983 | .4997 | .5011 | .5024 | .5038 |
| **3.2** | .5051 | .5065 | .5079 | .5092 | .5105 | .5119 | .5132 | .5145 | .5159 | .5172 |
| **3.3** | .5185 | .5198 | .5211 | .5224 | .5237 | .5250 | .5263 | .5276 | .5289 | .5302 |
| **3.4** | .5315 | .5328 | .5340 | .5353 | .5366 | .5378 | .5391 | .5403 | .5416 | .5428 |
| **3.5** | .5441 | .5453 | .5465 | .5478 | .5490 | .5502 | .5514 | .5527 | .5539 | .5551 |
| **3.6** | .5563 | .5575 | .5587 | .5599 | .5611 | .5623 | .5635 | .5647 | .5658 | .5670 |
| **3.7** | .5682 | .5694 | .5705 | .5717 | .5729 | .5740 | .5752 | .5763 | .5775 | .5786 |

Fig. 8.14

its *characteristic,* and the decimal is its *mantissa.* The reason for this is quite simple. When $x = a \cdot 10^n$ is written in scientific notation, then $\log x = \log a + n$ expresses $\log x$ as the sum of its mantissa and its characteristic. The mantissa of $\log x$ is determined by the first few digits of $x$, and the characteristic gives the location of the decimal point.

Even when $0 < x < 1$ it is useful to express $\log x$ as the sum of its characteristic and its mantissa. For example,

$$\log (0.0345) = \log [(3.45)10^{-2}] = \log 3.45 + (-2)$$

$$= 0.5378 - 2.$$

Here the mantissa is 0.5378 and the characteristic is $-2$. It is important that you guard yourself from any temptation to write $0.5378 - 2$ as $-2.5378$. These are two different numbers since $-2.5378 = -2 - 0.5378$, whereas $0.5378 - 2 = -2 + 0.5378$!

You will often be given $\log x$ and asked to find $x$. For example, if you are told that $\log x = 3.5403$, you observe that the mantissa of $\log x$ is 0.5403 and its characteristic is 3. Looking through the table of logarithms you next observe that $\log 3.47 = 0.5403$; so $x = (3.47)10^3 = 3{,}470$.

The table of logarithms in this book lists values of $x$ to three digits and values of $\log x$ to four digits. You must always bear in mind, though, that $\log x$ is usually an irrational number. When we write $\log 3 = 0.4771$, for example, we really mean that $\log x \approx 0.4771$ out to four places. There is a method called *linear interpolation* for estimating $\log x$ when $x$ has four significant digits in its numeral. Let's see how this works by looking at a few examples.

Suppose we are asked to compute $\log 3.228$. In Figure 8.15 we have represented that portion of the graph of $y = \log x$ that lies between $x = 3.22$ and $x = 3.23$, as well as the straight line connecting the points $P(3.22, 0.5079)$ and $Q(3.23, 0.5092)$. It should be noted, though, that we have distorted the scale to exaggerate the difference between these two graphs. In reality, they lie much closer to each other. The actual value of $\log 3.228$ is the ordinate of the point $R'$ on the graph of $y = \log x$ whose abscissa is 3.228. Linear interpolation asserts that we may use the ordinate of the point $R$ as an approximation of $\log 3.228$. We may compute the ordinate of $R$ by noting that $\triangle PRS$ is similar to $\triangle PQT$, so that $\overline{RS}/\overline{QT} = \overline{PS}/\overline{PT} = 8/10$. Thus $\overline{RS} = (0.8)(\overline{QT}) = (0.8)(0.5092 - 0.5079) = (0.8)(0.0013) = 0.00104 \approx 0.0010$. The ordinate of $R$ is $0.5079 + 0.0010 = 0.5089$. By linear interpolation, $\log 3.228 = 0.5089$.

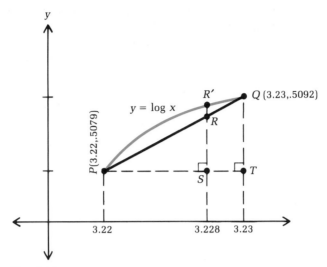

Fig. 8.15

These computations can be efficiently carried out by using the following format.

| x | log x |
|---|---|
| 3.23 | 0.5092 |
| 3.228 | |
| 3.22 | 0.5079 |

$$(0.8)(0.0013) = 0.00104 \approx 0.0010$$

Therefore, we can compute that log $3.228 = 0.5079 + 0.0010 = 0.5089$.

For our next example we want to find log 0.9187. Writing $0.9187 = (9.187)10^{-1}$, we see that we must use linear interpolation to approximate log 9.187.

| x | log x |
|---|---|
| 9.19 | 0.9633 |
| 9.187 | |
| 9.18 | 0.9628 |

$$(0.7)(0.0005) = 0.00035 \approx 0.0004$$

Thus, we have log $9.187 = 0.9628 + 0.0004 = 0.9632$, and log $0.9187 = 0.9632 - 1$.

Looking back over these examples, we see that the method of

linear interpolation assumes that if the fourth digit of $x$ is $k$, then the mantissa of log $x$ lies $k/10$ of the way between the entry for the first three digits of $x$ and the next entry in the table of logarithms. The process can be reversed and linear interpolation used to approximate $x$ if the mantissa of log $x$ does not happen to be in the table. Let's use interpolation to find $t$ if log $t = 1.7310$. Looking in the table of logarithms we see that $0.7310$ is not listed as a mantissa. In fact, it is between $0.7308$ and $0.7316$. Noting that $0.7310$ is one-fourth of the way from $0.7308$ to $0.7316$ and that $1/4 = 0.25 \approx 0.3$, we deduce that the fourth digit of $x$ must be 3. The following format is useful for computations of this kind.

| $x$ | $\log x$ | | |
|---|---|---|---|
| 5.39 | 0.7316 | | |
| ? | 0.7310 | 0.0002 | 0.0008 |
| 5.38 | 0.7308 | | |

$$\frac{0.0002}{0.0008} = \frac{1}{4} = 0.25 \approx 0.3$$

Thus, we establish that log $5.383 = 0.7310$ and log $53.83 = 1.7310$; so $t = 53.83$.

With a little practice, you should be able to do most interpolation problems in your head.

## Exercise set 8.6

*In the first ten problems, find each logarithm.*

$\left( \circ \right)$    **1**   log 2.38                        **2**   log 4

         **3**   log 8.6                         **4**   log 9.88

         **5**   log 35                          **6**   log 342,000

$\left( \circ \right)$    **7**   log 0.0091                  **8**   log 0.852

         **9**   log 0.0011                 **10**   log 0.000000101

*In problems 11 to 20, find $x$ if log $x$ is the given number. Express $x$ both in scientific and standard notation.*

$\left( \circ \right)$    **11**   0.3010       **12**   0.5441       **13**   2.7033       **14**   3.6096

       **15**   5.9566       **16**   2.9983       **17**   0.8768 $-$ 2       **18**   0.7825 $-$ 3

       **19**   0.0086 $-$ 1      **20**   0.6875 $-$ 2

*Use linear interpolation to find each logarithm.*

(∘)  **21** log 2.075   **22** log 3.542   **23** log 82.43   **24** log 996.7

**25** log 5,743   **26** log 49,560   **27** log 0.2978   **28** log 0.07126

**29** log 0.0001234   **30** log 0.0002003

*Now for some interpolation problems in which you are given log x and want to find x out to four digits. Express your answer in both scientific and standard notation.*

(∘)  **31** 0.1243   **32** 0.5530   **33** 2.9475   **34** 1.9579

(∘)  **35** 4.6556   **36** 3.9753   **37** 0.7421 − 1   **38** 0.7280 − 1

**39** 0.3798 − 4   **40** 0.2491 − 3

*We close with some problems that involve numbers that arise in some physical context. In each problem find the logarithm of the given number, using linear interpolation if necessary.*

(∘)  **41** The distance from the earth to the sun is approximately 92,900,000 miles.

**42** The distance from the earth to the moon is approximately 239,000 miles.

**43** The diameter of the earth is approximately 7,926 miles.

**44** The diameter of the moon is approximately 2,160 miles.

**45** The total land area of the earth is approximately 57,300,000 square miles.

**46** The total ocean area of the earth is approximately 139,500,000 square miles.

**47** 1 centimeter ≈ 0.000006214 miles

**48** 1 foot ≈ 0.0001894 miles

**49** 1 millimeter ≈ 0.003281 feet

**50** 1 micron ≈ 0.00003937 inches

(∘)

*Sample solutions for exercise set 8.6*

**1**  log 2.38 = 0.3766

**7**  log 0.0091 = log (9.1) $10^{-3}$ = 0.9590 − 3

**13**  From the table, if log $x$ = 2.7033, then $x$ = (5.05) 10 = 505.

**21**  Using the format introduced in the text, we have

| $x$ | $\log x$ |
|------|----------|
| 2.08 | 0.3181 |
| 2.075 | |
| 2.07 | 0.3160 |

$$(0.5)\,(0.0021) = 0.00105 \approx 0.0011.$$

Thus, $\log 2.075 = 0.3160 + 0.0011 = 0.3171$.

**31**

| $x$ | $\log x$ | | |
|------|----------|---|---|
| 1.34 | 0.1271 | | 0.0032 |
| ? | 0.1243 | 0.0004 | |
| 1.33 | 0.1239 | | |

$$\frac{0.0004}{0.0032} = \frac{1}{8} = 0.125 \approx 0.1$$

$$\log 1.331 = 0.1243$$

The answer is 1.331 or, in scientific notation, $(1.331)\,10^{0}$.

**37**

| $x$ | $\log x$ | | |
|------|----------|---|---|
| 5.53 | 0.7427 | | 0.0008 |
| ? | 0.7421 | 0.0002 | |
| 5.52 | 0.7419 | | |

$$\frac{0.0002}{0.0008} = 0.25 \approx 0.3$$

$$\log 5.523 \approx 0.7421$$

We want a number $x$ whose logarithm is $0.7421 - 1$. Hence, $x = (5.523)\,10^{-1} = 0.5523$.

**41** We are asked to compute $\log (92{,}900{,}000) = \log [(9.29)\,10^{7}] = 7.9680$.

## (8.7) (Computations with logarithms)

We mentioned earlier that logarithms were originally developed to aid in performing various numerical computations. Nowadays, such computations are usually done with the aid of a slide rule or a computer. Nonetheless, such devices are not always available, and for that reason you will find it helpful to learn the computational applications of logarithms. As you will soon see, logarithms can be used to transform the multiplication operation into addition, division into subtraction, and problems involving exponents into multiplication problems. The needed properties were discussed in Section 8.5, but for your convenience we will repeat them. Here $x_1, x_2 > 0; r \in R$.

$(\circ)$     LO.3   $\log x_1 x_2 = \log x_1 + \log x_2$

$(\circ)$     LO.4   $\log \dfrac{x_1}{x_2} = \log x_1 - \log x_2$

$(\circ)$     LO.5   $\log (x_1)^r = r(\log x_1)$

## Products

To illustrate how logarithms are used to compute products we consider $(34.8)(0.202)$. By LO.3, $\log [(34.8)(0.202)] = \log 34.8 + \log 0.202$, Hence

$$\left. \begin{array}{l} \log 34.8 \;=\; 1.5416 \\[4pt] \log 0.202 \;=\; 0.3054 - 1 \end{array} \right\} \text{ add}$$

$\log [(34.8)(0.202)] \;=\; 1.8470 - 1 = 0.8470.$

The final step is to find a number $x$ whose logarithm is 0.8470. From the table of logarithms we find that $x = 7.03$ is such a number. We deduce that $(34.8)(0.202) = 7.03$. Actual computation will show that $(34.8)(0.202) = 7.0296$. You must expect this kind of inexactness because, as we have stated, the logarithms in our table are actually approximations of irrational numbers.

## Quotients

To illustrate the division process we compute the quotient $Q = 880/6.2$. By LO.4, $\log Q = \log 880 - \log 6.2$. Consulting the table of logarithms we write

$$\left. \begin{array}{l} \log 880 = 2.9445 \\[4pt] \log 6.2 = 0.7924 \end{array} \right\} \text{ subtract}$$

$\log Q = 2.1521$

From the table of logarithms we see that $Q = (1.419) 10^2 = 141.9$.

A bit of care must be shown when dealing with divisors whose logarithms have a negative characteristic. To compute $S = 34.8/0.202$, we note that

$$\left. \begin{array}{l} \log 34.8 = 1.5416 \\ \\ \log 0.202 = 0.3054 - 1 \end{array} \right\} \text{ subtract}$$

When subtracting log 0.202 from log 34.8 we must remember to multiply log 0.202 by $-1$ and add. Thus log $S = 1.5416 - (0.3054 - 1) = 1.5416 - 0.3054 + 1 = 1.2362 + 1 = 2.2362$. Using linear interpolation we see that $S = (1.723)10^2 = 172.3$.

## Powers and roots

To compute $\sqrt{2.38}$ we note that by LO.5, log $\sqrt{2.38} = (1/2)(\log 2.38) = (1/2)(0.3766) = 0.1883$. Using linear interpolation we see that $\sqrt{2.38} = 1.543$. Similarly, $3^{4.1}$ may be computed by observing that log $3^{4.1} = (4.1)(\log 3) = (4.1)(0.4771) = 1.9561$; so by linear interpolation, $3^{4.1} = (9.038)10^1 = 90.38$.

## Alternate form of the characteristic

Sometimes, it is convenient to express the characteristic of a logarithm in a form other than the usual one. For example, let's consider the quotient $T = 23.5/88.1$. We know that log $T = \log 23.5 - \log 88.1 = 1.3711 - 1.9450 = -0.5739$. But this negative number does not express log $T$ as the sum of its mantissa and its characteristic; so we cannot directly compute $T$. To avoid this negative decimal, it is convenient to write log 23.5 in the form $2.3711 - 1$ and proceed as follows:

$$\left. \begin{array}{l} \log 23.5 = 2.3711 - 1 \\ \\ \log 88.1 = 1.9450 \end{array} \right\} \text{ subtract}$$

$$\log T = 0.4261 - 1$$

$$T = (2.668)10^{-1} = 0.2668.$$

For our second example we compute $\sqrt[3]{0.6}$. We know that log $\sqrt[3]{0.6} = (1/3)(\log 0.6) = (1/3)(0.7782 - 1) = 0.2594 - (1/3)$. But here again we have a logarithm that is not expressed as the sum of its mantissa and its characteristic. The difficulty stems from the fact that the characteristic of log 0.6 was not divisible by 3. This can be avoided by writing log $0.6 = 2.7782 - 3$ (or 5.7782

$- 6$, $8.7782 - 9$, and so on) since 3 (or 6 or 9) is divisible by 3. Proceeding in this fashion we see that $\log \sqrt[3]{0.6} = (1/3)(\log 0.6) = (1/3)(2.7782 - 3) = 0.9261 - 1$. Now that we have $\log \sqrt[3]{0.6}$ expressed as the sum of its mantissa and characteristic, by linear interpolation we find $\sqrt[3]{0.6} = (8.436)10^{-1} = 0.8436$.

## Additional examples

Until now, the examples we have considered were fairly routine. Logarithms, however, are useful for more complicated computational problems. Let's begin with $P = (65.1)(.0239)/0.8181$. What is needed here is a systematic application of LO.3 and LO.4, together with linear interpolation to find $\log 0.8181$.

$$\left.\begin{array}{l} \log 65.1 = 1.8136 \\ \log 0.0239 = 0.3784 - 2 \end{array}\right\} \text{add}$$

$$\left.\begin{array}{l} \log (65.1)(0.0239) = 2.1920 - 2 \\ \log 0.8181 = 0.9129 - 1 \end{array}\right\} \text{subtract}$$

$$\log P = 1.2791 - 1 = 0.2791$$

$$P = 1.901$$

Now let's compute $Q = (\sqrt[5]{.0327})/(86.9)(2.3)$. Proceeding systematically, we have that

$$\log \sqrt[5]{0.0327} = \frac{1}{5}(\log 0.0327) = \frac{1}{5}(0.5145 - 2) = \frac{1}{5}(3.5145 - 5)$$

$$= 0.7029 - 1$$

and

$$\left.\begin{array}{l} \log 86.9 = 1.9390 \\ \log 2.3 = 0.3617 \end{array}\right\} \text{add}$$

$$\log (86.9)(2.3) = 2.3007.$$

Thus

$$\left.\begin{array}{l} \log \text{numerator} = 2.7029 - 3 \\ \log \text{denominator} = 2.3007 \end{array}\right\} \text{subtract}$$

$$\log Q = 0.4022 - 3$$

$$Q = (2.525)10^{-3}.$$

In connection with this example you should note that log numerator was written in the form $2.7029 - 3$ in order to avoid a negative decimal in log $Q$; furthermore, log 0.0327 was written as $3.5145 - 5$ so that log $\sqrt[5]{0.0327}$ would be the sum of its mantissa and characteristic.

Finally, let us find the value of $R = \sqrt[3]{[(52.8)\,(3.4)^2]/824}$. We know that log $(3.4)^2 = 2(\log 3.4) = 2(0.5315) = 1.0630$; so

$$
\left.\begin{array}{rl}
\log 52.8 \;=\; & 1.7226 \\
\log (3.4)^2 \;=\; & 1.0630
\end{array}\right\} \text{ add}
$$

$$
\begin{array}{rl}
\overline{\log (52.8)(3.4)^2 \;=\;} & 2.7856
\end{array}
$$

$$
\left.\begin{array}{rl}
\;=\; & 3.7856 - 1 \\
\log 824 \;=\; & 2.9159
\end{array}\right\} \text{ subtract}
$$

$$
\begin{array}{rl}
\overline{\log \text{ radicand } \;=\;} & 0.8697 - 1.
\end{array}
$$

Hence log $R = (1/3)(0.8697 - 1) = (1/3)(2.8697 - 3) = 0.9566 - 1$ and $R = 0.905$.

## An application

Logarithms are useful in problems involving compound interest, which is interest that is periodically added to the principal and then earns interest itself. To see how this works, let's suppose you have \$1,000 invested at 8% interest compounded quarterly. The rate of 8% is the annual rate of interest. Since three months is one-fourth of a year, the rate for a three-month period (a *quarter*) would be $(1/4)(8\%) = 2\%$. At the end of three months your account would be credited with $(1,000)(0.02) = \$20$, and you would have \$1,020 on deposit. At this point the extra \$20 starts earning interest, so that at the end of six months you would be credited with $(1,020)(0.02) = \$20.40$, and your new balance would be \$1,040.40. Indeed, at the end of each three-month period you would be credited with (amount on deposit) (0.02). If you have $D$ dollars on deposit, your new balance would be $D + 0.02D = (1.02)D$. A formula for finding the amount $A_n$ in your account after $n$ years would be $A_n = 1,000(1.02)^{4n}$. More generally, if $D$ dollars is invested at $i\%$ compounded $k$ times a year, then the amount on deposit after $n$ years is

$$
A_n = D\left(1 + \frac{i}{100k}\right)^{nk}
$$

In the example we have been discussing, the amount of money on deposit after 10 years is $A_{10} = 1,000\,(1.02)^{4 \cdot 10} = 1,000\,(1.02)^{40}$. Using logarithms to compute this, we observe that

$$\log 1.02 \;=\; 0.0086$$

$$\log (1.02)^{40} \;=\; 40(0.0086) = 0.3440$$

$$(1.02)^{40} \;=\; 2.208$$

Hence $A_{10} = (1,000)(2.208) = \$2,208.$

## Exercise set 8.7

*Use logarithms to perform each of the following calculations. Then check your answer by actual computation.*

$\left(\circ\right)$    **1**   (2.3)(4.9)                  **2**   (73)(8.9)

      **3**   5/7                          **4**   17/13

*Use logarithms to perform each computation.*

**5**   (2,510)(693)               **6**   (837,000)(9,330)

**7**   $\dfrac{6.38}{3.15}$                    **8**   $\dfrac{88.3}{54,900}$

$\left(\circ\right)$   **9**   $\dfrac{3.89}{5.86}$              **10**   $\dfrac{0.00138}{0.893}$

$\left(\circ\right)$   **11**   $\sqrt[4]{0.03}$             **12**   $\sqrt[3]{0.28}$

     **13**   $\sqrt[5]{0.49}$             **14**   $\sqrt{0.2}$

**15**   $\dfrac{(55.3)(938)}{0.31}$        **16**   $\dfrac{(0.613)(0.092)}{31,200}$

**17**   $\dfrac{(313)^2(2.114)}{639.2}$      **18**   $\dfrac{(54,900)(0.00093)^3}{(3,131)^2}$

**19**   $\sqrt{(231)(3.04)(0.65)}$      **20**   $\sqrt[4]{(504)(25.8)(3090)}$

$\left(\circ\right)$   **21**   $\dfrac{\sqrt{(451)(59,000)}}{323}$      **22**   $\dfrac{625}{\sqrt[5]{(310)(22,500)}}$

     **23**   $\dfrac{\sqrt[3]{55.32}}{\sqrt[5]{7.124}}$        **24**   $\dfrac{\sqrt{78.36}}{\sqrt[3]{(3.1)(2.932)}}$

$\left(\circ\right)$   **25**   $3^{\sqrt{2}}$               **26**   $2^{\sqrt{3}}$

By definition, log $x$ is the unique number $y$ such that $10^y = x$. Thus when we say that log $3 = 0.4771$, we mean that $10^{0.4771} = 3$.

*Express each of the following as an integer.*

**27** $10^{0.9149}$

**28** $10^{0.6964}$

**29** $10^{1.0719}$

**30** $10^{3.9571}$

*Compute each of the following by means of logarithms. You will want to use the fact that log $\pi = 0.4971$ in some of these problems.*

$(\circ)$    **31**   What is the radius of a circle whose area is 6?

**32**   What is the radius of a circle whose area is 17/3?

**33**   The volume $V$ of a sphere is given by the formula $V = \dfrac{4}{3}\pi r^3$ where $r$ is its radius. The diameter of the earth is approximately 7,926 miles. Assuming that the earth is a sphere, determine its volume.

**34**   Assuming that the moon is a sphere with diameter 2,160 miles, use the formula in problem 33 to find its volume.

**35**   Use the formula of problem 33 to find the radius of a sphere whose volume is 60.

**36**   Use the formula of problem 33 to find the radius of a sphere whose volume is 238.

$(\circ)$    **37**   In the United States, distance is usually measured by a system in which 1 mile = 5,280 feet, 1 yard = 3 feet, and 1 foot = 12 inches. In most other parts of the world, the *metric system* is used, in which 1 kilometer = 1,000 meters and 1 meter = 100 centimeters. One can convert from one system to the other by using the fact that 1 inch = 2.54 centimeters. How many kilometers are there in a mile?

**38**   Using the data given in problem 37, how many meters are there in a yard?

**39**   Using the data of problem 37, how many inches are there in a meter?

**40**   The speed of light is approximately 186,300 miles per second. Use this figure to determine the length in miles of a *light year*, the distance traveled in one year by a beam of light.

**41**   The *period* of a simple pendulum is the time needed for one complete oscillation. The period $T$ in seconds of a simple pendulum of length $L$ feet is given by $T = 2\pi\sqrt{\dfrac{L}{g}}$ where $g = 32.16$. Find the period of a pendulum 4 feet long.

**42**   Use the formula of problem 41 to find the length of a pendulum whose period is 2 seconds. Incidentally, this is the length of a pendulum that you might use in a clock.

**43** The area $A$ of a triangle whose sides have lengths $a$, $b$, and $c$ is given by $A = \sqrt{s(s-a)(s-b)(s-c)}$ where $s = (1/2)(a+b+c)$. Compute the area of a triangle whose sides have lengths 1.35 inches, 2.78 inches, and 3 inches.

**44** The radius of the circle circumscribing a triangle whose sides have lengths $a, b,$ and $c$ is given by the formula $R = abc/[4\sqrt{s(s-a)(s-b)(s-c)}\,]$ where $s = (1/2)(a+b+c)$. Compute the radius of the circle circumscribing the triangle of problem 39.

**45** The proud parents of a baby put \$500 in trust for him at birth in an account that yields 6 percent compounded semiannually. How much money will be in the account on the child's twenty-first birthday?

**46** A man has invested \$9,000 in an account that pays 12 percent interest compounded quarterly. How much will be in the account at the end of 8 years?

$(\circ)$ **47** The *effective interest rate* corresponding to $i$ percent compounded $k$ times a year is defined to be the rate that gives the same yield when compounded annually. The effective interest rate $r$ can be computed from the formula $r = \left(1 + \dfrac{i}{100k}\right)^k - 1$. Determine the effective interest rate corresponding to 12 percent compounded quarterly. What is the effective interest rate corresponding to 12 percent compounded monthly?

**48** Using the formula of problem 45, what is the effective interest rate corresponding to 8 percent compounded quarterly?

$(\circ)$

*Sample solutions*
*for exercise set 8.7*

**1** Using logarithms to compute $(2.3)(4.9)$, we get

$$\left.\begin{array}{l} \log 2.3 \;=\; 0.3617 \\[4pt] \log 4.9 \;=\; 0.6902 \end{array}\right\} \text{add}$$

$$\overline{\phantom{xxxxxx}}$$

$\log (2.3)(4.9) \;=\; 1.0519.$

We must now use linear interpolation to obtain the answer.

$$\left.\begin{array}{l} \log 1.13 \;=\; 0.0531 \\[6pt] \left.\begin{array}{l} \log\; ? \;=\; 0.0519 \\[6pt] \log 1.12 \;=\; 0.0492 \end{array}\right]\;0.0027 \end{array}\right]\;0.0039$$

$$\frac{0.0027}{0.0039} \;=\; \frac{27}{39} = 0.69 \approx 0.7$$

$\log 1.12\,7 \approx 0.0519$

So $(2.3)(4.9) = 11.27$. To check the answer, actually perform the required computation.

$$4.9$$

$$2.3$$

$$\overline{\phantom{xx}}$$

$$147$$

$$98$$

$$\overline{\phantom{xx}}$$

$$11.27$$

9 $\left.\begin{array}{l}\log 3.89 = 1.5899 - 1 \\[4pt] \log 5.86 = 0.7679\end{array}\right\}$ subtract

$$\overline{\phantom{xxxxxx}}$$
$$0.8220 - 1$$

Using linear interpolation, we see that

$\left.\begin{array}{l}\log 6.64 = 0.8222 \\[8pt] \left.\begin{array}{l}\log\ ? = 0.8220 \\[8pt] \log 6.63 = 0.8215\end{array}\right] 0.0005 \\ \end{array}\right] 0.0007$

$$\frac{0.0005}{0.0007} = \frac{5}{7} \approx 0.7$$

$$\log 6.637 = 0.8220$$

Therefore, $3.89/5.86 = 0.6637$.

11  We have that $\log 0.03 = 0.4771 - 2$; so $\log \sqrt[4]{0.03} = (1/4)(0.4771 - 2) = (1/4)(2.4771 - 4) = 0.6193 - 1$. It follows that $\sqrt[4]{0.03} = 0.4162$.

21 $\left.\begin{array}{l}\log 451 = 2.6542 \\[4pt] \log 59{,}000 = 4.7709\end{array}\right\}$ add

$$\overline{\phantom{xxxxxxxx}}$$
$$\log (451)(59{,}000) = 7.4251$$
$$\overline{\phantom{xxxxxxxxxxxx}}$$
$$\log \sqrt{(451)(59{,}000)} = \frac{1}{2}(7.4251) = 3.7126$$

At this point we have

$\left.\begin{array}{l}\log \sqrt{(451)(59{,}000)} = 3.7126 \\[4pt] \log 323 = 2.5092\end{array}\right\}$ subtract

$$\overline{\phantom{xxxxxxx}}$$
$$\log \text{quotient} = 1.2034.$$

Using linear interpolation, we see that the answer is 15.97.

25  Back in Section 8.3 we showed you how to define $3^{\sqrt{2}}$ and told you that you would soon have a way of computing such numbers. Logarithms

can be used very nicely for this sort of thing. To see this, note that $\log 3^{\sqrt{2}}$ $= \sqrt{2}\,(\log 3) = \sqrt{2}\,(0.4771)$. We may as well use logarithms to perform this multiplication; so we note that $\log 2 = 0.3010$, $\log \sqrt{2} = (1/2)(0.3010)$ $= 0.1505$. Thus

$$\left. \begin{array}{r} \log \sqrt{2} \;\; = \;\; 0.1505 \\[4pt] \log 0.4771 \;\; = \;\; 0.6786 - 1 \end{array} \right\} \; \text{add}$$

$$\log\,(\sqrt{2}\,)(0.4771) \;\; = \;\; 0.8291 - 1$$

$$(\sqrt{2}\,)(0.4771) \;\; = \;\; 0.6747.$$

This says that $\log 3^{\sqrt{2}} = 0.6747$; so $3^{\sqrt{2}} = 4.728$.

**31** The formula for the area $A$ of a circle with radius $r$ is $A = \pi r^2$. We know that $6 = \pi r^2$; so $r^2 = 6/\pi$ and $r = \sqrt{6/\pi}$. Using logarithms, we see that

$$\left. \begin{array}{r} \log 6 \;\; = \;\; 0.7782 \\[4pt] \log \pi \;\; = \;\; 0.4971 \end{array} \right\} \; \text{subtract}$$

$$\log \frac{6}{\pi} \;\; = \;\; 0.2811$$

$$\log \sqrt{\frac{6}{\pi}} \;\; = \;\; \frac{1}{2}(0.2811) = 0.1406.$$

Thus $r = 1.382$.

**37** There are $(5{,}280)(12)$ inches in a mile. Since 1 inch $= 2.54$ centimeters, there are $(5{,}280)(12)(2.54)$ centimeters in a mile. Since 1 kilometer $= 10^5$ centimeters, the answer is $(5{,}280)(12)(2.54)(10^{-5})$.

$$
\begin{array}{rr}
\log 5{,}280 = & 3.7226 \\[4pt]
\log 12 = & 1.0792 \\[4pt]
\log 2.54 = & 0.4048 \\[4pt]
\log 10^{-5} = & -5.0000 \\
\hline
\log \text{ product} = & 0.2066
\end{array}
$$

We have shown that 1 mile $= 1.609$ kilometers.

**47** The effective interest rate corresponding to 12 percent compounded quarterly is given by $r = [1 + (12/400)]^4 - 1 = (1.03)^4 - 1$. Noting that $\log\ 1.03 = 0.0128$, we see that $\log\ 1.03^4 = 4(0.0128) = 0.0512$. Hence $(1.03)^4 = 1.125$, and $r = 1.125 - 1 = 0.125$. Thus the effective rate is 12.5 percent. For monthly compounding, we have $r = (1.01)^{12} - 1$, which produces an effective rate of 12.6 percent.

## (8.8) (Exponential equations)

*exponential equation*

If an equation contains a variable in one or more exponents, it is called an *exponential equation.* To solve the exponential equation $7^x = 11$, we take common logarithms of both members to obtain $\log 7^x = \log 11$. By law LO.5 of Section 8.5, $\log 7^x = x(\log 7)$; so $x(\log 7) = \log 11$, and $x = (\log 11/\log 7) = (1.0414/0.8451) = 1.232$.

It is important that you realize that when we write $\log 11/\log 7$, we mean that the logarithms are to be divided, *not* subtracted. This is important because $(\log 11/\log 7) \neq \log 11 - \log 7$. To see this, we use the fact that $(\log 11/\log 7) = 1.232$; so we need only observe that $\log 11 - \log 7 = 1.0410 - 0.8451 = 0.1963 \neq 1.232$.

To convince you that the foregoing technique really works, let's apply it to the exponential equation $4^x = 8$, whose solution set is clearly $\left\{ \dfrac{3}{2} \right\}$. Taking common logarithms we have $\log 4^x = \log 8$. By law LO.5,

$$x(\log 4) = \log 8$$

$$x = \frac{\log 8}{\log 4} = \frac{0.9031}{0.6021} = \frac{3}{2}$$

as we would expect. Again, it is important to note that this is an actual division process, and *not* subtraction. In other words, we are after the number $\log 8/\log 4$, which is not equal to $\log 8 - \log 4$. Indeed, $\log 8/\log 4 = 1.5$, whereas $\log 8 - \log 4 = \log (8/4) = \log 2 = 0.3010$.

For a slightly more complicated example, consider $8^{2x+5} = 13$. Taking logarithms of both members will produce $\log 8^{2x+5} = \log 13$. By law LO.5, $(2x + 5) \log 8 = \log 13$; so

$$(2x)(\log 8) + 5 \log 8 = \log 13$$

$$(2x)(\log 8) = \log 13 - 5 \log 8$$

$$x = \frac{\log 13 - 5 \log 8}{2(\log 8)}$$

Substituting $\log 13 = 1.1139$, $\log 8 = 0.9031$ we have

$$x = \frac{1.1139 - 5(0.9031)}{2(0.9031)} = \frac{1.1139 - 4.5155}{1.8062} = \frac{-3.4016}{1.8062}$$

$$= -1.883.$$

Exponential equations can also be used to compute the value of logarithms with a base other than 10. To evaluate $y = \log_6 14$, one writes the equivalent exponential equation $6^y = 14$. Taking common logarithms of both members will produce

$$\log 6^y = \log 14$$

$$y(\log 6) = \log 14$$

$$y = \frac{\log 14}{\log 6} = \frac{1.1461}{0.7782} = 1.4728.$$

This shows that $\log_6 14 = 1.4728$.

The same technique can be used to develop a formula for computing $\log_a x$ from $\log_b x$. The first step is to write the equation $y = \log_a x$ in the equivalent exponential form

$$x = a^y.$$

Taking logarithms with base $b$ now produces

$$\log_b x = \log_b a^y = y(\log_b a),$$

whence

$$y = \frac{\log_b x}{\log_b a}.$$

Replacing $y$ with $\log_a x$ in this last equation shows that

**(1)** $\log_a x = \dfrac{\log_b x}{\log_b a}.$

In particular, if $b = 10$ then (1) takes on the form

**(2)** $\log_a x = \dfrac{\log x}{\log a}.$

This shows that if a table of common logarithms is available, then logarithms with *any* base can be computed.

*natural logarithms*

In addition to common logarithms, another type of logarithm is of special interest. These are called *natural logarithms,* which are logarithms with base $e$ where $e$ is an irrational number whose value is approximately 2.71828. An explanation of how natural logarithms arise properly belongs in a calculus book, but it is of interest that they can be computed by means of (2). Indeed,

**(3)** $\log_e x = \dfrac{\log x}{\log e} = \dfrac{\log x}{0.4343} = 2.303 \log x$

since $\log e = 0.4343$ and $1/0.4343 = 2.303$. For example, $\log_e 10 = 2.303 \log 10 = 2.303$, while $\log_e 2 = (2.303)(0.3010) = 0.6932$.

The associated exponential function $f(x) = e^x$ is also important. For example, when a bank says that it is offering $i\%$ interest *compounded continuously*, it means that if $D$ dollars are placed on deposit, then at the end of $x$ years there will be

$$A(x) = De^{(i/100)x}$$

dollars in the account. For a practical application, assume that $100 is deposited in an account that earns 6% interest compounded continuously and that you wish to determine the amount that will be on deposit at the end of 2 years and the length of time for the account to triple in value. To answer the first item, note that $A(x) = 100e^{0.06x}$; so $A(2) = 100e^{0.12}$. Therefore,

$$\log A(2) = \log 100e^{0.12}$$

$$= \log 100 + \log e^{0.12}$$

$$= 2 + .12 \log e.$$

Using the fact that $\log e = 0.4343$, we see that $\log A(2) = 2 + (0.12)(0.4343) = 2 + 0.0521 = 2.0521$, from which it follows that $A(2) = (1.127)10^2 = 112.7$. At the end of 2 years there will be $112.70 on deposit. To find the length of time required for the deposit to triple, the equation $300 = 100e^{0.06x}$ must be solved. Simplifying, and then taking common logarithms will result in the equations

$$3 = e^{0.06x}$$

$$\log 3 = \log e^{0.06x} = (0.06x)\log e;$$

so $x = \log 3 / [(0.06) \log e] = 0.4771 / [(0.06)(0.4343)] = 18.31$. It would take 18.31 years for the account to triple in value.

In closing, we consider an application to biology. In Section 8.3 we considered a bacteria culture that started with 10,000 bacteria and after 3 days had a count of 2,160,000. We noted that the number of bacteria at the end of $x$ days is given by the function $N(x) = (10,000)6^x$. Now suppose you have been asked to determine the time at which there were 75,000 bacteria in the culture. This amounts to solving the exponential equation $75,000 = (10,000)6^x$, or equivalently, $7.5 = 6^x$. Thus

$$\log 7.5 = \log 6^x = x(\log 6)$$

$$x = \frac{\log 7.5}{\log 6} = \frac{0.8751}{0.7782} \approx 1.12.$$

It would take approximately 1.12 days for there to be 75,000 bacteria in the culture.

## Exercise set 8.8

*Solve each of the following exponential equations, rounding off your answer to the nearest hundredth.*

($\circ$)   **1**  $4^x = 5$                    **2**  $9^x = 2$                    **3**  $7^x = \dfrac{1}{2}$

    **4**  $2^x = \dfrac{3}{5}$                    **5**  $3^{4x+3} = 5$                    **6**  $12^{8x-1} = 3$

($\circ$)   **7**  $20^{3-2x} = 5$            **8**  $8^{1-3x} = 6$            **9**  $6^{x+1} = 8^{x-2}$

    **10**  $2^{2x+7} = 5^{x-4}$        **11**  $3^{2x-1} = 4^{3-x}$        **12**  $4^{x+3} = 6^{3-x}$

*Evaluate each of the following logarithms, rounding off your answer to the nearest thousandth.*

($\circ$)   **13**  $\log_7 3$                **14**  $\log_3 12$                **15**  $\log_2 25$

    **16**  $\log_6 3$                **17**  $\log_e 6$                **18**  $\log_e 4$

    **19**  $\log_e \sqrt{e}$          **20**  $\log_e \sqrt[3]{e}$

*The remaining problems involve some type of exponential equation.*

**21**  The interest on a savings account is 8 percent compounded continuously. If \$1,000 is deposited in the account, how much will be in the account at the end of 10 years? How long will it take the account to double in value?

**22**  Suppose \$1,000 is deposited in an account that earns 8 percent interest compounded quarterly. How long will it take for the account to double in value? Round your answer up to the next quarter year.

($\circ$)   **23**  In problem 31 of Exercise set 8.3, how long would it take for the culture to double in size?

**24**  In problem 32 of Exercise set 8.3, how long would it take for the culture to double in size?

**25**  The amount of a certain radioactive element present at time $x$ seconds is given by the function $A(x) = A_0 e^{-0.06x}$. What is its half-life?

**26**  The half-life of radium is 1,600 years. If the amount of radium present at the end of $x$ years is given by $A(x) = A_0 e^{-kx}$, find $k$.

($\circ$)   **27**  In problem 35 of Exercise set 8.3, how long would it take for the substance to reach a temperature of $140°$ F.?

**28**   In problem 36 of Exercise set 8.3, how long would it take for the object to reach a temperature of 200° F.?

**29**   Recall (problem 37 of Exercise set 8.3) that the atmospheric pressure $p(x)$ in inches of mercury at an altitude of $x$ miles above sea level is given approximately by $p(x) = 30.0(0.8)^x$. What altitude would yield a barometric pressure of 21.0 inches of mercury?

**30**   Recall (problem 38 of Section 8.3) that when a wound is healing, an approximation for the area $A(x)$ remaining unhealed after $x$ days is $A(x) = A_0 b^x$. If one-third of a wound is healed in 8 days, how long would it take for it to be half-healed?

(∘)  **31**   Radioactive carbon (carbon 14) has a half-life of approximately 5,600 years. This fact is frequently used by scientists to date objects that were once alive. The technique is based on the fact that a living organism maintains a constant ratio of radioactive carbon to stable carbon in its tissues. When the organism dies, the radioactive carbon experiences exponential decay, with the function $A(x) = A_0 b^x$ giving the amount of radioactive carbon present $x$ years after its death. A piece of charcoal is found to have only 14 percent of its original radioactive carbon. Approximately how old is the charcoal?

**32**   In the late 1940s and the 1950s a number of manuscripts were found near the northwestern corner of the Dead Sea. These are commonly called the Dead Sea Scrolls. One manuscript found in 1952 in a valley called Wadi Murabbaat contained 79.8 percent of its original radioactive carbon. Approximately when was it written?

(∘)

*Sample solutions*
*for exercise set 8.8*

**3**   Taking common logarithms will produce

$$\log 7^x = \log \frac{1}{2}$$

$$x(\log 7) = \log \frac{1}{2}$$

$$x = \frac{\log \dfrac{1}{2}}{\log 7} = \frac{-0.3010}{0.8451} \approx -0.36.$$

**9**   Again we take common logarithms and solve for $x$.

$$\log 6^{x+1} = \log 8^{x-2}$$

$$(x + 1)\log 6 = (x - 2)\log 8$$

$$x \log 6 + \log 6 = x \log 8 - 2 \log 8$$

$$x \log 8 - x \log 6 = \log 6 + 2 \log 8$$

$$x = \frac{\log 6 + 2 \log 8}{\log 8 - \log 6}$$

Using $\log 6 = 0.7782$ and $\log 8 = 0.9031$ shows

$$x = \frac{0.7782 + 2(0.9031)}{0.9031 - 0.7782} = \frac{2.5844}{0.1249} \approx 20.69.$$

**13**  We must solve the exponential equation $7^x = 3$. Alternately, by equation (2) of the text, we note that

$$\log_7 3 = \frac{\log 3}{\log 7} = \frac{0.4771}{0.8451} \approx 0.565.$$

**23**  From the sample solution to problem 31 of Exercise set 8.3, the growth function for this particular culture is $N(x) = 5,000(81^x)$ where $x$ is the time in days. We are asked to find $x$ so that $N(x) = 10,000$. This amounts to solving the equation $10,000 = 5,000(81^x)$, or $2 = 81^x$. Thus $\log 81^x = \log 2$; so

$$x = \frac{\log 2}{\log 81} = \frac{0.3010}{1.9085} \approx 0.16.$$

It will take 0.16 days for the culture to double in size.

**27**  From the sample solution to problem 35 of Exercise set 8.3, the function giving the temperature at the end of $x$ hours is $T(x) = 70 + 100(0.09)^x$. We want

$$140 = 70 + 100(0.09)^x$$

$$70 = 100(0.09)^x$$

$$(0.09)^x = \frac{70}{100} = 0.7;$$

so

$$\log (0.09)^x = \log 0.7$$

$$x(\log 0.09) = \log 0.7$$

$$x = \frac{\log 0.7}{\log 0.09} = \frac{0.8451 - 1}{0.9542 - 2} = \frac{-0.1549}{-1.0458} \approx 0.15 \text{ hours.}$$

**31**  Since carbon 14 has a half-life of 5,600 years, we know that

$$\frac{1}{2}A_0 = A_0 b^{5,600}$$

$$\frac{1}{2} = b^{5,600};$$

so $b = (1/2)^{1/5,600}$. We must determine the value of $x$ for which

$$A_0 b^x = 0.14 A_0$$

$$b^x = 0.14.$$

Replacing $b$ with $(1/2)^{1/5,600}$ yields $(1/2)^{x/5,600} = 0.14$. We next raise both members to the 5,600 power, and then take logarithms.

$$\left(\frac{1}{2}\right)^x = (0.14)^{5,600}$$

$$\log\left(\frac{1}{2}\right)^x = \log(0.14)^{5,600}$$

$$x\left(\log\frac{1}{2}\right) = 5,600(\log 0.14)$$

$$x = \frac{(5,600)(\log 0.14)}{\log\dfrac{1}{2}} = \frac{(5,600)(0.1461 - 1)}{-0.3010}$$

$$= \frac{(5,600)(-0.8539)}{-0.3010} = 15,890$$

The charcoal is approximately 15,890 years old.

# (9)

## Systems of equations

### (9.1) (Systems of linear equations in two variables)

The solution of a problem will often involve more than one equation in several variables. When this happens, the equations are called *system of equations* a *system of equations*. The *solution set* of a system of equations is simply the set of solutions that are common to each of the equations in the system. In this section we shall discuss systems of two linear equations in the variables $x$ and $y$; these are systems of the form

$$\begin{cases} a_1 x + b_1 y = c_1 \\ a_2 x + b_2 y = c_2 \end{cases}$$

where $a_1$, $a_2$, $b_1$, $b_2$, $c_1$, and $c_2$ are real numbers with $a_1$ or $b_1$ different from zero and $a_2$ or $b_2$ different from zero. The graph of the solution set of the first equation in the system displayed above is a straight line $L_1$ and that of the second equation is a straight line $L_2$. The graph of the solution set of the system is the intersection of $L_1$ and $L_2$.

To illustrate the possibilities, we consider some examples. We begin with the system

$$\begin{cases} -2x + y = 3 \\ x + y = 3. \end{cases}$$

When we write these equations in slope-intercept form, they become

$$\begin{cases} y = 2x + 3 \\ y = -x + 3; \end{cases}$$

so the graphs of the equations are distinct straight lines, each having $y$-intercept 3. It follows that the solution set of the system consists of the single point (0,3). The graph of this system appears in Figure 9.1.

For our second example we consider the system

$$\begin{cases} x + y = 1 \\ 3x + 3y = 3. \end{cases}$$

*dependent system*

The second equation can be obtained by multiplying both members of the first equation by 3; so the equations have the same solution set. The solution set of this system is infinite, and it appears in Figure 9.2. In general, a system of $n$ equations in $n$ variables is called a *dependent system* if its solution set is infinite. For a system of two linear equations in two variables, this occurs when one equation is a multiple of the other.

Finally we consider the system

$$\begin{cases} 3x + 2y = 2 \\ 6x + 4y = 8. \end{cases}$$

Writing the equations in slope-intercept form produces the system

$$\begin{cases} y = -\dfrac{3}{2}x + 1 \\ \\ y = -\dfrac{3}{2}x + 2. \end{cases}$$

*inconsistent system*

Both lines have slope $-(3/2)$, and they are clearly distinct; so we have a pair of parallel lines. The solution set is the empty set, and the graph of this particular system is shown in Figure 9.3. A system of equations whose solution set is empty is called an *inconsistent system*.

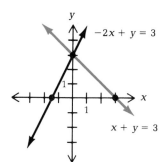

Fig. 9.1

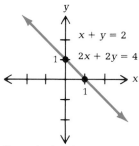

Dependent system

Fig. 9.2

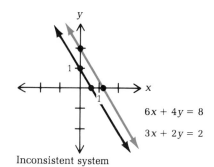

Inconsistent system

Fig. 9.3

In summary then, the system

$$(1) \quad \begin{cases} a_1 x + b_1 y = c_1 \\ a_2 x + b_2 y = c_2 \end{cases}$$

will have one of three possible solution sets:

**1** The solution set can consist of a single ordered pair. This occurs when the graphs of the two equations intersect at a single point.

**2** The system can be *dependent* in that it has an infinite solution set. This occurs when the two equations have the same graph.

**3** The system can be *inconsistent* in the sense that it has an empty solution set. Here the graphs of the equations are two parallel lines.

*method of substitution*

Since graphing is often tedious and sometimes inaccurate due to the fact that measurements can only be estimated, it is time to introduce algebraic methods for solving system (1). The first method we shall consider is called the *method of substitution*. To use it, you solve for $x$ or $y$ in one of the equations, and then substitute this value in the remaining equation in order to obtain an equation in a single variable. We illustrate the technique by solving the system

$$\begin{cases} 2x + 3y = 4 \\ 5x - y = -7. \end{cases}$$

Solving the second equation for $y$ produces $-y = -5x - 7$; so $y = 5x + 7$. When we replace $y$ with $5x + 7$ in the first equation, we obtain

$$2x + 3(5x + 7) = 4$$

$$2x + 15x + 21 = 4$$

$$17x = -17$$

$$x = -1.$$

To obtain the corresponding value of $y$, we note that $y = 5x + 7 = 5(-1) + 7 = -5 + 7 = 2$. The solution set is $\{(-1, 2)\}$. We check our work by substituting back into the original system as follows:

$$2(-1) + 3(2) = -2 + 6 = 4$$

$$5(-1) - 2 = -5 - 2 = -7.$$

The second method of solving the system

$$\begin{cases} a_1 x + b_1 y = c_1 \\ a_2 x + b_2 y = c_2 \end{cases}$$

of two linear equations in $x$ and $y$ is based on the fact that any solution of the given system is also a solution of the equation $k_1(a_1 x + b_1 y) + k_2(a_2 x + b_2 y) = k_1 c_1 + k_2 c_2$ where $k_1$, $k_2 \in R$. To see why this is true, we note that any solution $(s,t)$ of the original system must have the property that

$$a_1 s + b_1 t = c_1$$

$$a_2 s + b_2 t = c_2.$$

Hence, if we replace $a_1 s + b_1 t$ by $c_1$ and $a_2 s + b_2 t$ by $c_2$, we see that $k_1(a_1 s + b_1 t) + k_2(a_2 s + b_2 t) = k_1 c_1 + k_2 c_2$. This is precisely the statement that $(s,t)$ is a solution of the equation $k_1(a_1 x + b_1 y) + k_2(a_2 x + b_2 y) = k_1 c_1 + k_2 c_2$. The idea is to choose $k_1$ and $k_2$ so that the resulting equation will involve only one of the *elimination of variables* variables $x$ or $y$. For this reason, the method is called *elimination of variables*.

Let's illustrate this second technique by solving the system

$$\begin{cases} 2x + 3y = 4 \\ 5x - y = -7 \end{cases}$$

again. One approach is to observe that if three times the second equation is added to the first equation, the resulting equation will be free of $y$. Multiplying the second equation by $3$ produces $15x - 3y = -21$; so we have

$$2x + 3y = 4$$

$$\frac{15x - 3y = -21}{17x \qquad = -17}$$

$$x = -1.$$

If $(-1, y)$ is to be a solution of the system, it must be true that

$$2(-1) + 3y = 4$$

$$3y = 6$$

$$y = 2.$$

We obtain $\{(-1, 2)\}$ as the solution set, which agrees with the result obtained by the method of substitution.

As a further illustration, consider the system

$$\begin{cases} 4x - 9y = 21 \\ 6x + 6y = -1. \end{cases}$$

Using the method of substitution, we solve the second equation for $x$ and get $6x = -6y - 1$; so $x = -y - (1/6)$. Substituting this value of $x$ in the first equation, we see that

$$4\left(-y - \frac{1}{6}\right) - 9y = 21$$

$$-4y - \frac{4}{6} - 9y = 21$$

$$-13y = 21 + \frac{2}{3} = \frac{65}{3}$$

$$y = -\frac{5}{3}.$$

Hence $x = -y - \frac{1}{6} = -\left(-\frac{5}{3}\right) - \frac{1}{6} = \frac{5}{3} - \frac{1}{6} = \frac{3}{2}$. The solution set is $\left\{\left(\frac{3}{2}, -\frac{5}{3}\right)\right\}$. Then we substitute these values into the original system:

$$4\left(\frac{3}{2}\right) - 9\left(-\frac{5}{3}\right) = 6 + 15 = 21$$

$$6\left(\frac{3}{2}\right) + 6\left(-\frac{5}{3}\right) = 9 - 10 = -1.$$

If elimination of variables is used on this system, we can eliminate $x$ by adding 3 times the first equation to $-2$ times the second equation.

$$3(4x - 9y = 21) \qquad 12x - 27y = 63$$

$$-2(6x + 6y = -1) \qquad \underline{-12x - 12y = 12}$$

$$- 39y = 65$$

$$y = -\frac{65}{39} = -\frac{5}{3}.$$

If $\left(x, -\frac{5}{3}\right)$ is to be a solution, it must be true that

$$4x - 9\left(-\frac{5}{3}\right) = 21$$

$$4x + 15 = 21$$

$$4x = 6$$

$$x = \frac{3}{2}.$$

Again, the result agrees with the one obtained by the method of substitution.

Word problems can sometimes be solved by representing each of two unknown quantities by a different variable. You must remember, though, to translate the given data into the form of two independent equations.

**Problem:** The Ra-Lee Bicycle Shop sells three-speed and ten-speed bicycles. On a certain day they sell four of the three-speed models and six ten-speed bicycles for total sales of $1,270. The next day they take in $680 by selling five of the three-speed and two of the ten-speed models. What are the selling prices of the two models?

**Solution:** Let $x$ denote the selling price of a three-speed bike and $y$ the selling price of a ten-speed bike. The revenue taken in the first day must then be $4x + 6y$, and we are told that this is $1,270$. This gives us the equation $4x + 6y = 1,270$. Similarly, the next day's revenue of $5x + 2y$ is $680$. Thus, we have established the system

$$\begin{cases} 4x + 6y = 1,270 \\ 5x + 2y = 680 \end{cases}$$

to be solved. We proceed by adding $-3$ times the second equation to the first equation.

$$4x + 6y = \quad 1,270$$
$$\underline{-15x - 6y = -2,040}$$
$$-11x \qquad = -770$$
$$x = 70$$

To find $y$, we note that

$$4(70) + 6y = 1,270$$
$$280 + 6y = 1,270$$
$$6y = 990$$
$$y = 165.$$

The solution set is $\{(70, 165)\}$.

*Check:* $4(70) + 6(165) = 280 + 990 = 1,270$ and $5(70) + 2(165) = 350 + 330 = 680$.

**Problem:** You have $5,000 invested, part of it at 4% and part at 7%. If the yearly dividend totals $284, how much was invested at each amount?

**Solution:** Let $x$ denote the amount invested at 4% and $y$ the amount at 7%. The fact that you have invested a total of $5,000 can then be expressed by the equation $x + y = 5,000$. The income from the 4% investment is $0.04x$ and that from the 7% investment $0.07y$. Since the total dividend is $284, it must be true that $0.04x + 0.07y = 284$. Thus the system

$$\begin{cases} x + y = 5{,}000 \\ 0.04x + 0.07y = 284 \end{cases}$$

must be solved. We proceed by adding $-1$ times the second equation to $0.07$ times the first equation.

$$0.07x + 0.07y = \phantom{-}350$$
$$-0.04x - 0.07y = -284$$
$$\overline{\phantom{-}0.03x \phantom{- 0.07y} = \phantom{-}66}$$

$$x = \frac{66}{0.03} = \frac{6{,}600}{3} = 2{,}200$$

$y = 5{,}000 - 2{,}200 = 2{,}800.$

You have \$2,200 invested at 4% and \$2,800 at 7%.

Do you remember that the preceding example was also used in Section 4.5? There it was solved by introducing only one variable. The advantage of using two variables is that it is often easier to set up the equations, but then, of course, one must solve a system of two equations in two variables rather than a single equation in one variable.

**Problem:** A painter has contracted to paint a large room. The first day his helper started work at 8:00 A.M. and was joined by the painter an hour later. They worked straight through until 2:00 P.M., at which time the room was half-painted. The next morning, they both started work at 8:00, but the painter had to leave at 10:30 A.M. in order to start another job. The helper completed the job by staying on until 6:00 P.M., with an hour off for lunch. How long would it have taken each man if he had been working alone?

**Solution:** Let $x$ be the length of time it would take the helper and $y$ the length of time for the painter. The fraction of the job completed by the helper in one hour is $1/x$ and the fraction completed by the painter $1/y$. The first day, the helper worked 6 hours and the painter 5 hours. In 6 hours' time the helper completed $6(1/x) = 6/x$ of the job; in 5 hours' time the painter completed $5(1/y) = 5/y$ of the job. Since they completed half the job the first day, this tells us that $6/x + 5/y = 1/2$. The second day, the helper worked 9 hours, completing $9(1/x) = 9/x$ of the job. The painter worked 2-1/2 hours, completing $(5/2)(1/y) = 5/2y$ of the job. Since half the job was completed the second day, we know that $9/x +$

$5/2y = 1/2$. We must now solve the system of equations

$$\begin{cases} \dfrac{6}{x} + \dfrac{5}{y} = \dfrac{1}{2} \\[2ex] \dfrac{9}{x} + \dfrac{5}{2y} = \dfrac{1}{2}. \end{cases}$$

These are not linear equations, but the methods of this section can be applied if we treat $1/x$ and $1/y$ as the variables. Accordingly, we add *twice* the second equation to $-1$ times the first equation:

$$-\frac{6}{x} - \frac{5}{y} = -\frac{1}{2}$$

$$\frac{18}{x} + \frac{5}{y} = 1$$

$$\overline{\qquad\qquad\qquad}$$

$$\frac{12}{x} \qquad = \frac{1}{2}$$

$$x = 24.$$

Hence

$$\frac{6}{24} + \frac{5}{y} = \frac{1}{2}$$

$$\frac{5}{y} = \frac{1}{4}$$

$$y = 20.$$

It would take the helper 24 hours and the painter 20 hours if they were working alone.

$$\textit{Check: } \frac{6}{24} + \frac{5}{20} = \frac{1}{4} + \frac{1}{4} = \frac{1}{2} \quad \text{and} \quad \frac{9}{24} + \frac{5}{2 \cdot 20} = \frac{3}{8} + \frac{1}{8} = \frac{1}{2}.$$

## Exercise set 9.1

*In problems 1 to 8, sketch the graph of each system of equations. Determine from the graph whether each system is dependent or inconsistent or has a single point in its solution set. In the last case, determine the coordinates of the solution from the graph.*

$(\circ)$  **1** $\begin{cases} x - 2y = 4 \\ 8y - 4x = -8 \end{cases}$   **2** $\begin{cases} 3x - 4y = 5 \\ 8y - 6x + 10 = 0 \end{cases}$

**3** $\begin{cases} 2x + 3y = 6 \\ 12 - 4x - 6y = 0 \end{cases}$   **4** $\begin{cases} 2y - 5x = 10 \\ 4y = 10x \end{cases}$

**5** $\begin{cases} x + 2y = 2 \\ 3x - 2y = 6 \end{cases}$   **6** $\begin{cases} 4x + y = 1 \\ x + \dfrac{y}{4} = 1 \end{cases}$

**7** $\begin{cases} 3x - y = 3 \\ x - y = 3 \end{cases}$   **8** $\begin{cases} y - 2x = 4 \\ x + 2y = 8 \end{cases}$

*Use the method of substitution to solve each of the following systems of equations.*

$(\circ)$  **9** $\begin{cases} 3x - y = 16 \\ 4x - 9y = 6 \end{cases}$   **10** $\begin{cases} x + 5y = 11 \\ 2x + 3y = 1 \end{cases}$

**11** $\begin{cases} 3x + 4y = -1 \\ 9x + 2y = 2 \end{cases}$   **12** $\begin{cases} 12x - 5y = 2 \\ 8x - 15y = -1 \end{cases}$

**13** $\begin{cases} 6x - y = 5 \\ 21x + 2y = 3 \end{cases}$   **14** $\begin{cases} 13x + 7y = 1 \\ 4y - 3x = 3 \end{cases}$

*Use elimination of variables to solve each of the following systems of equations.*

$(\circ)$  **15** $\begin{cases} 3x + 5y = -2 \\ 7x - 4y = -36 \end{cases}$   **16** $\begin{cases} -4x + 6y = -6 \\ -9x + 7y = -20 \end{cases}$

**17** $\begin{cases} 2x - 7\overset{\cdot}{y} = 8 \\ 5x - 3y = \dfrac{11}{2} \end{cases}$   **18** $\begin{cases} 4x - 3y = 7 \\ 6x - 10y = \dfrac{26}{3} \end{cases}$

**19** $\begin{cases} 3x + 4y = 5 \\ 4x + 7y = 2 \end{cases}$   **20** $\begin{cases} 7x + 3y = 8 \\ 3x + 7y = 8 \end{cases}$

*Use a system of two equations in two variables to solve each of the following word problems.*

**21**  The difference of two numbers is 5. Twice the smaller plus the larger equals 32. What are the numbers?

**22**  The sum of two numbers is 10. The larger is two less than twice the smaller. Find the numbers.

$(\circ)$  **23**  The sum of the digits of a certain two-digit integer is 6. Interchanging the digits increases the number by 18. What is the number?

**24**  A man buys three shirts and two ties at a certain clothing store, and it costs him $30. His friend buys four shirts and one tie, spending $35. Assuming that all shirts were at the same price and that all ties were at the same price, what was the price of each?

**25** A man has $12,500 to invest, some of it at 4 percent and some at 7 percent. How much should be invested at each rate in order to produce a yearly income of $740?

**26** A man puts 20 gallons of gasoline in his car. Some of it was premium gas worth 62.9¢ per gallon, and some a low-lead mixture worth 57.9¢ per gallon. If the total bill was $12.23, how many gallons of each did he buy?

**27** Last week, a plumber and his helper each worked 40 hours. Their total joint pay was $520. This week, the plumber worked 36 hours and his helper 32 hours, and they earned a total of $448. Find the hourly salary of each.

**28** The Sno-King Company sells two types of snowmobiles, the deluxe Sno-Jet and the economy Sno-Kub model. During January, they sold 90 Sno-Jets and 76 Sno-Kubs for a gross revenue of $97,900. During February, they sold 65 Sno-Jets and 80 Sno-Kubs, taking in $80,750. Find the sales price of each type of snowmobile.

**29** The Nutte Haus sells a deluxe nut mixture for $1.30 per pound and an economy blend for 95¢ per pound. One day they sold $165.75 worth of nuts. If the prices had been 5¢ per pound lower on both grades, the figure would have been $158.50. How many pounds of each were sold?

**30** A man owns stock in two companies, Sno-King Co., and the Squirrel Nut Co. Sno-King stock is worth $110 per share and Squirrel Nut $75 per share. Right now, his investments are valued at $10,700. If Sno-King dropped to $100 per share and Squirrel rose to $80, his investments would be worth $10,200. How many shares of each does he own?

( ∘ )  **31** An airplane flying at its normal speed on a recent 630-mile trip encountered a head wind, and the trip took 1 hour and 45 minutes. On the return trip, there was a tail wind half as strong as the earlier head wind, and the trip took 1 hour and 30 minutes. What was the speed of the plane, and how strong was the head wind?

**32** On a recent canoe trip, it took 45 minutes to travel 12 miles downstream and 90 minutes for the return trip. What was the (actual) speed of the canoe, and how fast was the current?

**33** During a recent storm the basement of the library was flooded. A pump was started, and after an hour's time a second pump was added. It took a total of 4-1/2 hours to clear the water out of the basement. If both pumps had been used right from the start, the basement would have been drained in 18 minutes' less time. How long would it have taken for the original pump to drain the basement if it had been used alone? How long for the second pump?

**34** When mowing the family's lawn, John uses a power mower and his brother a riding mower. On July 1, John worked 30 minutes and his brother 100 minutes to get the lawn mowed. On July 8, John worked 2 hours and his brother 40 minutes to complete the same job. On July 15, John mowed the lawn himself, and on July 22 his brother mowed the lawn himself. How long did it take to mow the lawn on the fifteenth and the twenty-second?

**1** These equations comprise an inconsistent system. Graphed, they look like this.

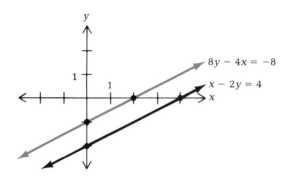

**9** Solve for $y$ in the first equation thus: $y = 3x - 16$. Substitute this value of $y$ in the second equation:

$$4x - 9(3x - 16) = 6$$

$$4x - 27x + 144 = 6$$

$$-23x = -138$$

$$x = 6.$$

Then $y = 3(6) - 16 = 18 - 16 = 2$. The solution set is $\{(6, 2)\}$.

*Check:* $3 \cdot 6 - 2 = 18 - 16 = 2$ and $4 \cdot 6 - 9 \cdot 2 = 24 - 18 = 6$.

**15** Add four times the first equation to five times the second equation.

$$4(3x + 5y = -2)$$
$$5(7x - 4y = -36)$$

$$12x + 20y = -8$$
$$\underline{35x - 20y = -180}$$
$$47x \phantom{+ 20y} = -188$$
$$x = -4$$

To obtain the value of $y$, note that if $(-4, y)$ is to be a solution of the system, then

$$3(-4) + 5y = -2$$

$$-12 + 5y = -2$$

$$5y = 10$$

$$y = 2.$$

The solution set $= \{(-4, 2)\}$.

*Check:* $3(-4) + 5(2) = -12 + 10 = -2$ and $7(-4) - 4(2) = -28 - 8 = -36$

**23** Let $x$ denote the tens digit and $y$ the units digit. The number is then $10x + y$. Interchanging the two digits will produce the number $10y + x$. We are given the information that this new number is 18 more than the original number; so

$$(10y + x) - (10x + y) \ = \ 18$$
$$10y + x - 10x - y \ = \ 18$$
$$9y - 9x \ = \ 18$$
$$y - x \ = \ 2.$$

We also have that $x + y = 6$; so we must solve the system of equations

$$\begin{cases} y - x = 2 \\ y + x = 6. \end{cases}$$

Adding the two equations produces $2y = 8$; so $y = 4$. But then $x = 6 - y = 6 - 4 = 2$. The number we are after is 24.

*Check:* The sum of the digits of 24 is 6. If the digits are interchanged, the resulting number is 42; furthermore, $42 - 24 = 18$.

**31** Let $x$ denote the speed of the airplane and $y$ the velocity of the head wind. Then the ground speed of plane on trip out $= x - y$ and the ground speed of plane on return trip $= x + y/2$. On the trip out, it took $7/4$ hours to cover 630 miles. Using the fact that *distance* $= (rate)(time)$ we see that

$$630 = (x - y)\left(\frac{7}{4}\right).$$

Similarly, the return trip took $3/2$ hours to cover the same distance; so

$$630 = \left(x + \frac{y}{2}\right)\left(\frac{3}{2}\right).$$

If the first equation is multiplied by $4/7$, we obtain $x - y = 360$. Multiplying the second equation by $2/3$ produces $x + (y/2) = 420$. Thus we must solve the system of equations

$$\begin{cases} x - y = 360 \\ x + \dfrac{y}{2} = 420. \end{cases}$$

Subtracting the first of these equation from the second one, we have $(3/2)y = 60$; so $y = 40$. Then $x = y + 360 = 40 + 360 = 400$. The plane was traveling at 400 mph, and the head wind had a force of 40 mph.

*Check:* Ground speed on the trip was $400 - 40 = 360$ mph and $(7/4)(360) = 7 \cdot 90 = 630$. Ground speed coming back was $400 + 20 = 420$ mph and $(3/2)(420) = 3 \cdot 210 = 630$.

## (9.2) (Systems in two variables—solution by determinants)

In this section we develop a third method of solving a system of two linear equations in two variables. Before doing so, however, it is necessary to introduce the concept of a *second-order determinant.* The determinant of the square array

*second-order*

*determinant*

$$a \quad b$$

$$c \quad d$$

of real numbers $a, b, c, d$ is denoted by the symbol

$$\begin{vmatrix} a & b \\ c & d \end{vmatrix}$$

and is defined to be the *number $ad = bc$.* For example,

$$\begin{vmatrix} 1 & 2 \\ 3 & 4 \end{vmatrix} = 1 \cdot 4 - 3 \cdot 2 = 4 - 6 = -2.$$

Such determinants are called *second-order* determinants because the array of numbers

$$a \quad b$$

$$c \quad d$$

has two rows and two columns. The following diagram may help you to remember the definition of a second order determinant:

$$\begin{vmatrix} a & b \\ c & d \end{vmatrix} = \quad \overset{a \qquad b}{\underset{c \qquad d}{\times}} \quad = ad - bc.$$

Here are some more examples:

$$\begin{vmatrix} 1 & -5 \\ 2 & 7 \end{vmatrix} = 7 - (-10) = 7 + 10 = 17,$$

$$\begin{vmatrix} -1 & 3 \\ 2 & -6 \end{vmatrix} = 6 - 6 = 0.$$

Let's see what this has to do with the solution of the system of linear equations

$$(1) \quad \begin{cases} a_1 x + b_1 y = c_1 \\ a_2 x + b_2 y = c_2. \end{cases}$$

Using elimination of variables, we may solve for $x$ by adding $b_2$ times the first equation to $-b_1$ times the second equation as follows:

$$a_1 b_2 x + b_1 b_2 y = c_1 b_2$$
$$\underline{-a_2 b_1 x - b_1 b_2 y = -c_2 b_1}$$
$$(a_1 b_2 - a_2 b_1)x = c_1 b_2 - c_2 b_1.$$

If $D = \begin{vmatrix} a_1 & b_1 \\ a_2 & b_2 \end{vmatrix} = a_1 b_2 - a_2 b_1 \neq 0$, we see that

$$x = \frac{c_1 b_2 - c_2 b_1}{D} = \frac{\begin{vmatrix} c_1 & b_1 \\ c_2 & b_2 \end{vmatrix}}{D}.$$

A similar argument shows that

$$y = \frac{c_2 a_1 - c_1 a_2}{D} = \frac{\begin{vmatrix} a_1 & c_1 \\ a_2 & c_2 \end{vmatrix}}{D}.$$

*Cramer's rule*

The foregoing formulas for $x$ and $y$ are known as *Cramer's rule*. It becomes a little easier to remember this rule if you note that in each denominator, $D$ is the determinant of the array

$$\begin{array}{cc} a_1 & b_1 \\ a_2 & b_2 \end{array}$$

formed by the coefficients of $x$ and $y$ in the original equations. The numerator of the equation for $x$ is the determinant of the array obtained by replacing the coefficients of $x$ by $c_1$ and $c_2$; the numer-

**331 (∘) Systems in two variables—solution by determinants**

ator of the equation for $y$ is the determinant of the array obtained by replacing the coefficients of $y$ by $c_1$ and $c_2$. To sum it all up, if you are given the system of linear equations

**(1)**
$$\begin{cases} a_1 x + b_1 y = c_1 \\ a_2 x + b_2 y = c_2 \end{cases}$$

and if $D = \begin{vmatrix} a_1 & b_1 \\ a_2 & b_2 \end{vmatrix} \neq 0$, then the system has the unique solution

(∘)

$$x = \frac{\begin{vmatrix} c_1 & b_1 \\ c_2 & b_2 \end{vmatrix}}{D} \quad \text{and} \quad y = \frac{\begin{vmatrix} a_1 & c_1 \\ a_2 & c_2 \end{vmatrix}}{D}.$$

To illustrate Cramer's rule, consider the system

$$\begin{cases} 2x + 3y = 4 \\ 5x - y = -7 \end{cases}$$

that was solved in Section 9.1 by substitution as well as by elimination of variables. Here

$$D = \begin{vmatrix} 2 & 3 \\ 5 & -1 \end{vmatrix} = 2(-1) - 5 \cdot 3 = -2 - 15 = -17;$$

so

$$x = \frac{\begin{vmatrix} 4 & 3 \\ -7 & -1 \end{vmatrix}}{-17} = \frac{-4 - 3(-7)}{-17} = \frac{-4 + 21}{-17} = \frac{+17}{-17} = -1$$

and

$$y = \frac{\begin{vmatrix} 2 & 4 \\ 5 & -7 \end{vmatrix}}{-17} = \frac{-14 - 20}{-17} = \frac{-34}{-17} = 2.$$

In order to apply Cramer's rule to (1), it is essential that the determinant $D \neq 0$. To see what happens if $D = 0$, we consider three cases. In the first instance, $\begin{vmatrix} c_1 & b_1 \\ c_2 & b_2 \end{vmatrix} \neq 0$. We add $b_2$ times the first equation to $-b_1$ times the second equation:

$$a_1 b_2 x + b_1 b_2 y = c_1 b_2$$

$$\underline{-a_2 b_1 x - b_1 b_2 y = -c_2 b_1}$$

$$(a_1 b_2 - a_2 b_1)x = c_1 b_2 - c_2 b_1$$

**(2)** $0 \cdot x = c_1 b_2 - c_2 b_1$

Any solution of system (1) must also be a solution of equation (2). But (2) has no solution since $c_1 b_2 - c_2 b_1 \neq 0$. It follows that system (1) is inconsistent.

In the second instance, $\begin{vmatrix} a_1 & c_1 \\ a_2 & c_2 \end{vmatrix} \neq 0$. Adding $a_2$ times the first equation to $-a_1$ times the second equation produces

**(3)** $0 \cdot y = c_1 a_2 - c_2 a_1$.

Any solution of (1) must be a solution of (3); so here too we have an inconsistent system.

Finally, we consider $\begin{vmatrix} c_1 & b_1 \\ c_2 & b_2 \end{vmatrix} = 0$ and $\begin{vmatrix} a_1 & c_1 \\ a_2 & c_2 \end{vmatrix} = 0$. The equation $b_2(a_1 x + b_1 y) - b_1(a_2 x + b_2 y) = b_2 c_1 - c_2 b_1$ is then an *identity* in that both members reduce to 0. This identity may be rewritten as

$$b_1(a_2 x + b_2 y) - c_2 b_1 = b_2(a_1 x + b_2 y) - c_1 b_2$$

$$b_1(a_2 x + b_2 y - c_2) = b_2(a_1 x + b_2 y - c_1).$$

If $b_1 \neq 0$, then $a_2 x + b_2 y - c_2 = (b_2/b_1)(a_1 x + b_1 y - c_1)$. It follows from this that the second equation of system (1) is a nonzero multiple of the first equation, and we have a dependent system. If $b_1 = 0$ then $a_1 \neq 0$, and a similar argument shows that system (1) is dependent here, too.

Here is a summary of the foregoing discussion. Given the system

**(1)** $\begin{cases} a_1 x + b_1 y = c_1 \\ a_2 x + b_2 y = c_2 \end{cases}$

if $D = \begin{vmatrix} a_1 & b_1 \\ a_2 & b_2 \end{vmatrix} \neq 0$ then Cramer's rule may be used to solve for

$x$ and $y$. If $D = 0$ there are two possibilities.

**(a)** If $\begin{vmatrix} c_1 & b_1 \\ c_2 & b_2 \end{vmatrix} \neq 0$ or $\begin{vmatrix} a_1 & c_1 \\ a_2 & c_2 \end{vmatrix} \neq 0$, the system is inconsistent.

**(b)** If $\begin{vmatrix} c_1 & b_1 \\ c_2 & b_2 \end{vmatrix} = 0$ and $\begin{vmatrix} a_1 & c_1 \\ a_2 & c_2 \end{vmatrix} = 0$, the system is dependent.

For example, consider the system of equations

$$\begin{cases} 2x - 5y = 3 \\ -4x + 10y = -6. \end{cases}$$

Here $D = \begin{vmatrix} 2 & -5 \\ -4 & 10 \end{vmatrix} = 20 - 20 = 0$, $\begin{vmatrix} 3 & -5 \\ -6 & 10 \end{vmatrix} = 30 - 30 = 0$, and

$\begin{vmatrix} 2 & 3 \\ -4 & -6 \end{vmatrix} = -12 - (-12) = 0$; so the system is dependent. On the

other hand, the system

$$\begin{cases} 3x - 2y = 1 \\ 6x - 4y = 3 \end{cases}$$

is inconsistent since $D = \begin{vmatrix} 3 & -2 \\ 6 & -4 \end{vmatrix} = -12 - (-12) = 0$ and $\begin{vmatrix} 1 & -2 \\ 3 & -4 \end{vmatrix}$

$= -4 - (-6) = -4 + 6 = 2 \neq 0$.

## Exercise set 9.2

In problems 1 to 8, evaluate each determinant.

**(∘)** **1** $\begin{vmatrix} 1 & 5 \\ -3 & 2 \end{vmatrix}$  **2** $\begin{vmatrix} 4 & -1 \\ -2 & 3 \end{vmatrix}$  **3** $\begin{vmatrix} 2 & -6 \\ -1 & 4 \end{vmatrix}$  **4** $\begin{vmatrix} -3 & 2 \\ -5 & 1 \end{vmatrix}$

**5** $\begin{vmatrix} 2 & 6 \\ -1 & -3 \end{vmatrix}$  **6** $\begin{vmatrix} \dfrac{1}{2} & -\dfrac{1}{3} \\ \dfrac{1}{4} & \dfrac{5}{6} \end{vmatrix}$  **7** $\begin{vmatrix} \dfrac{1}{5} & \dfrac{3}{4} \\ \dfrac{3}{10} & \dfrac{1}{2} \end{vmatrix}$  **8** $\begin{vmatrix} 4 & -6 \\ \dfrac{1}{3} & -\dfrac{1}{2} \end{vmatrix}$

In problems 9 to 20, use Cramer's rule to solve problems 9 to 20 of Exercise set 9.1.

Classify each of the following systems according to whether it is dependent, inconsistent, or has a solution set containing a single point. In the last case, use Cramer's rule to determine the solution set.

21 $\begin{cases} 6x + 8y = 3 \\ 9x + 12y = 2 \end{cases}$  22 $\begin{cases} 3x - 5y = 8 \\ -6x + 10y = -16 \end{cases}$

23 $\begin{cases} \dfrac{2}{3}x - \dfrac{3}{4}y = 1 \\ 8x - 9y = 12 \end{cases}$  24 $\begin{cases} \dfrac{2}{5}x + \dfrac{1}{4}y = 1 \\ 8x + 5y = 10 \end{cases}$

25 $\begin{cases} 2x - 3y = 6 \\ 3x - y = -2 \end{cases}$  26 $\begin{cases} \dfrac{1}{2}x - \dfrac{2}{3}y = 1 \\ 3x - 4y = 6 \end{cases}$

(∘) 27 $\begin{cases} 2x + \sqrt{2}\,y = 3\sqrt{2} \\ -3x + 4\sqrt{2}\,y = \dfrac{1}{\sqrt{2}} \end{cases}$  28 $\begin{cases} \sqrt{5}\,x + \dfrac{y}{2} = 6 \\ x - \dfrac{y}{\sqrt{20}} = \dfrac{4\sqrt{5}}{5} \end{cases}$

29 $\begin{cases} x - \sqrt{3}\,y = \sqrt{6} \\ \sqrt{3}\,x - 3y = 3\sqrt{2} \end{cases}$  30 $\begin{cases} \sqrt{2}\,x + 2y = 4 \\ x + \sqrt{2}\,y = 2 \end{cases}$

(∘)

*Sample solutions for exercise set 9.2*

1 $\begin{vmatrix} 1 & 5 \\ -3 & 2 \end{vmatrix} = 1 \cdot 2 - (-3) \cdot 5 = 2 - (-15) = 2 + 15 = 17$

9 $D = \begin{vmatrix} 3 & -1 \\ 4 & -9 \end{vmatrix} = -27 - 4(-1) = -27 + 4 = -23$

$x = \dfrac{\begin{vmatrix} 16 & -1 \\ 6 & -9 \end{vmatrix}}{-23} = \dfrac{-144 - 6(-1)}{-23} = \dfrac{-138}{-23} = 6$

$y = \dfrac{\begin{vmatrix} 3 & 16 \\ 4 & 6 \end{vmatrix}}{-23} = \dfrac{18 - 64}{-23} = \dfrac{-46}{-23} = 2.$

Solution set $\{(6, 2)\}$

15 $D = \begin{vmatrix} 3 & 5 \\ 7 & -4 \end{vmatrix} = -12 - 35 = -47$

$x = \dfrac{\begin{vmatrix} -2 & 5 \\ -36 & -4 \end{vmatrix}}{-47} = \dfrac{8 - (-36)5}{-47} = \dfrac{8 + 180}{-47} = \dfrac{188}{-47} = -4$

$$y = \frac{\begin{vmatrix} 3 & -2 \\ 7 & -36 \end{vmatrix}}{-47} = \frac{-108 - 7(-2)}{-47} = \frac{-108 + 14}{-47} = \frac{-94}{-47} = -2$$

Solution set = $\{(-4, 2)\}$

21  The system is inconsistent.

$$D = \begin{vmatrix} 6 & 8 \\ 9 & 12 \end{vmatrix} = 72 - 72 = 0$$

$$\begin{vmatrix} 3 & 8 \\ 2 & 12 \end{vmatrix} = 36 - 16 = 20 \neq 0$$

27  $$D = \begin{vmatrix} 2 & \sqrt{2} \\ -3 & 4\sqrt{2} \end{vmatrix} = 8\sqrt{2} + 3\sqrt{2} = 11\sqrt{2}$$

$$x = \frac{\begin{vmatrix} 3\sqrt{2} & \sqrt{2} \\ \dfrac{1}{\sqrt{2}} & 4\sqrt{2} \end{vmatrix}}{11\sqrt{2}} = \frac{24 - 1}{11\sqrt{2}} = \frac{23}{11\sqrt{2}} = \frac{23\sqrt{2}}{22}$$

$$y = \frac{\begin{vmatrix} 2 & 3\sqrt{2} \\ -3 & \dfrac{1}{\sqrt{2}} \end{vmatrix}}{11\sqrt{2}} = \frac{\sqrt{2} + 9\sqrt{2}}{11\sqrt{2}} = \frac{10\sqrt{2}}{11\sqrt{2}} = \frac{10}{11}$$

Solution set = $\left\{ \left( \dfrac{23\sqrt{2}}{11}, \dfrac{10}{11} \right) \right\}$

# (9.3) (Systems of linear equations in three variables)

At this point, you should have a pretty firm grasp of the ways to solve a system of two linear equations in two variables. Now we turn to systems of three linear equations in three variables, with good reason. Many practical situations give rise to such systems; it's that simple!

You should recall that an ordered pair of real numbers is simply a pair $(a,b)$ in which we keep track of the order in which the numbers appear. Similarly, an *ordered triple* of real numbers is a triple $(a,b,c)$ in which we keep track of which number appears first, which second, and which third. Thus we want $(1,2,3) = (1,2,3)$, but $(1,2,3) \neq (1,3,2)$, $(1,2,3) \neq (3,1,2)$, and so on. Ordered pairs of real numbers were introduced because a *solution* of a linear equation $ax + by = c$ in the variables $x$ and $y$ is an *ordered pair* $(x_0,y_0)$ of real numbers such that $ax_0 + by_0 = c$. An ordered pair of real numbers is needed because it is necessary to substitute for both $x$ and $y$ in the equation.

All right then! Let's take a linear equation

$ax + by + cz = d$

in the three variables $x$, $y$, and $z$. A *solution* of such an equation is an ordered triple $(x_0,y_0,z_0)$ of real numbers such that with $x = x_0$, $y = y_0$, and $z = z_0$ the equation is satisfied; in other words, such that $ax_0 + by_0 + cz_0$ really does equal $d$. For example, the triple $(1,-3,-4)$ is a solution of $5x - 2y + z = 7$ because $5(1) - 2(-3) + (-4) = 5 + 6 - 4 = 7$; the triple $(1,2,3)$ is *not* a solution because $5(1) - 2(2) + 3 = 5 - 4 + 3 = 4 \neq 7$.

To *solve* the system of linear equations

$$\begin{cases} a_1 x + b_1 y + c_1 z = d_1 \\ a_2 x + b_2 y + c_2 z = d_2 \\ a_3 x + b_3 y + c_3 z = d_3 \end{cases}$$

is to find all ordered triples that are solutions to all three equations. Thus we must determine the set intersection $\{(x,y)|a_1 x + b_1 y + c_1 z = d_1\} \cap \{(x,y)|a_2 x + b_2 y + c_2 z = d_2\} \cap \{(x,y)|a_3 x + b_3 y + c_3 z = d_3\}$. This is equivalent to finding the intersection of three planes; so if we tried to solve this system by graphing, we would need three-dimensional graphs. Because that is impractical, we will devote our attention to algebraic techniques.

Given a system of three linear equations in the variables $x$, $y$, $z$, the idea is to use substitution or elimination of variables to obtain a new system of two linear equations in two variables whose solution set contains all solutions to the original system. The techniques of Section 9.1 can then be used to solve the new system; the value of the remaining variable is obtained by substituting back into the original system. To see this procedure at work, consider the system

$$(1) \begin{cases} 3x - 4y + 2z = 25 \\ 2x - 5y + 3z = 28 \\ 7x - y + 3z = 35. \end{cases}$$

By subtracting four times the third equation from the first equation and five times the third equation from the second equation, we obtain the new system

$$(2) \begin{cases} -25x - 10z = -115 \\ -33x - 12z = -147 \end{cases}$$

of two linear equations in two variables. Arguments similar to those used in connection with the elimination-of-variables method of Section 9.1 now show that any solution of system (1) is also a solution of system (2). To solve (2), add two-fifths of the first equation to $-(1/3)$ times the second equation:

$$\frac{2}{5}(-25x - 10z = -115) \qquad\qquad 10x + 4z = \phantom{-}46$$

$$-\frac{1}{3}(-33x - 12z = -147) \qquad\qquad \underline{-11 \phantom{x}- 4z = -49}$$
$$\phantom{-\frac{1}{3}(-33x - 12z = -147)} \qquad\qquad \phantom{-11 -} -x \phantom{- 4z} = -3$$

$$x = 3.$$

Hence, $30 + 4z = 46$, $4z = 16$, and $z = 4$. The next step is to substitute $x = 3$ and $z = 4$ back into one of the equations in the original system in order to obtain a value of $y$:

$$7x - y + 3z = 35$$
$$7 \cdot 3 - y + 3 \cdot 4 = 35$$
$$-y + 33 = 35$$
$$y = -2.$$

Thus, the only possible solution of system (1) is $(3, -2, 4)$. We verify that this is indeed the solution by substituting it back into the system of equations:

$$3(3) - 4(-2) + 2(4) = 9 + 8 + 8 = 25,$$

$$2(3) - 5(-2) + 3(4) = 6 + 10 + 12 = 28,$$

$$7(3) - (-2) + 3(4) = 21 + 2 + 12 = 35.$$

Hence, the solution set is $\{(3, -2, 4)\}$.

Here is a summary of the method for solving a system of three linear equations in three variables.

**Step 1:** Select one of the equations. By adding suitable multiples of this equation to the remaining equations, obtain a system of two equations in two variables. (Alternately, you may solve the selected equation for one of the variables, and then substitute this value in the remaining equations to arrive at a system of two equations in two variables.)

**Step 2:** Solve the system of two equations in two variables that you obtained in Step 1. Any of the methods described earlier in the chapter will do.

**Step 3:** At the conclusion of Step 2 you will have produced numerical values for two of the variables. Substitute back into one of the original equations to obtain the value of the remaining variable.

**Step 4:** Check your answer!

If the original system has a unique solution, this procedure will get it for you.

On the other hand (just as in the two-variable case), there are systems of three linear equations in three variables that do not have a unique solution. If such a system has an infinite solution set, it is a *dependent system;* if it has no solution, it is an *inconsistent system.* We now illustrate these possibilities.

*dependent system*

*inconsistent system*

Consider the system

$$\text{(3)} \quad \begin{cases} 5x - y + 6z = 7 \\ 2x + y + 3z = 1 \\ 11x - 5y + 12z = 19. \end{cases}$$

Solving the first equation for $y$ produces $y = 5x + 6z - 7$. By substituting this expression in the remaining equations, the system

$$\text{(4)} \quad \begin{cases} 2x + (5x + 6z - 7) + 3z = 1 \\ 11x - 5(5x + 6z - 7) + 12z = 19 \end{cases}$$

is produced. Simplifying both equations, we may write

$$\text{(5)} \quad \begin{cases} 7x + 9z = 8 \\ -14x - 18z = -16. \end{cases}$$

Since the second equation of this last system is a multiple of the first equation, we deduce that system (5) is dependent. Direct substitution now shows that any solution of system (5) is also a solution of (3); so (3) is also a dependent system. This can also be seen from the fact that the third equation in system (3) can be expressed as three times the first equation plus $-2$ times the second equation.

We now have a look at the system

$$(6) \quad \begin{cases} 5x + 5y + 4z = 1 \\ x - 13y - 2z = 6 \\ 2x - y + z = 2. \end{cases}$$

By subtracting four times the third equation from the first equation and adding two times the third equation to the second equation, we obtain the new system

$$(7) \quad \begin{cases} -3x + 9y = -7 \\ 5x - 15y = 10 \end{cases}$$

which is equivalent to

$$(8) \quad \begin{cases} x - 3y = \dfrac{7}{3} \\ x - 3y = 2. \end{cases}$$

Any solution of system (6) must clearly be a solution of system (8). Since (8) is inconsistent, it follows that (6) must also be inconsistent.

In closing, we consider a word problem whose solution involves a system of three linear equations in three variables.

**Problem:** A group of people spent $56.00 for tickets at a theater that charges $2.50 for adults, $1.50 for students, and $1.00 for children. Had they arrived before 6:00 P.M., they could have taken advantage of the matinee rates of $1.50 for adults, $1.25 for students, and $1.00 for children, thereby saving $18.00. If there had been two more adults in the group and the tickets had been bought in advance, the special group rate of 75¢ for children and $1.25 for all others could have been used. Then, the admission would have been $35.50. How many adults, students, and children were in the original group?

**Solution:** Let $x$, $y$, and $z$, respectively, denote the number of adults, students, and children in the original group. If all costs are expressed in terms of dollars, the original admission of $56.00 leads to the equation

$$\frac{5}{2}x + \frac{3}{2}y + z = 56,$$

since (*number of adults*)(5/2) + (*number of students*)(3/2) + (*number of children*)(1) = total admission. Similarly, the matinee admission data can be expressed in the equation

$$\frac{3}{2}x + \frac{5}{4}y + z = 56 - 18.$$

For the group admission plan, there would be $x + 2$ adults; so the equation becomes

$$\frac{5}{4}(x + 2) + \frac{5}{4}y + \frac{3}{4}z = \frac{71}{2}.$$

Thus, we must solve the system

$$\begin{cases} \dfrac{5}{2}x + \dfrac{3}{2}y + z = 56 \\[2mm] \dfrac{3}{2}x + \dfrac{5}{4}y + z = 38 \\[2mm] \dfrac{5}{4}x + \dfrac{5}{4}y + \dfrac{3}{4}z + \dfrac{5}{2} = \dfrac{71}{2} \end{cases}$$

of three linear equations in three variables. This system is clearly equivalent to the system

$$\begin{cases} 5x + 3y + 2z = 112 \\ 6x + 5y + 4z = 152 \\ 5x + 5y + 3z = 132. \end{cases}$$

If three times the third equation is subtracted from five times the first equation and the third equation is subtracted from the second, we obtain the new system

$$\begin{cases} 10x + z = 164 \\ x + z = 20 \end{cases}$$

of two linear equations in two variables. Subtracting the second equation from the first of this new system, we see that $9x = 144$, so that $x = 16$. It is immediately apparent that $z = 20 - x = 4$. Since $5x + 3y + 2z = 112$, we must have

$$5(16) + 3y + 2(4) = 112$$

$$3y + 88 = 112$$

$$3y = 24$$

$$y = 8.$$

There were 16 adults, 8 students, and 4 children in the original group.

*Check:* When we substitute these values into the situations in the original problem, we get $(16)(\$2.50) + (8)(\$1.50) + (4)(\$1.00) = \$40 + \$12 + \$4 = \$56$; $(16)(\$1.50) + (8)(\$1.25) + (4)(\$1.00) = \$24 + \$10 + \$4 = \$38$; and $(18)(\$1.25) + (8)(\$1.25) + (4)(\$.75) = \$22.50 + \$10.00 + \$3.00 = \$35.50$.

## *Exercise set 9.3*

*Determine the solution set of each of the following systems of linear equations. If the system is dependent or inconsistent, state that fact.*

(∘) 1  $\begin{cases} 4x + 5y - z = -12 \\ -3x + y + 2z = 1 \\ 2x + y - 4z = -19 \end{cases}$
  2  $\begin{cases} -2x + 7y - 3z = -1 \\ 2x - 3y + 4z = 7 \\ 6x + 2y + 5z = 18 \end{cases}$

3  $\begin{cases} 4x - 3y + 3z = 11 \\ 8x + 6y + 5z = 13 \\ -3x + 2y - z = 3 \end{cases}$
  4  $\begin{cases} 3x + 5y - 4z = -11 \\ -2x - 8y + 5z = 19 \\ -x + 2y - 3z = -13 \end{cases}$

5  $\begin{cases} 5x + 4y - 14z = 7 \\ -3x + 2y + 14z = -2 \\ 2x + 2y - 6z = 7 \end{cases}$
  6  $\begin{cases} 6x - 9y + 3z = 3 \\ 9x - 12y - 18z = 6 \\ -15x - 6y + 6z = -36 \end{cases}$

7  $\begin{cases} 10x - 8y + 2z = 46 \\ -4x + 6y + 6z = -20 \\ 24x + 12y - 6z = 6 \end{cases}$
  8  $\begin{cases} 6x - 5y - 5z = -8 \\ 3x - 2y + \dfrac{1}{2}z = 3 \\ -8x + 7y + 12z = 22 \end{cases}$

(∘) 9  $\begin{cases} 4x - 3y + 2z = 1 \\ 9x - 2y + 3z = 2 \\ 3x - 7y + 3z = 1 \end{cases}$
  10  $\begin{cases} 2y + z = 2 \\ x + 2z = 6 \\ x - 2y + z = 0 \end{cases}$

11  $\begin{cases} 3x + 2y + 4z = 7 \\ -y + z = -1 \\ x + 2z = 1 \end{cases}$
  12  $\begin{cases} 3x + 6z = 10 \\ x + 2y = 6 \\ x - y + 3z = 2 \end{cases}$

*Solve each of the remaining problems by forming an appropriate system of three equations in three variables.*

**13** Three numbers add up to 45. The first number minus the second number is 13, and the second number is four less than twice the third. What are the numbers?

**14** The sum of three numbers is 6. The second is three times the third, and the first number plus twice the second number is zero. What are the numbers?

( ∘ ) **15** Mildred Myser finds a bag containing quarters, dimes, and nickels. There are 23 coins, and their total value is $3.40. If there were twice as many quarters as dimes, how many of each were there?

**16** I have $120. It is in the form of $10, $5, and $1 bills. I have as many fives as I have ones, and the number of fives is two less than twice the number of tens. How many bills of each type do I have?

**17** The Ridem Bike Shop sells three-speed, five-speed, and ten-speed bicycles. They sold 40 three-speed bicycles in June, 45 in July, and 30 in August. The five-speed bike sales were 15 in June, 9 in July, and 12 in August. The corresponding figures for ten-speed bikes was 18, 30, and 20. If the total sales figure for June was $5,619, for July $6,705, and for August $4,980, find the selling price of each of the three types of bicycles.

**18** A plumber has a helper and an apprentice. During a certain week they each work 40 hours, and they earn a grand total of $680. The following week, the plumber works 40 hours, his helper 36 hours, and the apprentice 32 hours; the payroll totals $636. If the apprentice earns $2 less per hour than the helper, what are their hourly rates?

**19** A man owns 70 shares of stock. There are three types of stock involved, some worth $10 per share, some $6, and some $4. The total value of his holdings is $470. If the $10 stock increased to $11, the $6 to $7, and the $4 dropped to $3, the stock would be worth $510. How many shares of each does he own?

**20** An electrician has two helpers, Kilo and Watt. On a recent wiring job, each of the three worked 8 hours, but they managed to complete only half the job. The next day Kilo worked 12 hours and Watt 8 hours to complete another third of the job. The electrician and Watt finished up the remainder in 4 hours' time on the third day. How long would it have taken each man to do the wiring job if he been working alone?

**21** Find a parabola $y = ax^2 + bx + c$ that contains the points (1,7), (3,5), and (−1,1).

**22** At 9:00 A.M. a certain river is 28 feet below flood stage. At 10:00 A.M. it is 18 feet below flood stage, and at 11:00 A.M. it is 10 feet below flood stage. The flood control officials are trying to predict the duration and extent of any potential flood. One method they decide to use is to assume that the height $y$ in feet of the river above flood stage is given by a formula of the form $y = ax^2 + bx + c$. Here $x$ measures the time in hours starting at 9:00 A.M. A negative value of $y$ corresponds to the river's being below flood stage and a positive value to its being above flood stage. If these assumptions are correct, what values should $a$, $b$, and $c$ have? When will the river reach flood stage? When will it crest, and how high will it be above flood stage then?

**1** Subtract 5 times the second equation from the first equation, and then subtract the second from the third to arrive at the system

$$\begin{cases} 19x - 11z = -17 \\ 5x - 6z = -20. \end{cases}$$

Any of the methods developed earlier in the chapter may now be used to solve this new system; let us choose elimination of variables. We subtract 19 times the second equation from 5 times the first equation to obtain an equation in $z$ alone:

$$95x - 55z = -85$$

$$\underline{95x - 114z = -380}$$

$$-59z = -295$$

$$z = 5.$$

Hence

$$5x - 6z = -20$$

$$5x - 6(5) = -20$$

$$5x - 30 = -20$$

$$5x = 10$$

$$x = 2.$$

Substituting $x = 2$, $z = 5$ in the second equation of the original system, we see that $y = 1 + 3x - 2z = 1 + 6 - 10 = -3$. The solution set is consequently $\{(2, -3, 5)\}$.

*Check:* $4(2) + 5(-3) - 5 = 8 - 15 - 5 = -12$ and $-3(2) + (-3) + 2(5) = -6 - 3 + 10 = 1$; $2(2) + (-3) - 4(5) = 4 - 3 - 20 = -19$

**9** Let us solve the first equation for $z$ and then substitute this value in the remaining equations. We then have

$$2z = 1 - 4x + 3y$$

$$z = \frac{1}{2} - 2x + \frac{3}{2}y.$$

Carrying out the substitution produces

$$\begin{cases} 9x - 2y + \dfrac{3}{2} - 6x + \dfrac{9}{2}y = 2 \\[2mm] 3x - 7y + \dfrac{3}{2} - 6x + \dfrac{9}{2}y = 1 \end{cases}$$

which is equivalent to

$$\begin{cases} 3x + \dfrac{5}{2}y = \dfrac{1}{2} \\[2mm] -3x - \dfrac{5}{2}y = -\dfrac{1}{2}. \end{cases}$$

This shows that we have a dependent system. The solution set is infinite.

**15**  Let $x$ denote the number of quarters, $y$ the number of dimes, and $z$ the number of nickels. The fact that there are 23 coins may be expressed by the equation $x + y + z = 23$; that their value is \$3.40 becomes $25x + 10y + 5z = 340$; and finally, that there are twice as many quarters as dimes is $x = 2y$. We must therefore solve this system

$$\begin{cases} x + y + z = 23 \\ 25x + 10y + 5z = 340 \\ x = 2y. \end{cases}$$

Substituting $x = 2y$ in the first two equations, we see that

$$\begin{cases} 3y + z = 23 \\ 60y + 5z = 340 \end{cases}$$

which becomes

$$\begin{cases} 3y + z = 23 \\ 12y + z = 68 \end{cases}$$

when we take one-fifth of the second equation. Thus $9y = 45$, $y = 5$, $z = 23 - 15 = 8$, and $x = 2y = 10$. There must be 10 quarters, 5 dimes, and 8 nickels.

*Check:* \$2.50 + \$.50 + \$.40 = \$3.40

## (9.4) (Systems in three variables—solution by determinants)

Our goal in this section is to extend Cramer's rule to systems of three linear equations in three variables. We first define the determinant of a square array

$$
\begin{array}{ccc}
a_1 & b_1 & c_1 \\
a_2 & b_2 & c_2 \\
a_3 & b_3 & c_3
\end{array}
$$

of real numbers. Such a determinant is called a *third-order determinant* because it is defined on an array of real numbers containing three rows and three columns. The determinant of the array above is the real number given by the rule

$$
\begin{vmatrix}
a_1 & b_1 & c_1 \\
a_2 & b_2 & c_2 \\
a_3 & b_3 & c_3
\end{vmatrix} = a_1 b_2 c_3 + b_1 c_2 a_3 + c_1 a_2 b_3 - a_3 b_2 c_1 - b_3 c_2 a_1 - c_3 a_2 b_1.
$$

There is an easy pictorial way to keep track of this rule. Rewrite the first two columns of the determinant, draw in the indicated arrows, and then multiply as shown by the arrows, prefixing each product by the sign indicated at the tip of the arrow:

$$
= a_1 b_2 c_3 + b_1 c_2 a_3 + c_1 a_2 b_3 - a_3 b_2 c_1 - b_3 c_2 a_1 - c_3 a_2 b_1.
$$

We illustrate by computing the determinants

$$
\begin{vmatrix}
1 & 2 & 3 \\
4 & 5 & 6 \\
7 & 8 & 9
\end{vmatrix} \quad \text{and} \quad
\begin{vmatrix}
-1 & 4 & 3 \\
0 & 1 & 3 \\
2 & -1 & 0
\end{vmatrix}
$$

for which we have

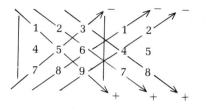

$$= (1)(5)(9) + (2)(6)(7) + (3)(4)(8) - (7)(5)(3) - (8)(6)(1) - (9)(4)(2)$$
$$= 45 + 84 + 96 - 105 - 48 - 72 = 0;$$

$$= (-1)(1)(0) + (4)(3)(2) + (3)(0)(-1) - (2)(1)(3) - (-1)(3)(-1) - (0)(0)(3)$$
$$= 0 + 24 + 0 - 6 - 3 - 0 = 15.$$

Now let's consider the system

$$\begin{cases} a_1 x + b_1 y + c_1 z = d_1 \\ a_2 x + b_2 y + c_2 z = d_2 \\ a_3 x + b_3 y + c_3 z = d_3 \end{cases}$$

of three linear equations in the variables $x$, $y$, and $z$. The statement of Cramer's rule is entirely analogous to the two-variable case and can even be proved in much the same way. It's a tedious procedure, however, and not particularly illuminating; so we'll omit it. Here, then, is the statement of Cramer's rule for the preceding system.

If $D = \begin{vmatrix} a_1 & b_1 & c_1 \\ a_2 & b_2 & c_2 \\ a_3 & b_3 & c_3 \end{vmatrix} \neq 0$, then the system has a unique solution given by the formula

$$x = \frac{\begin{vmatrix} d_1 & b_1 & c_1 \\ d_2 & b_2 & c_2 \\ d_3 & b_3 & c_3 \end{vmatrix}}{D} \quad y = \frac{\begin{vmatrix} a_1 & d_1 & c_1 \\ a_2 & d_2 & c_2 \\ a_3 & d_3 & c_3 \end{vmatrix}}{D} \quad z = \frac{\begin{vmatrix} a_1 & b_1 & d_1 \\ a_2 & b_2 & d_2 \\ a_3 & b_3 & d_3 \end{vmatrix}}{D}.$$

You should notice that the numerator for $x$ is the determinant of the array formed by replacing the first column of $D$ by $d_1$, $d_2$, $d_3$. Similarly, if the second column of $D$ is replaced by $d_1$, $d_2$, $d_3$, the resulting determinant is the numerator for $y$, and if the third is, the resulting determinant is the numerator for $z$.

We illustrate the use of Cramer's rule by considering the system

$$\begin{cases} x - 3z = -11 \\ 2x + 5y = 16 \\ x - y + 2z = 0. \end{cases}$$

In this case

$$D = \begin{vmatrix} 1 & 0 & -3 \\ 2 & 5 & 0 \\ 1 & -1 & 2 \end{vmatrix} \begin{matrix} 1 & 0 \\ 2 & 5 \\ 1 & -1 \end{matrix}$$

$$= (1)(5)(2) + (0)(0)(1) + (-3)(2)(-1) - (1)(5)(-3) - (-1)(0)(1) - (2)(2)(0)$$

$$= 10 + 0 + 6 - (-15) - 0 - 0 = 10 + 6 + 15 = 31.$$

Letting $N_x$, $N_y$, $N_z$ denote the numerators for $x$, $y$, $z$, we have

$$N_x = \begin{vmatrix} -11 & 0 & -3 \\ 16 & 5 & 0 \\ 0 & -1 & 2 \end{vmatrix} \begin{matrix} -11 & 0 \\ 16 & 5 \\ 0 & -1 \end{matrix}$$

$$= (-11)(5)(2) + (0)(0)(0) + (-3)(16)(-1)$$

$$- (0)(5)(-3) - (-1)(0)(-11) - (2)(16)(0)$$

$$= -110 + 0 + 48 - 0 - 0 - 0 = -62;$$

$$N_y = \begin{vmatrix} 1 & -11 & -3 \\ 2 & 16 & 0 \\ 1 & 0 & 2 \end{vmatrix} \begin{matrix} 1 & -11 \\ 2 & 16 \\ 1 & 0 \end{matrix}$$

$$= (1)(16)(2) + (-11)(0)(1) + (-3)(2)(0)$$

$$- (1)(16)(-3) - (0)(0)(1) - (2)(2)(-11)$$

$$= 32 - 0 - 0 - (-48) - 0 - (-44) = 32 + 48 + 44 = 124;$$

$$= (1)(5)(0) + (0)(16)(1) + (-11)(2)(-1)$$

$$- (1)(5)(-11) - (-1)(16)(1) - (0)(2)(0)$$

$$= 0 + 0 + 22 - (-55) - (-16) - 0 = 22 + 55 + 16 = 93.$$

Hence $x = N_x/D = -62/31 = -2$, $y = N_y/D = 124/31 = 4$, and $z = N_z/D = 93/31 = 3$. The solution set is $\{(-2, 4, 3)\}$.

*Check:* $-2 - 3(3) = -2 - 9 = -11$; $2(-2) + 5(4) = -4 + 20 = 16$; $-2 - 4 + 2(3) = -2 - 4 + 6 = 0$.

In case the determinant $D$ of the array formed by the coefficients of the variables of a system of three linear equations in three variables is zero, the system must be either dependent or inconsistent. Though we omit the proof, the details are summarized following. Given the system

$$\begin{cases} a_1 x + b_1 y + c_1 z = d_1 \\ a_2 x + b_2 y + c_2 z = d_2 \\ a_3 x + b_3 y + c_3 z = d_3 \end{cases}$$

if $D = \begin{vmatrix} a_1 & b_1 & c_1 \\ a_2 & b_2 & c_2 \\ a_3 & b_3 & c_3 \end{vmatrix} \neq 0$, then Cramer's rule may be used to solve for $x$, $y$, and $z$. If $D = 0$, there are two possibilities.

**(a)** If $\begin{vmatrix} d_1 & b_1 & c_1 \\ d_2 & b_2 & c_2 \\ d_3 & b_3 & c_3 \end{vmatrix} \neq 0$, $\begin{vmatrix} a_1 & d_1 & c_1 \\ a_2 & d_2 & c_2 \\ a_3 & d_3 & c_3 \end{vmatrix} \neq 0$, or $\begin{vmatrix} a_1 & b_1 & d_1 \\ a_2 & b_2 & d_2 \\ a_3 & b_3 & d_3 \end{vmatrix} \neq 0$, the system is inconsistent.

**(b)** If all three determinants listed above are zero, the system is dependent.

For example, the system

$$\begin{cases} 3x - 2y + 3z = 5 \\ x + 3y = 2 \\ 2x - 5y + 3z = 4 \end{cases}$$

is inconsistent since $D = \begin{vmatrix} 3 & -2 & 3 \\ 1 & 3 & 0 \\ 2 & -5 & 3 \end{vmatrix} = 27 - 0 - 15 - 18 - 0 -$

$(-6) = 0$, whereas $\begin{vmatrix} 5 & -2 & 3 \\ 2 & 3 & 0 \\ 4 & -5 & 3 \end{vmatrix} = 45 - 0 - 30 - 36 - 0 - (-12)$

$= -9 \neq 0$.

## Exercise set 9.4

In problems 1 to 6, evaluate each determinant.

(○)  1  $\begin{vmatrix} 1 & 3 & 0 \\ 2 & 1 & -1 \\ 0 & 1 & 5 \end{vmatrix}$      2  $\begin{vmatrix} 0 & 1 & -4 \\ 2 & 1 & -3 \\ 0 & 4 & -2 \end{vmatrix}$

3  $\begin{vmatrix} 1 & -3 & 5 \\ 2 & 3 & 1 \\ 2 & -6 & 10 \end{vmatrix}$      4  $\begin{vmatrix} 2 & 0 & 1 \\ -4 & 0 & 2 \\ 1 & 0 & 3 \end{vmatrix}$

5  $\begin{vmatrix} 1 & 0 & -2 \\ 0 & 5 & 2 \\ 2 & 5 & 2 \end{vmatrix}$      6  $\begin{vmatrix} 2 & 3 & -1 \\ 0 & 1 & 2 \\ 4 & 6 & -2 \end{vmatrix}$

Use Cramer's rule to solve each of the following systems of equations.

(○)  7  $\begin{cases} x + z = 6 \\ x + y = -2 \\ y + z = 2 \end{cases}$      8  $\begin{cases} x + y + z = 4 \\ y + z = 2 \\ x - z = 1 \end{cases}$

9  $\begin{cases} x + y - 2z = 5 \\ x + 3y - z = 11 \\ x + 2z = 2 \end{cases}$      10  $\begin{cases} 2x - y = -7 \\ 3x + y + z = -15 \\ -x + 2y + z = 0 \end{cases}$

11  $\begin{cases} 3x + z = 0 \\ 4x + y - z = 4 \\ 2x - y + 2z = -1 \end{cases}$      12  $\begin{cases} x + y + z = 1 \\ 3y + 4z = 3 \\ x + y - z = 0 \end{cases}$

Classify the following systems according to whether they are dependent or inconsistent or have a solution set that contains a single point. In the last case, use Cramer's rule to determine the solution set.

$(\circ)$    13    $\begin{cases} x + 3z = 5 \\ 2x + y + 2z = 7 \\ 4x + 3y = 11 \end{cases}$      14    $\begin{cases} 2x - y - z = 1 \\ x + y + 2z = 0 \\ 5x - y = 2 \end{cases}$

15    $\begin{cases} 2x + y - 4z = 2 \\ 4x + y + 4z = 9 \\ 3x + y = 3 \end{cases}$      16    $\begin{cases} 5x - y + z = 0 \\ 2x + 3z = 4 \\ x + 3y - 6z = 5 \end{cases}$

17    $\begin{cases} x + 2y = 3 \\ x - 3z = 0 \\ x + y + 2z = 1 \end{cases}$      18    $\begin{cases} 3x + 2z = 0 \\ 3y + z = 1 \\ x + y + z = 1 \end{cases}$

19    $\begin{cases} x + y - 4z = -5 \\ 2x - y + z = 2 \\ x + y - 2z = -2 \end{cases}$      20    $\begin{cases} 2x - y + z = -2 \\ 4x - y - z = 0 \\ -x + 2y + z = 3 \end{cases}$

$(\circ)$

*Sample solutions*
*for exercise set 9.4*

1

$$\begin{vmatrix} 1 & 3 & 0 \\ 2 & 1 & -1 \\ 0 & 1 & 5 \end{vmatrix}\begin{matrix} 1 & 3 \\ 2 & 1 \\ 0 & 1 \end{matrix}$$

$= (1)(1)(5) + (3)(-1)(0) + (0)(2)(1) - (0)(1)(0) - (1)(-1)(1) - (5)(2)(3)$

$= 5 + 0 + 0 - 10 - (-1) - 30 = 5 + 1 - 30 = -24$

7    We begin by evaluating the determinant $D$ of the array formed by the coefficients of the variables:

$$D = \begin{vmatrix} 1 & 0 & 1 \\ 1 & 1 & 0 \\ 0 & 1 & 1 \end{vmatrix}\begin{matrix} 1 & 0 \\ 1 & 1 \\ 0 & 1 \end{matrix}$$

$= (1)(1)(1) + (0)(0)(0) + (1)(1)(1) - (0)(1)(1) - (1)(0)(1) - (1)(1)(0)$

$= 2.$

Using the fact that

$$\begin{vmatrix} 6 & 0 & 1 \\ -2 & 1 & 0 \\ 2 & 1 & 1 \end{vmatrix} = 6 + 0 - 2 - 2 - 0 - 0 = 2,$$

$$\begin{vmatrix} 1 & 6 & 1 \\ 1 & -2 & 0 \\ 0 & 2 & 1 \end{vmatrix} = -2 + 0 + 2 - 0 - 0 - 6 = -6,$$

**351 ($\circ$) Systems in three variables—solution by determinants**

$$\begin{vmatrix} 1 & 0 & 6 \\ 1 & 1 & -2 \\ 0 & 1 & 2 \end{vmatrix} = 2 + 0 + 6 - 0 - (-2) - 0 = 10,$$

we see that $x = 2/2 = 1$, $y = -6/2 = -3$, and $z = 10/2 = 5$; so the solution set is $\{(1, -3, 5)\}$.

*Check:* $1 + 5 = 6$; $1 - 3 = -2$; and $-3 + 5 = 2$.

13  The system is dependent.

$$D = \begin{vmatrix} 1 & 0 & 3 \\ 2 & 1 & 2 \\ 4 & 3 & 0 \end{vmatrix} = 0 + 0 + 18 - 12 - 6 - 0 = 0$$

$$\begin{vmatrix} 5 & 0 & 3 \\ 7 & 1 & 2 \\ 11 & 3 & 0 \end{vmatrix} = 0 + 0 + 63 - 33 - 30 - 0 = 0$$

$$\begin{vmatrix} 1 & 5 & 3 \\ 2 & 7 & 2 \\ 4 & 11 & 0 \end{vmatrix} = 0 + 40 + 66 - 84 - 22 - 0 = 0$$

$$\begin{vmatrix} 1 & 0 & 5 \\ 2 & 1 & 7 \\ 4 & 3 & 11 \end{vmatrix} = 11 + 0 + 30 - 20 - 21 - 0 = 0.$$

# (9.5) (Systems involving second-degree equations)

Up to this point we have been discussing systems of linear equations. In this final section of the chapter we briefly consider systems of two equations in two variables in which at least one of the equations is quadratic. Two basic methods will be used to solve such a system: substitution, and an analogue of elimination of variables. If it is possible to solve either equation for one variable in terms of the other, then the method of substitution may be used to solve the system. The idea is to substitute the expression obtained for the given variable in the remaining equation, thereby obtaining an equation in a single variable. Hopefully, this new equation can be solved! We illustrate the technique by considering a few examples.

For our first example, consider the system

(1) $\begin{cases} x^2 + 4y^2 = 4 \\ \sqrt{3}x - 2y = 2 \end{cases}$

Since the second equation of this system is linear, it can be solved for $y$ in terms of $x$:

$$2y = \sqrt{3}x - 2$$

$$y = \frac{\sqrt{3}}{2}x - 1.$$

The next step is to replace $y$ with $(\sqrt{3}/2)x - 1$ in the first equation of (1) in order to obtain an equation in a single unknown:

(2) $x^2 + 4\left(\frac{\sqrt{3}}{2}x - 1\right)^2 = 4.$

By its construction, any solution of system (1) must also be a solution of (2); so we now simplify (2) and solve it:

$$x^2 + 4\left(\frac{3}{4}x^2 - \sqrt{3}x + 1\right) = 4$$

$$x^2 + 3x^2 - 4\sqrt{3}x + 4 = 4$$

$$4x^2 - 4\sqrt{3}x = 0$$

$$4x(x - \sqrt{3}) = 0$$

$$x = 0 \text{ or } x = \sqrt{3}.$$

The equation $y = (\sqrt{3}/2)x - 1$ may now be used to find the corresponding values of $y$. When $x = 0$, $y = (\sqrt{3}/2) \cdot 0 - 1 = -1$, and when $x = \sqrt{3}$, $y = (\sqrt{3}/2)\sqrt{3} - 1 = (3/2) - 1 = 1/2$. This shows that the solution set of system (1) is contained in $\{(0, -1), (\sqrt{3}, 1/2)\}$. The final step is to check these solutions by actually substituting them back into the equations of (1).

*Check:* When $x = 0$, $y = -1$, we have $0^2 + 4(-1)^2 = 0 + 4 = 4$ and $\sqrt{3} \cdot 0 - 2(-1) = 0 + 2 = 2$. With $x = \sqrt{3}$, $y = 1/2$ we may write $(\sqrt{3})^2 + 4(1/2)^2 = 3 + 1 = 4$ and $\sqrt{3}\sqrt{3} - 2(1/2) = 3 - 1 = 2$. We have now established that the solution set of (1) is $\{(0, -1), (\sqrt{3}, 1/2)\}$. The graphs of the equations in system (1) are displayed in Figure 9.4.

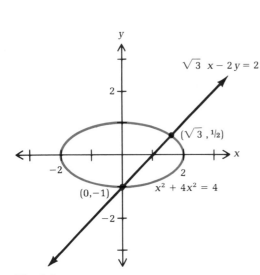

Fig. 9.4

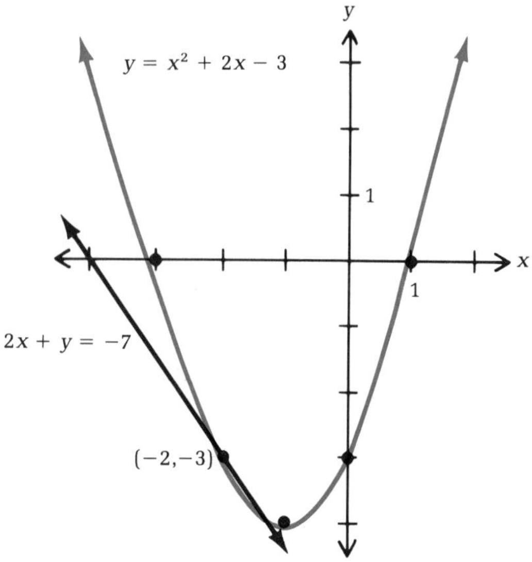

Fig. 9.5

Next we consider the system

$$(3) \quad \begin{cases} y = x^2 + 2x - 3 \\ 2x + y = -7. \end{cases}$$

From the second equation, we have $y = -2x - 7$. Substituting this expression for $y$ in the first equation produces $-2x - 7 = x^2 + 2x - 3$,

$$x^2 + 4x + 4 = 0$$

$$(x + 2)^2 = 0$$

$$x = -2.$$

When $x = -2$, $y = -2x - 7 = -2(-2) - 7 = 4 - 7 = -3$. Noting that $(-2)^2 + 2(-2) - 3 = 4 - 4 - 3 = -3$ and $2(-2) + (-3) = -4 - 3 = -7$, we deduce that the solution set of system (3) is $\{(-2, -3)\}$. See Figure 9.5 for graphs of the equations in system (3).

Our next example is the system

$$(4) \quad \begin{cases} y^2 - 4x^2 = 4 \\ y = x + 1. \end{cases}$$

The method of substitution produces the equation $(x + 1)^2 - 4x^2 = 4$; so

$$x^2 + 2x + 1 - 4x^2 = 4$$

$$3x^2 - 2x + 3 = 0$$

$$x = \frac{2 \pm \sqrt{4 - 36}}{2 \cdot 3} = \frac{2 \pm \sqrt{-32}}{6} = \frac{1 \pm 2i\sqrt{2}}{3}.$$

When $x = \dfrac{1 + 2i\sqrt{2}}{3}$, $y = x + 1 = \dfrac{4 + 2i\sqrt{2}}{3}$ and when $x = \dfrac{1 - 2i\sqrt{2}}{3}$, $y = \dfrac{4 - 2i\sqrt{2}}{3}$. Noting that $\left(\dfrac{4 \pm 2i\sqrt{2}}{3}\right)^2 -$

$$4\left(\frac{1 \pm 2i\sqrt{2}}{3}\right)^2 = \frac{8 \pm 16i\sqrt{2}}{9} - 4\left(\frac{-7 \pm 4i\sqrt{2}}{9}\right) = \frac{8}{9} - \frac{-28}{9}$$

$$= \frac{36}{9} = 4,$$ we conclude that the solution set of system (4) is

$$\left\{ \left(\frac{1 + 2i\sqrt{2}}{3}, \frac{4 + 2i\sqrt{2}}{3}\right), \left(\frac{1 - 2i\sqrt{2}}{3}, \frac{4 - 2i\sqrt{2}}{3}\right) \right\}.$$ Points

in the plane have ordered pairs of real numbers for their coordinates. The fact that the solution set of (4) consists of ordered pairs of nonreal complex numbers is an algebraic expression of the fact that the graphs of the two equations in the system do not intersect (see Figure 9.6).

When faced with a system of two quadratic equations in two variables, an appropriate method of solution is an adaptation of the method of elimination of variables. The idea is to observe that if $E_1$ and $E_2$ are the equations, then any solution of the system formed by $E_1$ and $E_2$ is also a solution of the equation

**(5)** $k_1 E_1 + k_2 E_2 \quad (k_1, k_2 \in R).$

One chooses $k_1$, $k_2$ so that equation (5) contains a single variable. To see how this is done, consider the system

**(6)** $\begin{cases} x^2 + y^2 = 5 \\ y^2 - 2x^2 = 2. \end{cases}$

This particular system can be solved by the method of substitution, but our purpose is to illustrate elimination of variables. If the second equation in system (6) is subtracted from the first equation, the result is

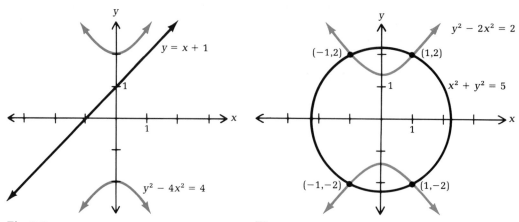

Fig. 9.6    Fig. 9.7

**(7)** $3x^2 = 3$.

Any solution of (6) is a solution of (7); so any solution of (6) must be a solution of the new system

**(8)** $\begin{cases} x^2 + y^2 = 5 \\ 3x^2 = 3. \end{cases}$

But (8) can easily be solved by observing that $x = \pm 1, (\pm 1)^2 + y^2 = 5$, $1 + y^2 = 5$, $y^2 = 4$; so $y = \pm 2$. It follows that the solution set of system (6) is contained in $\{(1, 2), (1, -2), (-1, 2), (-1, -2)\}$. Noting that $(\pm 2)^2 - 2(\pm 1)^2 = 4 - 2 = 2$, we deduce that this actually is the solution set of (6). The system is pictured in Figure 9.7.

Finally, we consider the system

**(9)** $\begin{cases} 3x^2 + 28y^2 = 75 \\ 6x^2 + 56y^2 - 8y - 4x = 130. \end{cases}$

The method of substitution would produce a disaster if you tried it on this system. Notice, though, that if twice the first equation is subtracted from the second equation, the resulting equation is $-8y - 4x = -20$. Any solution of the original system is, therefore, a solution of the new system

**(10)** $\begin{cases} 3x^2 + 28y^2 = 75 \\ -8y - 4x = -20. \end{cases}$

But (10) can be solved by substitution. Solving for $y$ in the equation $-8y - 4x = -20$ shows that $y = (20 - 4x)/8 = (5 - x)/2$. Replacing $y$ with $(5 - x)/2$ in the equation $3x^2 + 28y^2 = 75$, we have

$$3x^2 + 28\left(\frac{5 - x}{2}\right)^2 = 75$$

$$3x^2 + \frac{28}{4}(x^2 - 10x + 25) = 75$$

$$3x^2 + 7x^2 - 70x + 175 = 75$$

$$10x^2 - 70x + 100 = 0$$

$$x^2 - 7x + 10 = 0$$

$$(x - 2)(x - 5) = 0$$

$$x = 2 \text{ or } x = 5.$$

When $x = 2$, $y = (5 - x)/2 = (5 - 2)/3 = 3/2$, and when $x = $ , $y = (5 - 5)/2 = 0$. Noting that

$$3(2)^2 + 28\left(\frac{3}{2}\right)^2 = 3 \cdot 4 + \frac{28 \cdot 9}{4} = 12 + 63 = 75,$$

$$6(2)^2 + 56\left(\frac{3}{2}\right)^2 - 8\left(\frac{3}{2}\right) - 4(2) = 6 \cdot 4 + \frac{56 \cdot 9}{4} - \frac{24}{2} - 8$$

$$= 24 + 126 - 12 - 8 = 130,$$

$$3(5)^2 + 28(0)^2 = 3 \cdot 25 = 75,$$

$$6(5)^2 + 56(0)^2 - 8(0) - 4(5) = 6 \cdot 25 - 20 = 150 - 20 = 130,$$

we conclude that the solution set of (9) is $\{(2, 3/2), (5, 0)\}$.

In summary, then, if you are given a system of second-degree equations in two variables, you may replace either equation with the result of adding a multiple of that equation to a multiple of the other equation. The solution set of the new system contains all solutions of the original system and, hopefully, is easier to solve.

## Exercise set 9.5

*Find the solution set of each of the following systems of equations. In problems 1 to 10, sketch the graphs of the equations in each system.*

( ∘ ) **1** $\begin{cases} y = x^2 - 2x + 3 \\ y + 2x = 4 \end{cases}$     **2** $\begin{cases} y = x^2 - 5x - 4 \\ y + x - 1 = 0 \end{cases}$

**3** $\begin{cases} x^2 + y^2 = 10 \\ y + 2x = 5 \end{cases}$     **4** $\begin{cases} x^2 + y^2 = 25 \\ y - 7x + 25 = 0 \end{cases}$

**5** $\begin{cases} \dfrac{x^2}{4} + \dfrac{y^2}{9} = 1 \\ \dfrac{x}{2} + \dfrac{y}{3} = 1 \end{cases}$     **6** $\begin{cases} x^2 + 9y^2 = 18 \\ x = 3y \end{cases}$

**7** $\begin{cases} y^2 - 4x^2 = 12 \\ y + 2x = 6 \end{cases}$     **8** $\begin{cases} 3y^2 - 5x^2 = 7 \\ 6y - 5x = 7 \end{cases}$

**9** $\begin{cases} y = x^2 + 4x + 4 \\ 2x - y = 1 \end{cases}$     **10** $\begin{cases} x^2 + y^2 = 16 \\ x - y = 6 \end{cases}$

( ∘ ) **11** $\begin{cases} x^2 + y^2 = 10 \\ y^2 - x^2 = 8 \end{cases}$     **12** $\begin{cases} x^2 + y^2 = 17 \\ 2x^2 - y^2 = 31 \end{cases}$

**13** $\begin{cases} 2x^2 - y^2 = 9 \\ x^2 - 3y^2 = 7 \end{cases}$     **14** $\begin{cases} y^2 - 3x^2 = 4 \\ y^2 - 2x^2 = 5 \end{cases}$

**15** $\begin{cases} 4x^2 - y^2 = 5 \\ 2y^2 - x^2 = 4 \end{cases}$     **16** $\begin{cases} 2x^2 - 3y^2 = 7 \\ 3x^2 - y^2 = 14 \end{cases}$

( ∘ ) **17** $\begin{cases} 3x^2 + 3y^2 - 4y - 2x = 20 \\ x^2 + y^2 = 10 \end{cases}$     **18** $\begin{cases} 2x^2 + 2y^2 - 2x + 14y = 0 \\ x^2 + y^2 = 25 \end{cases}$

**19** $\begin{cases} 2x^2 - 8y^2 - 6y - 3x = 6 \\ x^2 - 4y^2 = 12 \end{cases}$     **20** $\begin{cases} 3x^2 - 5y^2 + 8x - 8y = 3 \\ 10y^2 - 6x^2 = 10 \end{cases}$

Substitution can also be used for a system like

$$\begin{cases} x^2 + 4y^2 = 5 \\ xy = 1. \end{cases}$$

Even though both equations are of the second degree, the second equation can be solved for $y$ to obtain $y = 1/x$. Now replace $y$ with $1/x$ in the first equation and solve for $x$:

$$x^2 + \frac{4}{x^2} = 5$$

$$x^4 + 4 = 5x^2$$

$$x^4 - 5x^2 + 4 = 0$$

$$(x^2 - 4)(x^2 - 1) = 0$$

$$(x + 2)(x - 2)(x + 1)(x - 1) = 0$$

$$x = 2, -2, 1, \text{ or } -1.$$

Since $y = 1/x$, the solution set of the system is contained in $\{(2, 1/2), (-2, -1/2), (1, 1), (-1, -1)\}$. Since $(\pm 2)^2 + 4(\pm 1/2)^2 = 4 + 4\cdot(1/4) = 4 + 1 = 5$, $(\pm 1)^2 + 4(\pm 1)^2 = 1 + 4 = 5$, this must be the solution set.

*Use the foregoing technique where needed to solve the following systems of equations.*

21 $\begin{cases} x^2 - y^2 = 8 \\ xy = 3 \end{cases}$
22 $\begin{cases} x^2 + 16y^2 = 20 \\ xy = 2 \end{cases}$

23 $\begin{cases} 4x^2 + 16y^2 = 8 \\ y = \dfrac{1}{2x} \end{cases}$
24 $\begin{cases} 3x^2 - 4y^2 = 11 \\ xy = 1 \end{cases}$

$\left(\circ\right)$ 25 $\begin{cases} 2y^2 - x^2 = 1 \\ 6y^2 - 2xy - 3x^2 = 1 \end{cases}$
26 $\begin{cases} 6x^2 - y^2 = 2 \\ 2y^2 + 3xy - 12x^2 = 2 \end{cases}$

27 $\begin{cases} x^2 + 4y^2 = 13 \\ 2x^2 - 8xy + 8y^2 = 2 \end{cases}$
28 $\begin{cases} 2x^2 + y^2 = 12 \\ 8x^2 - 5xy + 4y^2 = 28 \end{cases}$

The real number system was enlarged to the system of complex numbers in order to provide negative numbers with a square root. As a matter of fact, the techniques of this section can be used to find a square root for *any* complex number. To illustrate this, we find a complex number $x + yi$ whose square is $-3 - 4i$. Since $(x + yi)^2 = x^2 - y^2 + 2xyi = -3 - 4i$, this amounts to finding real solutions for the system of equations

$$\begin{cases} x^2 - y^2 = -3 \\ 2xy = -4. \end{cases}$$

From the second equation, $y = -(2/x)$; so we have

$$x^2 - \left(-\frac{2}{x}\right)^2 = -3$$

$$x^2 - \frac{4}{x^2} = -3$$

$$x^4 - 4 = -3x^2$$

$$x^4 + 3x^2 - 4 = 0$$

$$(x^2 + 4)(x^2 - 1) = 0$$

$$(x^2 + 4)(x + 1)(x - 1) = 0$$

$$x = 1, \text{ or } x = -1.$$

(Why are the roots $x = \pm 2i$ discarded?) Since $y = -(2/x)$, we must have $y = -2$ when $x = 1$ and $y = 2$ when $x = -1$. Either of the numbers

$1 - 2i$ or $-1 + 2i$ has the property we seek.

*Check:* $(1 - 2i)^2 = 1 - 4i + 4i^2 = -3 - 4i$

*In problems 29 to 32, find a complex number whose square is the given complex number.*

**29**  $-8 + 6i$                    **30**  $15 - 8i$

**31**  $5 - 12i$                    **32**  $-7 + 24i$

*Solve each of the remaining problems by introducing an appropriate system of two equations in two variables.*

$(\circ)$  **33**  The hypotenuse of a right triangle is 13 inches, and its area is 30 square inches. What are the lengths of its remaining sides?

**34**  The diagonal of a rectangle is 5 feet and its area is 12 square feet. What are its dimensions?

**35**  The sum of the reciprocals of two numbers is $4/15$ and the sum of the numbers is 16. What are the numbers?

**36**  The sum of the reciprocals of two numbers is $5/24$ and the product of the numbers is 96. Find the numbers.

**37**  The area $A$ of a trapezoid with parallel sides $a$, $b$ and altitude $h$ is given by $A = (h/2)(a + b)$. One parallel side of a trapezoid is 2 inches longer than the other, and its altitude is the sum of the lengths of the two parallel sides. If its area is 50 square inches, find the lengths of its parallel sides.

**38**  The sum of the squares of the digits of a certain two-digit number is 40. Interchanging the digits will increase the number by 36. What is the number?

**39**  A group of hikers is going to climb Tame Mountain. It is a 100-mile trip from their lodge to the mountain and then 16 miles up a rather easy trail to the summit. It takes them a total of 10 hours to drive to the mountain and climb the trail. They are able to descend from Tame Mountain three times faster than they climbed, but they drive 10 mph slower on the way back to the lodge. The return trip took a total of 5 hours and 10 minutes. What was the average speed of their car on the way to the mountain and their average climbing speed?

**40**  A man invested a total of $14,000, some of it in Fund A and some in Fund B. That year, his return from Fund A was $480 and from Fund B $240. At the end of the year, he sold a total of $2,000 of his investments. The dividend from Fund A was down 1 percent the following year, and his return was $350. The dividend rate from Fund B doubled that same year, and his return was $400. Determine the rate of return of both funds in the year of purchase.

1  Solve the second equation for $y$ to obtain $y = -2x + 4$. Replace $y$ with $-2x + 4$ in the first equation and solve for $x$.

$$-2x + 4 = x^2 - 2x + 3$$
$$0 = x^2 - 1$$
$$x = 1, -1.$$

When $x = 1$, $y = -2 + 4 = 2$; when $x = -1$, $y = -2(-1) + 4 = 6$. Noting that $1^2 - 2(1) + 3 = 1 - 2 + 3 = 2$, $(-1)^2 - 2(-1) + 3 = 1 + 2 + 3 = 6$, it follows that the solution set of the system is $\{(1, 2), (-1, 6)\}$. Here is a sketch of the graphs of the equations in the system.

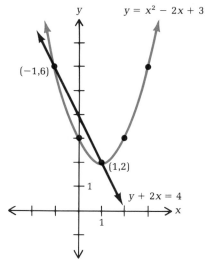

11     $\quad x^2 + y^2 = 10 \qquad\qquad\qquad x^2 + y^2 = 10$

$$\underline{-x^2 + y^2 = 8} \qquad\qquad\quad \underline{-x^2 + y^2 = 8}$$
$$2y^2 = 18 \qquad\qquad\quad\; 2x^2 \qquad = 2$$
$$y^2 = 9 \qquad\qquad\qquad\quad x^2 = 1$$
$$y = \pm 3 \qquad\qquad\qquad\quad x = \pm 1.$$

$$(\pm 1)^2 + (\pm 3)^2 = 1 + 9 = 10$$

$$-(\pm 1)^2 + (\pm 3)^2 = -1 + 9 = 8$$

The solution set is $\{(1, 3), (1, -3), (-1, 3), (-1, -3)\}$.

**17**  Subtract 3 times the second equation from the first equation.

$$3x^2 + 3y^2 - 4y - 2x = 20$$

$$\underline{3x^2 + 3y^2 \qquad\qquad = 30}$$

$$-4y - 2x\; = -10$$

Form the new system

$$\begin{cases} x^2 + y^2 = 10 \\ -4y - 2x = -10 \end{cases}$$

whose solution set contains all solutions of the original system. Solve for $x$ in the second equation of the new system, obtaining $x = 5 - 2y$. Replace $x$ with $5 - 2y$ in the first equation, and solve for $y$.

$$(5 - 2y)^2 + y^2 \;=\; 10$$

$$4y^2 - 20y + 25 + y^2 \;=\; 10$$

$$5y^2 - 20y + 25 \;=\; 10$$

$$5y^2 - 20y + 15 \;=\; 0$$

$$y^2 - 4y + 3 \;=\; 0$$

$$(y - 3)(y - 1) \;=\; 0$$

$$y \;=\; 3, \text{ or } y = 1.$$

When $y = 3$, $x = 5 - 6 = -1$; when $y = 1$, $x = 5 - 2 = 3$. This shows the solution set is contained in $\{(-1, 3), (3, 1)\}$. Noting that $3(-1)^2 + 3(3)^2 - 4(3) - 2(-1) = 3 + 27 - 12 + 2 = 20$, $3(3)^2 + 3(1)^2 - 4(1) - 2(3) = 27 + 3 - 4 - 6 = 20$, and $(-1)^2 + (3)^2 = 10 = (3)^2 + (1)^2$, we deduce that the solution set is, indeed, $\{(-1, 3), (3, 1)\}$.

**25**  Subtract 3 times the first equation from the second equation.

$$6y^2 - 2xy - 3x^2 = 1$$

$$\underline{6y^2 \qquad\quad - 3x^2 = 3}$$

$$-2xy \qquad\quad = -2$$

$$xy = 1$$

Form the new system

$$\begin{cases} 2y^2 - x^2 = 1 \\ xy = 1. \end{cases}$$

Solve by substitution:

$$2\left(\frac{1}{x}\right)^2 - x^2 = 1$$

$$2 - x^4 = x^2$$

$$x^4 + x^2 - 2 = 0$$

$$(x^2 + 2)(x^2 - 1) = 0$$

$$x = \pm i\sqrt{2} \text{ or } \pm 1.$$

Since $y = 1/x$, when $x = \pm 1$, $y = \pm 1$ and when $x = i\sqrt{2}$, $y = -(i\sqrt{2}/2)$, $x = -i\sqrt{2}$, $y = i\sqrt{2}/2$. The natural assertion is that the solution set ought to be

$$\left\{(1, 1), (-1, -1), \left(i\sqrt{2}, -\frac{i\sqrt{2}}{2}\right), \left(-i\sqrt{2}, \frac{i\sqrt{2}}{2}\right)\right\}.$$

*Check:* $2(\pm 1)^2 - (\pm 1)^2 = 2 - 1 = 1$; $6(\pm 1)^2 - 2(\pm 1)(\pm 1) - 3(\pm 1)^2 = 6 - 2$
$- 3 = 1$; $\quad 2\left(\mp\dfrac{i\sqrt{2}}{2}\right)^2 - (\pm i\sqrt{2})^2 = 2\left(-\dfrac{1}{2}\right) - (-2) = -1 + 2 = 1$;

$6\left(\mp\dfrac{i\sqrt{2}}{2}\right)^2 - 2(\pm i\sqrt{2})\left(\mp\dfrac{i\sqrt{2}}{2}\right) - 3(\pm i\sqrt{2})^2 = 6\left(-\dfrac{1}{2}\right) -$
$2(1) - 3(-2) = -3 - 2 + 6 = 1.$

**33**  Letting $x$ and $y$ denote the lengths of the sides of the triangle, we know from the theorem of Pythagoras that $x^2 + y^2 = 169$. Since the area of the triangle is $(1/2)xy$, we must solve the system of equations

$$\begin{cases} x^2 + y^2 = 169 \\[2mm] \dfrac{1}{2}xy = 30. \end{cases}$$

Since $y = 60/x$, it must be true that

$$x^2 + \left(\frac{60}{x}\right)^2 = 169$$

$$x^4 + 3{,}600 = 169x^2$$

$$x^4 - 169x^2 + 3{,}600 = 0$$

$$(x^2 - 25)(x^2 - 144) = 0$$

$$x = \pm 5, \pm 12.$$

The sides have lengths 5 inches and 12 inches.

*Check:* $5^2 + 12^2 = 25 + 144 = 169$ and $(1/2)(5)(12) = 30$.

# (10)

## *Sequences and series*

### (10.1) (Basic terminology)

The notion of a *sequence* of numbers is extremely useful in mathematics. A *finite sequence* of real numbers is simply a finite list of such numbers. For example, the list 2, 4, 6, 8, 10 is a finite sequence of numbers. The numbers 2, 4, 6, 8, and 10 are the *terms* of the sequence, with 2 being the *first term,* 4 the *second term,* and so on. A more formal definition follows.

(∘)    *A finite sequence of real numbers is a function whose domain is the first n natural numbers for some $n \in N$.*

The example we gave can be thought of in this light by defining a function $s$ by the rule $s(1) = 2$, $s(2) = 4$, $s(3) = 6$, $s(4) = 8$, $s(5) = 10$. The domain of $s$ is $\{1, 2, 3, 4, 5\}$, and the original list of numbers can be recaptured from $s$ by writing $s(1)$, $s(2)$, $s(3)$, $s(4)$, $s(5)$ in that order. Thus the list of numbers associated with the function $s(n) = 1/n^2 + 1$ $(n = 1, 2, 3, 4)$ is $1/1^2 + 1$, $1/2^2 + 1$, $1/3^2 + 1$, $1/4^2 + 1$, which can be written more simply as $1/2$, $1/5$, $1/10$, $1/17$. In short, a *finite sequence* can either be thought of as a finite list of real numbers $s_1, s_2, \ldots, s_n$ or as a function whose domain is $\{1, 2, \ldots, n\}$ for some $n \in N$.

*infinite sequence*

Similarly, an *infinite sequence* may be viewed either as an *infinite* list of real numbers or as a function whose domain is the set of *all* natural numbers. The function $s(n) = n/n + 1$ $(n \in N)$ is an example of an infinite sequence. The associated list of numbers is $1/2, 2/3, 3/4, 4/5, \ldots, n/n + 1, \ldots$ with the understanding that $s(1)$ is written first, $s(2)$ second, and so on. The term $n/n + 1$

*nth term   general term*

is called the *nth term* or the *general term* of this sequence. The terms of the sequence are obtained by successively replacing $n$ with $1, 2, 3, \ldots$ in the expression for the general term. On the other hand, if the general term of a sequence is given to you, it is very easy to write the sequence as a function. For example, the sequence $2, 4, 6, 8, \ldots, 2n, \ldots$ becomes $s(n) = 2n$ $(n \in N)$ in the form of a function.

A sequence is often defined by simply specifying its general term $s_n$. It is even customary to use the notation $s_n$ in place of $s(n)$ when thinking of sequences as functions instead of lists of numbers. It is important, though, to realize that knowing the first few terms of a sequence does not allow you to determine the general term. This may easily be seen from a concrete example. Consider the sequences whose general term $s_n$ is given following:

**(1)** $s_n = 2n$

**(2)** $s_n = 2n + (n - 1)(n - 2)(n - 3)$

**(3)** $s_n = 2n + n(n - 1)(n - 2)(n - 3)$

**(4)** $s_n = 2n + \dfrac{(n - 1)(n - 2)(n - 3)}{n}.$

The first three terms of each sequence

**(1)** $2, 4, 6, 8, 10, \ldots, s_n, \ldots$

**(2)** $2, 4, 6, 14, 34, \ldots, s_n, \ldots$

**(3)** $2, 4, 6, 32, 130, \ldots, s_n, \ldots$

**(4)** $2, 4, 6, \dfrac{19}{2}, \dfrac{74}{5}, \ldots, s_n, \ldots$

are all the same—$2, 4, 6$—but each sequence has a different general term. In other words, if you are given the infinite sequence $2, 4,$

6, . . . , it could be *any* of the sequences above, or possibly even some other sequence.

Up to this point we have been considering sequences whose general term was given by a formula. This is not always possible to do. Consider the sequence whose $n$th term is your weight (to the nearest pound) at 8:00 A.M. on the $n$th day of the current year. This clearly defines a finite sequence, but there is no formula for its $n$th term. As another example, consider the sequence whose $n$th term is the $n$th digit to the right of the decimal point in the decimal expansion of $\pi$. Since $\pi = 3.14159. . .$ , we have $s_1 = 1$, $s_2 = 4$, $s_3 = 1$, and so forth; but, once more, there is no formula for the general term of this sequence.

## Exercise set 10.1

*Write out each of the following finite sequences in functional form.*

$(\circ)$    **1**   $9, -2, 5$            **2**   $6, -4, 3, 8, 5$

      **3**   $1, 1, 1, 1, 1$        **4**   $3, 1, 3, 1, 3, 1$

      **5**   $\dfrac{1}{2}, \dfrac{2}{3}, \dfrac{3}{4}, \dfrac{4}{5}$       **6**   $\dfrac{9}{4}, \dfrac{16}{9}, \dfrac{25}{16}$

*Write each of the following finite sequences as a list of real numbers.*

$(\circ)$    **7**   $s(n) = n^2 + n \ (n = 1, 2, 3)$

      **8**   $s(n) = \dfrac{1}{n} \ (n = 1, 2, 3, 4)$

      **9**   $s(n) = \dfrac{1}{2n + 3} \ (n = 1, 2, 3, 4, 5)$

      **10**   $s(n) = \dfrac{n + 1}{n^2} \ (n = 1, 2, 3)$

      **11**   $s(n) = (-1)^n \ (n = 1, 2, 3, 4)$

      **12**   $s(n) = 4 \ (n = 1, 2, 3, 4, 5, 6)$

*Write out the first five terms of each of the following sequences.*

$(\circ)$    **13**   $s_n = \dfrac{1}{2n}$          **14**   $s_n = 3n - 1$

      **15**   $s_n = \dfrac{1}{n^2}$          **16**   $s_n = \dfrac{n - 1}{n}$

**17** $s_n = 2 + 3n$          **18** $s_n = 2n - 4$

**19** $s_n = \dfrac{n^2}{n+1}$          **20** $s_n = 1 + (-1)^n$

*In problems 21 to 26, you are given the general term for two distinct sequences, as well as a positive integer k. Verify that the first k terms of the sequences are identical but the $(k + 1)$st terms differ.*

$(\circ)$  **21**  $s_n = 3n; s_n = 3n + (n-1)(n-2)(n-3); k = 3$

**22**  $s_n = 2^n; s_n = 2^n + (n-1)(n-2)(n-3); k = 3$

**23**  $s_n = 2n + 1; s_n = 2n + 1 + (n-1)(n-2)(n-3)(n-4); k = 4$

**24**  $s_n = n^3; s_n = n^3 + (n-1)(n-2)(n-3)(n-4); k = 4$

**25**  $s_n = n^2; s_n = 2n^2 - 3n + 2; k = 2$

**26**  $s_n = 3n; s_n = n^2 + 2; k = 2$

*In the remaining problems you are given the first few terms of an infinite sequence. Find two distinct possible formulas for the general term $s_n$.*

$(\circ)$  **27**  $5, 10, 15, \ldots$          **28**  $4, 8, 12, \ldots$

**29**  $1, 4, 9, 16, \ldots$          **30**  $1, 3, 5, 7, \ldots$

$(\circ)$

*Sample solutions for exercise set 10.1*

**1**  $s(1) = 9, s(2) = -2, s(3) = 5$

**7**  We substitute $n = 1, 2, 3$ in the formula for $s(n)$, and we get $s(1) = 1^2 + 1 = 2$, $s(2) = 2^2 + 2 = 6$, and $s(3) = 3^2 + 3 = 12$.

**13**  With $n = 1, 2, 3, 4, 5$ we have $s_1 = \dfrac{1}{2 \cdot 1} = \dfrac{1}{2}$, $s_2 = \dfrac{1}{2 \cdot 2} = \dfrac{1}{4}$, $s_3 = \dfrac{1}{2 \cdot 3} = \dfrac{1}{6}$, $s_4 = \dfrac{1}{2 \cdot 4} = \dfrac{1}{8}$, and $s_5 = \dfrac{1}{2 \cdot 5} = \dfrac{1}{10}$. Thus, the first five terms are $\dfrac{1}{2}, \dfrac{1}{4}, \dfrac{1}{6}, \dfrac{1}{8}$, and $\dfrac{1}{10}$.

**21**  The first four terms of the first sequence are 3, 6, 9, 12. For the second sequence,

$s_1 = 3 \cdot 1 + (1 - 1)(1 - 2)(1 - 3) = 3 + 0 = 3$

$s_2 = 3 \cdot 2 + (2 - 1)(2 - 2)(2 - 3) = 6 + 0 = 6$

$s_3 = 3 \cdot 3 + (3 - 1)(3 - 2)(3 - 3) = 9 + 0 = 9$

$s_4 = 3 \cdot 4 + (4 - 1)(4 - 2)(4 - 3) = 12 + 6 = 18.$

The first three terms of both sequences are 3, 6, 9; their fourth terms differ.

27 There is no single correct answer to this problem, though a reasonable answer is $s_n = 5n$; $s_n = 5n + (n - 1)(n - 2)(n - 3)$.

# (10.2) (Arithmetic sequences)

arithmetic sequence

An *arithmetic sequence* is one that has the property that the difference between successive terms remains constant. Thus, 2, 5, 8, 11, 14; $-4$, $-2$, 0, 2, 4, 6, 8; and 5, 0, $-5$, $-10$, $-15$ are all examples of *finite arithmetic sequences. Infinite arithmetic sequences* are those of the form

finite arithmetic
sequences
infinite arithmetic
sequences

$$s_n = c + (n - 1)d \qquad (c, d \text{ fixed real numbers}).$$

For example, if $s_n = 2 + (n - 1)3$, one has

$$s_1 = 2 + (1 - 1)3 = 2 + 0 = 2,$$

$$s_2 = 2 + (2 - 1)3 = 2 + 3 = 5,$$

$$s_3 = 2 + (3 - 1)3 = 2 + 6 = 8.$$

Notice, in fact, that

$$
\begin{array}{lll}
s_{n+1} & = 2 + [n + 1) - 1]3 & = 2 + 3n \\
s_n & = 2 + (n - 1)3 & = 2 + 3n - 3 \\
\hline
s_{n+1} - s_n & & = 3.
\end{array}
$$

Thus, if you are given the value of any term of the sequence, you add 3 to get the value of the next term.

Of course, this holds more generally. If $s_n = c + (n - 1)d$, then

$$s_1 = c + (1 - 1)d = c + 0 = c,$$

$$s_2 = c + (2 - 1)d = c + d,$$

$$s_3 = c + (3 - 1)d = c + 2d.$$

Indeed,

$$
\begin{array}{lll}
s_{n+1} & = c + [n + 1) - n]d & = c + nd \\
s_n & = c + (n - 1)d & = c + nd - d \\
\hline
s_{n+1} - s_n & & = d.
\end{array}
$$

Suppose, on the other hand, that you are given an arithmetic sequence $-3, 1, 5, 9, \ldots$ and asked to find $c, d$ so that $s_n = c + (n-1)d$ for all $n \in N$. Clearly, $c = s_1 = -3$ and $d = s_2 - s_1 = 1 - (-3) = 1 + 3 = 4$. Thus $s_n = -3 + (n-1)4$. For obvious reasons, if $s_n = c + (n-1)d$, the number $c$ is called the *first term* of the sequence and $d$ the *difference between successive terms*. In Section 10.1 we pointed out that the first few terms of a sequence do not determine the general term. The knowledge that we had an arithmetic sequence allowed us to find $s_n$ in the present situation.

Associated with any arithmetic sequence $s_n = c + (n-1)d$ is an *arithmetic series* $s_1 + s_2 + \ldots + s_n + \ldots$. Suppose you are asked to find the sum $S_k = s_1 + s_2 + \ldots + s_k$ of the first $k$ terms of this series. We begin with the case where $s_n = n$; so $S_k = 1 + 2 + \ldots + k$ = the sum of the first $k$ positive integers. Now $S_k$ may be written two different ways as

$$\left. \begin{array}{l} S_k = 1 \quad\quad + 2 \quad\quad + 3 \quad\quad + \ldots + k \\[4pt] S_k = k \quad\quad + (k-1) \; + (k-2) + \ldots + 1 \end{array} \right\} \text{ add, so that}$$

$$2S_k = (k+1) \; + (k+1) \; + (k+1) + \ldots + (k+1)$$

$$= k(k+1).$$

This leads us to the following formula.

$(\circ)$    AM.1   $1 + 2 + \ldots + k = \dfrac{k(k+1)}{2}$

We now consider the general situation where $s_n = c + (n-1)d$ and $S_k = s_1 + s_2 + \ldots + s_k$. Here

$$S_k = (c) + (c + d) + (c + 2d) + \ldots + [c + (k-1)d]$$

$$= kc + [d + 2d + \ldots + (k-1)d]$$

$$= kc + [1 + 2 + \ldots + (k-1)]d.$$

If AM.1 is written with $k - 1$ in place of $k$, it becomes $1 + 2 + \ldots + (k-1) = \dfrac{(k-1)k}{2}$.

Hence, $S_k$ may be written as

$$S_k = kc + \frac{(k-1)k}{2}d$$

or, finally, in the following form.

$(\circ)$    AM.2   $S_k = \dfrac{k}{2}[2c + (k-1)d]$

Using the fact that $s_1 = c, s_k = c + (k-1)d$, this can also be written in the following useful form.

$(\circ)$    AM.3   $S_k = \dfrac{k}{2}(s_1 + s_k)$

To illustrate the use of these formulas, let's consider some examples. First, we compute the sum of the first eight terms of the arithmetic series whose general term is $5n - 2$. Here $s_1 = 3$, $s_2 = 5 \cdot 2 - 2 = 8$; so $c = 3$ and $d = s_2 - s_1 = 8 - 3 = 5$. Hence, $S_8 = (8/2)[2 \cdot 3 + (8 - 1)5] = 4(6 + 35) = 4(41) = 164$. On the other hand, AM.3 can be used to show that $1 + 3 + 5 + 7 + 9 = (5/2)(1 + 9) = (5/2)(10) = 25$. The idea is to observe that there are five terms, the first term is $1$, and the last term is $9$.

**Problem:** Mr. M. E. Lucky has won the first prize in a slogan contest. The terms of the prize call for ten annual payments, but he may choose between the following options: (1) equal payments of $6,500, or (2) $4,000 for the first payment and an annual increase of $600 thereafter. Which option leads to the larger actual prize?

**Solution:** Mr. Lucky would receive 10($6,500) = $65,000 if he accepts the first option. With option 2, the payments form an arithmetic sequence with 4,000 the first term and 600 the difference between successive terms. Here we must compute $S_{10}$ when $c = 4,000$ and $d = 600$. By AM.2,

$$S_{10} = \dfrac{10}{2}(2 \cdot 4{,}000 + 9 \cdot 600) = 5(8{,}000 + 5{,}400)$$

$$= 5(13{,}400) = 67{,}000.$$

The second option pays $67,000; so it leads to the higher actual prize.

**Problem:** An 18% interest rate compounded monthly (with no interest for the first 30 days) is fairly standard for certain types of revolving charge accounts. What is the actual interest paid on a $120 purchase that is paid for in 12 monthly payments of $10 plus interest? How about 24 payments of $5 plus interest?

**Solution:** An 18% annual interest rate translates to 1-1/2% per month. There is no interest on the first payment; the interest on the second payment is $(0.015)(110)$. At the time the last payment is made, there is a $10.00 balance, so that the interest charge is $(0.015)(10)$. Since the interest charge decreases by a fixed amount each month, and since there are eleven interest payments, we are after the sum of the first eleven terms of a certain arithmetic series. By AM.3, $S_{11} = (11/2)[(0.015)(110) + (0.015)(10)] = (11/2)(1.80) = \$9.90$. For 24 months, the interest on the second payment is $(0.015)(115)$ and that on the last payment $(0.015)(5)$. There are 23 interest payments, so that we want $S_{23} = (23/2)[(0.015)(115) + (0.015)(5)] = (23/2)(1.80) = \$20.70$.

Up to this point our examples have all involved finding the sum of a certain arithmetic series. On occasion one needs to determine $n$ so that the sum of the first $n$ terms of a given arithmetic series has a certain fixed value. The final example serves to illustrate this type of problem.

**Problem:** Each Sunday Morton Miserly buries some money in a tin can in his back yard. The first week he buries $10, the second week $10.50, and each week thereafter he increases the amount to be buried by fifty cents. How many weeks will it take for him to accumulate $1,183?

**Solution:** If $s_n$ denotes the amount buried the $n$th week, then $s_1$, $s_2, \ldots, s_n, \ldots$ forms an arithmetic sequence. Clearly, $c = 10$ and $d = \dfrac{1}{2}$. We must determine $n$ so that

$$S_n = \frac{n}{2}\left[2 \cdot 10 + \frac{n-1}{2}\right] = 1{,}183,$$

$$10n + \frac{n^2 - n}{4} = 1{,}183,$$

$$40n + n^2 - n = 4{,}732,$$

$$n^2 + 39n - 4{,}732 = 0,$$

which factors as $(n - 52)(n + 91) = 0$; so $n = 52$ or $n = -91$. Since a negative number of weeks would make no sense as an answer, we deduce that it would take 52 weeks for him to save \$1,183.

$$\text{Check: } S_{52} = \frac{52}{2}\left(2\cdot 10 + \frac{51}{2}\right) = 26\left(20 + \frac{51}{2}\right) = 26\left(\frac{91}{2}\right)$$

$$= 13(91) = 1,183.$$

## Exercise set 10.2

In problems 1 to 8 you are given the first two terms of an arithmetic sequence. Write down the next two terms and find a formula for the general term.

( ∘ )  **1**  $-2, 1, \ldots$          **2**  $-9, -5, \ldots$

     **3**  $11, 8, \ldots$          **4**  $25, 16, \ldots$

     **5**  $\dfrac{1}{2}, \dfrac{3}{2}, \ldots$          **6**  $\dfrac{1}{3}, 2, \ldots$

     **7**  $2, -\dfrac{1}{2}, \ldots$          **8**  $1 + \sqrt{2}, 1, \ldots$

Here are some problems that require finding a value of $s_n$ for some natural number $n$.

( ∘ )  **9**  Find the ninth term of the arithmetic sequence $1, 4, \ldots$.

    **10**  Find the sixth term of the arithmetic sequence $-40, -26, \ldots$.

    **11**  Find the fifteenth term of the arithmetic sequence $10, 7, \ldots$.

    **12**  Find the eighth term of the arithmetic sequence $14, 8, \ldots$.

Suppose you are given $s_3 = 11$, $s_8 = 31$ in an arithmetic sequence and that you are asked to find a formula for the general term. You know that for suitable numbers $c$ and $d$,

$$s_n = c + (n - 1)d.$$

Thus

$$s_8 = c + 7d = 31$$

$$s_3 = c + 2d = 11,$$

which is a system of two linear equations in the variables $c$ and $d$. Subtracting the second equation from the first produces $5d = 20$; so $d = 4$. It follows that $c = 11 - 2d = 11 - 8 = 3$, and $s_n = 3 + (n - 1)4$.

*The foregoing example should provide you with the techniques you will need for problems 13 to 16.*

( ○ )  **13** Find $s_n$ if the third term of an arithmetic sequence is $-7$ and the tenth term is 14.

**14** Find $s_n$ if the fifth term of an arithmetic sequence is $-3$ and the twelfth term is 32.

**15** Find $s_1$ if the sixth term of an arithmetic sequence is 26 and the eleventh term is 6.

**16** Find $s_1$ if the fifth term of an arithmetic sequence is 20 and the ninth term is $-4$.

*Use AM.2 or AM.3 of the text to find the sum of each of the following arithmetic series.*

( ○ )  **17** $4 + 9 + 14 + 19$

**18** $6 + 8 + 10 + 12 + 14$

**19** $5 + (-1) + (-7) + (-13)$

**20** $20 + 14 + 8 + 2$

**21** Find $S_{10}$ if $s_n = 2n + 3$.

**22** Find $S_{14}$ if $s_n = 5n - 2$.

**23** Find $S_8$ for the arithmetic sequence $-6, -2, \ldots.$

**24** Find $S_{12}$ for the arithmetic sequence $-3, 1, \ldots.$

( ○ )  **25** Find the sum of the first 50 odd natural numbers.

**26** Find the sum of the first 40 even natural numbers.

**27** Find the sum of all the positive even integers less than 100.

**28** Find the sum of all odd, two-digit, natural numbers.

**29** During a special sales promotion, an automobile salesman received a bonus of $100 for the first car he sold, $115 for the second car, $130 for the third car, and a bonus increase of $15 per car for each car he sold thereafter. He sold 20 cars. What was his bonus?

**30** A $180 purchase is put on a revolving charge account whose interest is computed at the rate of 18 percent compounded monthly. If the account is paid off in eighteen payments of $10 plus interest, what is the total amount of interest?

**31** Mr. Poore started work at an annual salary of $4,000 and each year receives a $400 pay raise. At this rate, how many years will it take for his lifetime earnings to total $43,200?

**32** A builder wishes to arrange 376 bricks in a pile so that there will be 1 brick in the first row, 4 bricks in the second row, 7 in the third row, and so forth. Can this be done? If so, how many rows will it take, and how many bricks will be in the bottom row?

**1** The next two terms are 1 and 4. Since the first term is $-2$ and the difference between successive terms is $1 - (-2) = 3$, the general term is given by $s_n = -2 + (n-1)3$.

**9** Here, the first term is 1, and $d = 4 - 1 = 3$; so $s_n = 1 + (n-1)3$. In particular, $s_9 = 1 + 8 \cdot 3 = 25$.

**13** We know that $s_n = c + (n-1)d$. Hence

$$s_{10} = c + 9d = 14$$

$$s_3 = c + 2d = -7.$$

Subtracting, we have

$$7d = 21$$

$$d = 3,$$

$$c = -7 - 2d = -7 - 6 = -13.$$

Hence, $s_n = -13 + (n-1)3$.

**17** There are four terms and their sum is $S_4 = (4/2)(4 + 19) = 2 \cdot 23 = 46$.

**25** The odd natural numbers form an arithmetic sequence whose first term is 1 and in which the difference of successive terms is 2. Hence, $S_{50} = (50/2)[2 + (50 - 1)2] = 25(2 + 98) = 2{,}500$.

## (10.3) (Geometric sequences)

*geometric sequence*

*ratio*

A *geometric sequence* is one in which the quotient of successive terms remains constant. Given a geometric sequence $s_1, s_2, \ldots,$ $s_n, \ldots$ the number $r = s_{n+1}/s_n$ is called the *ratio* of the sequence. It has the property that $rs_n = s_{n+1}$, so that if any term is multiplied by the ratio, the result is the next term of the sequence. Examples of finite geometric sequences are: 1, 10, 100, 1,000; 1/2, $-3$, 18, $-108$; and 4, 2, 1, 1/2, 1/4, 1/8. Their ratios are, respectively, 10, $-6$, and 1/2.

Infinite geometric sequences are those of the form

$$s_n = ar^{n-1} \qquad (a, r \text{ fixed, nonzero numbers}).$$

Notice that $s_1 = ar^0 = a$ and $s_{n+1}/s_n = ar^n/ar^{n-1} = r$ so that $a$ is the first term and $r$ the ratio of such a sequence. Thus if $a = 1/8$ and $r = 4$, the resulting sequence is 1/8, 1/2, 2, 8, $\ldots$, $4^{n-1}/8$, $\ldots$. The first term of the sequence is 1/8, and each term (other than the first) is four times the preceding term.

Suppose, on the other hand, that you are given the first two terms of a geometric sequence and asked to determine the general term. This is accomplished from your knowledge that $s_n = ar^{n-1}$ where $a = s_1$ and $r = s_2/s_1$. For the geometric sequence 6, 4, ... , $a = 6$ and $r = 4/6 = 2/3$, so that $s_n = 6(2/3)^{n-1}$.

For an example of a geometric sequence we consider a woman whose starting salary with a certain company is $4,000 per year. Suppose she receives a 10% pay raise at the end of each year, and we are asked to determine her salary at the twenty-first year of employment. Her salary for the first year is $4,000, and for the second year it is $4,000 + (0.1)(4,000) = (1.1)(4,000)$. During the third year she earns $(1.1)(4,000) + (0.1)[(1.1)(4,000)] = (1.1)^2(4,000)$. If $s_n$ denotes her salary during her $n$th year of employment, then $s_1, s_2, \ldots, s_n, \ldots$ forms a geometric sequence whose first term is 4,000 and whose ratio is 1.1. We must compute $s_{21} = (4,000)(1.1)^{20}$.

$$\log 1.1 \quad = 0.0414$$

$$\log (1.1)^{20} = 0.8280$$

$$\log 4{,}000 \ = 3.6021$$

$$\overline{\log s_{21} \quad = 4.4301}$$

$$s_{21} \quad = (2.692)10^4.$$

Her salary would be $26,920.

*geometric series*

With every geometric sequence $s_n = ar_{n-1}$ there is associated a *geometric series* $a + ar + ar^2 + \ldots + ar^{n-1} + \ldots$. We next derive a formula for the sum $S_k$ of the first $k$ terms of such a series. We do this by observing that

$$S_k = a + ar + ar^2 + \ldots + ar^{k-2} + ar^{k-1}$$

$$rS_k = \quad ar + ar^2 + \ldots + ar^{k-2} + ar^{k-1} + ar^k$$

subtract

$$\overline{(1 - r)S_k = a \qquad\qquad\qquad\qquad\qquad - ar^k}$$

and finally that

(∘) **GM.1** $S_k = \dfrac{a - ar^k}{1 - r}$ $(r \neq 1)$

We illustrate the use of this formula by noting that if $s_n = 3 \cdot 2^{n-1}$, then $S_8 = s_1 + s_2 + \ldots + s_8$ is given by

$$S_8 = \frac{3 - 3 \cdot 2^8}{1 - 2} = \frac{3(1 - 256)}{-1} = 3 \cdot 255 = 865.$$

Using the fact that $s_k = ar^{k-1}$, GM.1 may also be written in the form

(∘)   GM.2 $S_k = \dfrac{a - rs_k}{1 - r}$   $(r \neq 1)$

Thus, to evaluate the sum $4 + 2 + 1 + 1/2 + 1/4 + 1/8$, we simply observe that this is the sum of the first six terms of a geometric series with first term $4$, ratio $1/2$, and sixth term $1/8$. By GM.2, we get

$$S_6 = \frac{4 - \dfrac{1}{2} \cdot \dfrac{1}{8}}{1 - \dfrac{1}{2}} = \frac{64 - 1}{16 - 8} = \frac{63}{8}.$$

Of course, if $r = 1$ the formula for $S_k$ becomes simply $S_k = ka$.

   Working through a problem will help you to understand the use of these formulas.

**Problem:** Due to such factors as air resistance and friction, a ball is rolling so that each second it travels nine-tenths as far as it did the preceding second. If it travels 4 feet the first second, how far will it go in 4 seconds?

**Solution:** We are after the sum of the first four terms of a geometric series whose first term is 4 and whose ratio is $9/10$. Hence

$$S_4 = \frac{4 - 4\left(\dfrac{9}{10}\right)^4}{1 - \dfrac{9}{10}} = \frac{4(10^4 - 9^4)}{10^4\left(1 - \dfrac{9}{10}\right)}$$

$$= \frac{4(10{,}000 - 6{,}561)}{10^3(10 - 9)} = \frac{4(3{,}439)}{1{,}000} = 13.756.$$

The ball will travel 13.756 feet in 4 seconds.

   And finally, have you wondered if there is some way to find the "sum" of an infinite geometric series

$$a + ar + ar^2 + \ldots + ar^{n-1} + \ldots .$$

To get a picture of what is involved, we begin by considering the series $1 + 1/2 + 1/4 + \ldots + (1/2)^{n-1} + \ldots$. Here

$$S_1 = 1$$

$$S_2 = \frac{1 - \left(\dfrac{1}{2}\right)^2}{1 - \dfrac{1}{2}} = \frac{1 - \dfrac{1}{4}}{1 - \dfrac{1}{2}} = \frac{4 - 1}{4 - 2} = \frac{3}{2} = 2 - \frac{1}{2}$$

$$S_3 = \frac{1 - \left(\dfrac{1}{2}\right)^3}{1 - \dfrac{1}{2}} = \frac{1 - \dfrac{1}{8}}{1 - \dfrac{1}{2}} = \frac{8 - 1}{8 - 4} = \frac{7}{4} = 2 - \frac{1}{4}$$

$$S_4 = \frac{1 - \left(\dfrac{1}{2}\right)^4}{1 - \dfrac{1}{2}} = \frac{1 - \dfrac{1}{16}}{1 - \dfrac{1}{2}} = \frac{16 - 1}{16 - 8} = \frac{15}{8} = 2 - \frac{1}{8}.$$

In general,

$$S_k = \frac{1 - \left(\dfrac{1}{2}\right)^k}{1 - \dfrac{1}{2}} = \frac{1 - \dfrac{1}{2^k}}{1 - \dfrac{1}{2}} = \frac{2^k - 1}{2^k - 2^{k-1}} = \frac{2^k - 1}{2^{k-1}} = 2 - \frac{1}{2^{k-1}}.$$

Now, as $k$ gets larger, the number $1/2^{k-1}$ gets closer and closer to zero. In fact, by taking $k$ large enough, $1/2^{k-1}$ can be made to get as close to zero as we please. This leads us to define the sum $1 + 1/2 + 1/4 + \ldots + (1/2)^{n-1} + \ldots = 2$.

We now suppose $s_n = ar^{n-1}$, so that by GM.2,

$$S_k = \frac{a - ar^k}{1 - r} = \frac{a}{1 - r} - \frac{ar^k}{1 - r}.$$

If $-1 < r < 1$, the term $ar^k/(1 - r)$ gets arbitrarily close to zero as $k$ gets larger and larger. Hence we define

---

$(\circ)$  GM.3  $a + ar + \ldots + ar^{n-1} + \ldots = \dfrac{a}{1 - r}$

---

In case $|r| \geq 1$, the term $ar^k/(1-r)$ does not get close to zero and there is no way to assign a value to the infinite sum $a + ar + ar^2 + \ldots + ar^{n-1} + \ldots$. A careful treatment of this topic involves the notion of a *limit* and properly belongs in a calculus text. Our goal in the foregoing discussion was to make GM.3 seem plausible to you and to give you some idea of how it might be established.

To illustrate the use of GM.3, we point out that the sum of the geometric series $9 + 3 + 1 + \ldots$ is $\dfrac{9}{1 - \dfrac{1}{3}} = \dfrac{27}{3 - 1} = \dfrac{27}{2}$. As a further illustration we ask you again to consider the problem of the rolling ball. Recall that the ball rolled 4 feet the first second, and during each second thereafter it rolled nine-tenths as far as it did the preceding second. Let's compute the total distance the ball would roll if it were allowed to continue rolling indefinitely. We want the sum of the infinite geometric series $4 + 4(9/10) + 4(9/10)^2 + \ldots + 4(9/10)^{n-1} + \ldots$, which is $4/[1 - (9/10)] = 40/(10 - 9) = 40$ feet.

Sums of infinite geometric series have an interesting application in connection with repeating decimals; these are decimal expansions that, after a finite number of decimal places, have an endlessly repeating group of digits. For example, we let a horizontal bar denote a repeating group of digits, and

$0.66\overline{6}$,

$0.45\overline{45}$,

$0.138523\overline{523}$

are all repeating decimals. Consider the problem of expressing a repeating decimal as the quotient of two integers. To illustrate the technique, consider the repeating decimal $0.45\overline{45}$, which can be written as $\dfrac{45}{10^2} + \dfrac{45}{10^4} + \dfrac{45}{10^6} + \ldots + \dfrac{45}{10^{2n}} + \ldots$. This is an infinite geometric series with $a = \dfrac{45}{10^2}$ and $r = \dfrac{1}{100}$. Its sum is $\dfrac{\dfrac{45}{100}}{1 - \dfrac{1}{100}}$

$= \dfrac{45}{100 - 1} = \dfrac{45}{99} = \dfrac{5}{11}$. Hence, $0.45\overline{45} = \dfrac{5}{11}$.

## Exercise set 10.3

*In problems 1 to 6 you are given the first two terms of a geometric sequence. Find the next three terms as well as the general term $s_n$.*

$\left(\circ\right)$

**1** $4, 8, \ldots$

**3** $9, 3, \ldots$

**5** $\dfrac{1}{2}, -2, \ldots$

**2** $5, 10, \ldots$

**4** $32, 16, \ldots$

**6** $\dfrac{1}{4}, -8, \ldots$

*Now for some problems that involve finding $s_n$ for some specific value of n.*

$\left(\circ\right)$ **7** Find the sixth term of the geometric sequence $6, 2, \ldots$.

**8** Find the eighth term of the geometric sequence $\dfrac{1}{2}, 2, \ldots$.

**9** The third term of a geometric sequence is $9/4$ and the sixth term is $243/256$. What is its first term?

**10** The fourth term of a geometric sequence is 8 and the seventh term is 512. Find the first term.

**11** The sixth term of a geometric sequence is 27 and the ratio is $-(1/3)$. Find the third term.

**12** The seventh term of a geometric sequence is 4,000 and the ratio is $1/10$. Find the fourth term.

**13** A man has a starting salary of \$7,000 per year and each year receives a 6 percent pay raise. What will his salary be during his seventh year of employment?

**14** A bartender has a habit of helping himself to an occasional drink and then refilling the bottle with colored water so no one will detect the loss. If he takes five 1-ounce drinks in this way from a full quart bottle and if no one else touches the bottle, how many ounces of whiskey will be left in the bottle?

*In problems 15 to 28, use the formulas GM.1 and GM.2 of the text to evaluate the sum of each finite geometric series.*

$\left(\circ\right)$ **15** $\dfrac{1}{2} + 1 + 2 + 4 + 8$

**17** $\dfrac{1}{2} - 1 + 2 - 4 + 8$

**19** $48 + 24 + 12 + 6$

$\left(\circ\right)$ **21** Find $S_8$ if $s_n = 3\left(\dfrac{2}{3}\right)^{n-1}$.

**16** $\dfrac{1}{3} + 1 + 3 + 9 + 27 + 81$

**18** $\dfrac{1}{3} - 1 + 3 - 9 + 27 - 81$

**20** $324 + 108 + 36 + 12 + 4$

**22** Find $S_{10}$ if $s_n = 4\left(\dfrac{5}{2}\right)^{n-1}$.

**23** Find $S_6$ if $s_n = 3(-4)^{n-1}$.　　　　**24** Find $S_7$ if $s_n = \dfrac{1}{4}(-6)^{n-1}$.

( ○ )　**25** A gambler trying desperately to recoup his losses decides to try doubling his bets. His first bet is $2, his second bet $4, and so forth. If he places eight bets and loses all eight of them, what are his losses?

**26** Two checkers players have agreed on the following stakes: at the end of the game, the winner gets 5¢ for the first man left on the board, 15¢ for the second, 45¢ for the third, and so on. How much money will be won if there are seven men left on the board?

**27** A chain letter is started by writing to four people and suggesting that they each send a copy of the letter to four other people. If the chain is unbroken when the sixth set of letters is mailed, how many letters have been sent out?

**28** During a recent auto trip a family covered 600 miles the first day. Due to fatigue and an increasing number of places to see, on each day thereafter they only cover four-fifths of the distance of the previous day. How far did they get in five days?

*In each of problems 29 to 38 you will be given an infinite geometric series. If the series has a sum, find it. If the series has no sum, state that.*

( ○ )　**29** $25 + 5 + 1 + \ldots$　　　　**30** $36 + 6 + 1 + \ldots$

**31** $25 - 5 + 1 - \ldots$　　　　**32** $36 - 6 + 1 - \ldots$

**33** $\dfrac{1}{4} + \dfrac{1}{2} + 1 + \ldots$　　　　**34** $\dfrac{1}{500} + \dfrac{1}{100} + \dfrac{1}{20} + \ldots$

**35** $9 + 6 + 4 + \ldots$　　　　**36** $16 + 6 + \dfrac{9}{4} + \ldots$

**37** $9 - 6 + 4 - \ldots$　　　　**38** $16 - 6 + \dfrac{9}{4} - \ldots$

( ○ )　**39** If a ball rebounds three-fourths as far as it falls, how far will it travel before coming to rest if it is dropped from a height of 10 feet?

**40** A ball dropped from a height of 8 feet rebounds seven-ninths as far as it falls. How far will it travel before coming to rest?

**41** The tip of a pendulum swings through an arc of 8 inches, and each arc thereafter is eleven-thirteenths the length of the preceding one. How far does the tip move before the pendulum comes to rest?

**42** You receive an inheritance that pays you $1,000 on your twenty-first birthday and thereafter an annual payment that is 80 percent of the preceding one. Approximately how much money can you expect from your inheritance if you live to a ripe old age?

**43** You are given a share of mining stock. The first year's dividend is $80 and each year the dividend is three-quarters that of the preceding year. Approximately how much can you expect to receive during your lifetime?

**44** Due to various factors such as friction and air resistance, a ball is rolling so that each second it travels seven-eighths as far as the preceding second. If it rolls 4 feet during the first second, how far will it roll before coming to rest?

*Express each of the following repeating decimals as the quotient of two integers.*

($\circ$)  **45** $0.\overline{3}$  **46** $0.\overline{4}$

**47** $0.1\overline{6}$  **48** $0.\overline{24}$

**49** $0.\overline{2}$  **50** $0.0\overline{5}$

**51** $4.1\overline{234}$  **52** $5.1\overline{600}$

($\circ$)

*Sample solutions for exercise set 10.3*

**1** Since $8/4 = 2$, the ratio of this sequence is 2. The next three terms are 16, 32, 64. Since the first term is 4, $s_n = 4 \cdot 2^{n-1} = 2^{n+1}$.

**7** Here $a = 6$, $r = 1/3$; so $s_6 = 6(1/3)^5 = 6/243 = 2/81$.

**15** There are five terms, with ratio 2, first term $1/2$, and fifth term 8. By GM.2,

$$S_5 = \frac{\dfrac{1}{2} - 2 \cdot 8}{1 - 2} = 16 - \frac{1}{2} = \frac{31}{2}.$$

**21** By GM.1,

$$S_8 = \frac{3 - 3\left(\dfrac{2}{3}\right)^8}{1 - \dfrac{2}{3}} = \frac{3 - \dfrac{256}{2{,}187}}{\dfrac{1}{3}} = 3\left(3 - \frac{256}{2{,}187}\right)$$

$$= 9 - \frac{256}{729} = \frac{6{,}305}{729}.$$

**25** His bets form a geometric sequence whose first term is 2 and whose ratio is 2. Hence

$$S_8 = \frac{2(1 - 2^8)}{1 - 2} = 2(2^8 - 1) = 2(256 - 1) = 510.$$

381 ($\circ$) **Geometric sequences**

He lost $510.

**29**  Here, $a = 25$ and $r = 1/5$; so the sum of the series is

$$\frac{25}{1 - \dfrac{1}{5}} = \frac{125}{5 - 1} = \frac{125}{4}.$$

**39**  The ball drops 10 feet and then rebounds $(3/4)(10) = 15/2$ feet. It then drops $15/2$ feet and rebounds $(3/4)(15/2) = 45/8$ feet. Following this, it drops $45/8$ feet and rebounds $(3/4)(45/8) = 135/32$ feet. The diagram may help you visualize what is happening.

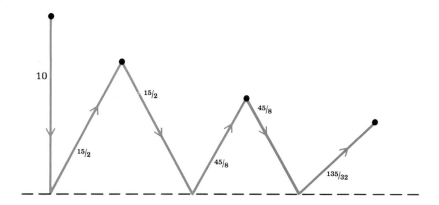

The distance traveled by the ball is $10 + 15 + 45/4 + 135/16 + \ldots$. If we temporarily ignore the initial 10-foot drop, we have a geometric series with $a = 15$ and $r = 3/4$. Its sum is $\dfrac{15}{1 - \dfrac{3}{4}} = \dfrac{60}{4 - 3} = 60$. We deduce that the ball traveled a total of $10 + 60 = 70$ feet before coming to rest.

**45**  $0.\bar{3} = \dfrac{3}{10} + \dfrac{3}{10^2} + \dfrac{3}{10^3} + \ldots$. Hence,

$$0.\bar{3} = \frac{\dfrac{3}{10}}{1 - \dfrac{1}{10}} = \frac{3}{10 - 1} = \frac{3}{9} = \frac{1}{3}.$$

## (10.4) (Binomial expansions)

When solving a quadratic equation, we made frequent use of the fact that $(x + y)^2 = x^2 + 2xy + y^2$. In this section we will develop a formula that enables us to express an *arbitrary* positive integral power of $x + y$ as a polynomial. Such an expression for $(x + y)^n$ ($n \in N$) is called a *binomial expansion* or a *binomial series*. For $n = 1, 2, 3, 4, 5$ the following expansions may be obtained by actually multiplying $x + y$ by itself an appropriate number of times:

$(x + y)^1 = x + y$

$(x + y)^2 = x^2 + 2xy + y^2$

$(x + y)^3 = x^3 + 3x^2y + 3xy^2 + y^3$

$(x + y)^4 = x^4 + 4x^3y + 6x^2y^2 + 4xy^3 + y^4$

$(x + y)^5 = x^5 + 5x^4y + 10x^3y^2 + 10x^2y^3 + 5xy^4 + y^5$

In each case, the listed observations can be made about the expansion of $(x + y)^n$:

**1** There are $n + 1$ terms.

**2** In each term the sum of the exponents of $x$ and $y$ is $n$.

**3** The exponent of $x$ decreases by 1 and that of $y$ increases by 1 as we move from any term to the next term.

**4** The first term in the expansion is $x^n$.

**5** The second term is $nx^{n-1}y$.

**6** The third term is $\dfrac{n(n-1)}{2} x^{n-2}y^2$   $(n > 1)$.

**7** The fourth term is $\dfrac{n(n-1)(n-2)}{3 \cdot 2} x^{n-3}y^3$   $(n > 2)$.

**8** The fifth term is $\dfrac{n(n-1)(n-2)(n-3)}{4 \cdot 3 \cdot 2} x^{n-4}y^4$   $(n > 3)$.

**9** The last term is $y^n$.

Now let $r$ be a positive integer. Since products of the form $r(r - 1)(r - 2)...(3)(2)$ keep coming up in connection with binomial expansions, we define $r!$ (read "*r factorial*") to be the product of the first $r$ positive integers. Thus $1! = 1$, $2! = 2 \cdot 1 = 2$, $3! = 3 \cdot 2 \cdot 1 = 3 \cdot 2 = 6$, $4! = 4 \cdot 3 \cdot 2 \cdot 1 = 4 \cdot 3 \cdot 2 = 24$, and so on. Once this

*binomial expansion*
*binomial series*

*"r factorial"*

is done, we may state the formula for the binomial expansion of $(x + y)^n$.

---

$(\circ)$    BI.1   $(x + y)^n = x^n + \dfrac{n}{1!} x^{n-1} y + \dfrac{n(n-1)}{2!} x^{n-2} y^2 +$

$$\dfrac{n(n-1)(n-2)}{3!} x^{n-3} y^3 + \ldots +$$

$$\dfrac{n(n-1)(n-2)\ldots(n-r+2)}{(r-1)!} x^{n-r+1} y^{r-1}$$

$$+ \ldots + y^n.$$

---

*binomial theorem*

*mathematical induction*

The validity of BI.1 is known as the *binomial theorem*. Its proof involves the principle of *mathematical induction* and is beyond the scope of our present discussion. We must, therefore, be content to see how the binomial theorem can be applied. For our first example we replace $x$ with $a$ and $y$ with 3 in BI.1 to obtain

$$(a + 3)^5 = a^5 + \frac{5}{1!} a^4 \cdot 3 + \frac{5 \cdot 4}{2!} a^3 \cdot 3^2 + \frac{5 \cdot 4 \cdot 3}{3!} a^2 \cdot 3^3$$

$$+ \frac{5 \cdot 4 \cdot 3 \cdot 2}{4!} a \cdot 3^4 + 3^5$$

$$= a^5 + (5 \cdot 3)a^4 + (10 \cdot 9)a^3 + (10 \cdot 27)a^2 + (5 \cdot 81)a + 243$$

$$= a^5 + 15a^4 + 90a^3 + 270a^2 + 405a + 243.$$

As a second example we observe that

$$(x - 2y)^6 = x^6 + \frac{6}{1!} x^5(-2y) + \frac{6 \cdot 5}{2!} x^4(-2y)^2 + \frac{6 \cdot 5 \cdot 4}{3!} x^3(-2y)^3$$

$$+ \frac{6 \cdot 5 \cdot 4 \cdot 3}{4!} x^2(-2y)^4 + \frac{6 \cdot 5 \cdot 4 \cdot 3 \cdot 2}{5!} x(-2y)^5 + (-2y)^6$$

$$= x^6 + (6)(-2)x^5 y + (15)(4)x^4 y^2 + (20)(-8)x^3 y^3$$

$$+ (15)(16)x^2 y^4 + (6)(-32)xy^5 + (1)(64)y^6$$

$$= x^6 - 12x^5 y + 60x^4 y^2 - 160x^3 y^3 + 240x^2 y^4$$

$$- 192xy^5 + 64y^6.$$

Finally, we point out that the formula for the $r$th term of the binomial expansion of $(x + y)^n$ reads thus.

$(\circ)$ BI.2 $\dfrac{n(n-1)...(n-r+2)}{(r-1)!} x^{n-r+1} y^{r-1}$

This formula is fairly easy to remember if you just keep the following items straight in your mind:

**1** The exponent of $y$ is $r-1$ and the denominator of the coefficient is $(r-1)!$.

**2** The exponent of $x$ is $n-(r-1)$.

**3** The numerator of the coefficient is $n(n-1)\ldots(n-r+2)$ where the last factor of the product is one greater than the exponent of $x$.

Let's find the fourth term of the binomial expansion of $(s-2t)^{12}$. The idea is to use BI.2 with $x=s$, $y=-2t$, $n=12$, and $r=4$. Items 1 to 3 above translate into the present situation as follows:

**1** The exponent of $(-2t)$ is 3 and the denominator of the coefficient is $3!$.

**2** The exponent of $s$ is $12-3=9$.

**3** The numerator of the coefficient is $(12)(11)(10)$. Note here that $10=9+1$.

We conclude that the fourth term is $[(12)(11)(10)/3!]\ s^9(-2t)^3 = (220)(-8)s^9\,t^3 = -1{,}760s^9\,t^3$.

## Exercise set 10.4

Write out the binomial expansion of each of the following.

$(\circ)$

**1** $(x+2)^4$         **2** $(t-4)^3$

**3** $(3w-2)^5$       **4** $(2s+1)^4$

**5** $\left(2x-\dfrac{y}{3}\right)^6$      **6** $\left(2z-\dfrac{v}{5}\right)^5$

**7** $\left(\dfrac{a}{3}+b^2\right)^3$      **8** $\left(\dfrac{c}{6}+\dfrac{1}{2}\right)^6$

Write out the first five terms of each of the following binomial expansions.

$(\circ)$

**9** $(x+y)^{18}$        **10** $(s+2t)^{14}$

**11** $\left(x-\dfrac{y}{3}\right)^{24}$      **12** $\left(s-\dfrac{t}{7}\right)^{21}$

**13** $(2a + b)^{13}$

**14** $\left(\dfrac{u}{3} + v\right)^9$

**15** $(h - 3)^{12}$

**16** $(k - 5)^{15}$

*Find the indicated term of each binomial series.*

$(\circ)$ **17** The sixth term of $(x + 2y)^{19}$

**18** The fourth term of $(s - 3)^{23}$

**19** The third term of $(a - 6)^{32}$

**20** The third term of $(b + 5)^{36}$

**21** The fourth term of $\left(t - \dfrac{1}{2}\right)^{28}$

**22** The seventh term of $\left(x + \dfrac{1}{3}\right)^{15}$

The binomial theorem can easily be used to estimate powers of numbers close to 1. For example, suppose you wish to find $(1.02)^8$ to the nearest hundredth. Using the fact that $(1.02)^8 = (1 + 0.02)^8$ you apply the binomial theorem as follows:

$$(1 + 0.02)^8 = 1^8 + \frac{8}{1!}(1)^7(0.02) + \frac{8 \cdot 7}{2!}(1)^6(0.02)^2$$

$$+ \frac{8 \cdot 7 \cdot 6}{3!}(1)^5(0.02)^3 + \ldots$$

$$= 1 + 8(0.02) + 28(0.0004) + 56(0.000008) + \ldots$$

$$= 1 + 0.16 + 0.0112 + 0.000448 + \ldots$$

$$\approx 1.17.$$

*Use the binomial theorem to estimate each of the following to the nearest hundredth.*

$(\circ)$ **23** $(1.03)^6$

**24** $(1.04)^7$

**25** $(0.98)^9$

**26** $(0.97)^{10}$

**27** $(1.01)^{18}$

**28** $(0.99)^{16}$

If $D$ dollars is on deposit in an account paying $r$ percent interest compounded $k$ times a year, then the balance at the end of $n$ years is $A_n = D\left(1 + \dfrac{r}{100k}\right)^{nk}$. The binomial theorem can be used to estimate the number $\left(1 + \dfrac{r}{100k}\right)^{nk}$ to any desired degree of accuracy.

(∘)  **29**  One thousand dollars is invested in an account paying 6 percent interest compounded monthly. Use the binomial theorem to compute the balance (to the nearest dollar) at the end of 2 years.

**30**  One hundred dollars is invested in an account paying 4 percent interest compounded quarterly. Use the binomial theorem to compute the balance (to the nearest dollar) at the end of 5 years.

**31**  The population of a town is expanding at the rate of 1 percent per year. With an initial population of 10,000, use the binomial theorem to determine the size of the town after 10 years.

**32**  Lumber production in the United States was $(7.43)10^9$ board feet one year. The production 2 years later was approximately 98-1/2 percent of this figure. If this trend were to continue (in other words, if the production for each year is 98-1/2 percent of the figure 2 years earlier) what would the production be 12 years later? The answer should be expressed in the form $a \cdot 10^9$ where $a$ is estimated to the nearest hundredth.

There are occasions when an infinite binomial series can be given a meaning. For example, it is true that with $|x| < 1$ and $m$ any real number, the infinite binomial expansion

$$(1 + x)^m = 1 + \frac{m}{1!} x + \frac{m(m-1)}{2!} x^2 + \frac{m(m-1)(m-2)}{3!} x^2$$

$$+ \ldots + \frac{m(m-1)\ldots(m-r+2)}{(r-1)!} x^{r-1} + \ldots$$

is valid in the sense that the difference between the left and right members of the expression gets closer and closer to 0 as more and more terms are added to the right member. Assuming this to be true, things like $\sqrt{0.98}$ can be computed to any desired degree of accuracy. Out to the nearest hundredth we have

$$\sqrt{0.98} = (1 - 0.02)^{1/2}$$

$$= 1 + \frac{\frac{1}{2}}{1!}(-0.02) + \frac{\frac{1}{2}\left(-\frac{1}{2}\right)}{2!}(-0.02)^2$$

$$+ \frac{\frac{1}{2}\left(-\frac{1}{2}\right)\left(-\frac{3}{2}\right)}{3!}(-0.02)^3 + \ldots$$

$$= 1 + \frac{1}{2}(-0.02) - \frac{1}{8}(0.0004) + \frac{1}{16}(-0.000008) + \ldots$$

$$= 1 - 0.01 - 0.00005 - \ldots$$

$$\approx 0.99.$$

Write out the first four terms of the binomial series expansions of each of the following:

(∘)   **33**   $(1 + x)^{1/3}$                           **34**   $(1 + x)^{3/2}$

         **35**   $(1 + x)^{-1}$                          **36**   $(1 + x)^{-2/3}$

Compute to the nearest hundredth.

(∘)   **37**   $\sqrt{1.08}$                                **38**   $\sqrt[3]{1.09}$

         **39**   $\dfrac{1}{0.98}$                             **40**   $\dfrac{1}{1.03}$

         **41**   $\sqrt[4]{0.80}$                             **42**   $(0.96)^{3/2}$

(∘)

*Sample solutions for exercise set 10.4*

**1**   $(x + 2)^4 \;=\; x^4 + \dfrac{4}{1!}x^3 \cdot 2 + \dfrac{4 \cdot 3}{2!}x^2 \cdot 2^2 + \dfrac{4 \cdot 3 \cdot 2}{3!}x \cdot 2^3 + 2^4$

$\qquad\qquad\quad\; = x^4 + (4 \cdot 2)x^3 + (6 \cdot 4)x^2 + (4 \cdot 8)x + 16$

$\qquad\qquad\quad\; = x^4 + 8x^3 + 24x^2 + 32x + 16$

**9**   $(x + y)^{18} \;=\; x^{18} + \dfrac{18}{1!}x^{17}y + \dfrac{18 \cdot 17}{2!}x^{16}y^2 + \dfrac{18 \cdot 17 \cdot 16}{3!}x^{15}y^3$

$\qquad\qquad\qquad +\; \dfrac{18 \cdot 17 \cdot 16 \cdot 15}{4!}x^{14}y^4 + \ldots$

$\qquad\qquad\; = x^{18} + 18x^{17}y + 153x^{16}y^2 + 816x^{15}y^3$

$\qquad\qquad\qquad +\; 3{,}060x^{14}y^4 + \ldots.$

**17**   You want the sixth term of $(x + 2y)^{19}$. The exponent of $2y$ is 5 and the denominator of the coefficient is 5!. The exponent of $x$ must, therefore, be $19 - 5 = 14$, and the numerator of the coefficient is $(19)(18)(17)(16)(15)$. Hence, the term is

$$\dfrac{(19)(18)(17)(16)(15)}{5!}x^{14}(2y)^5 = 372{,}096x^{14}y^5.$$

**23**   $(1 + 0.03)^6 \;=\; 1^6 + \dfrac{6}{1!}(1^5)(0.03) + \dfrac{6 \cdot 5}{2!}(1^4)(0.03)^2$

$\qquad\qquad\qquad +\; \dfrac{6 \cdot 5 \cdot 4}{3!}(1^3)(0.03)^3 + \ldots$

$\qquad\qquad\; = 1 \; + \; 6(0.03) + 15(0.0009) + 20(0.000027) + \ldots$

$$= 1 + 0.18 + 0.0135 + 0.00054 + \ldots$$

$$\approx 1.19.$$

29  We must compute $A_2 = 1{,}000\left(1 + \dfrac{6}{1{,}200}\right)^{24} = 1{,}000(1 + 0.005)^{24}$ to the nearest dollar. By the binomial theorem,

$$(1 + 0.005)^{24} = 1 \quad + 24(0.005) + \frac{24 \cdot 23}{2!}(0.005)^2$$

$$+ \frac{24 \cdot 23 \cdot 22}{3!}(0.005)^3 + \ldots$$

$$= 1 \quad + 0.120 + (276)(0.000025)$$

$$+ (2{,}024)(0.000000125) + \ldots$$

$$= 1 \quad + 0.120 + 0.0069 + 0.000253 \approx 1.127.$$

Hence, $A_2 = (1{,}000)(1.127) = \$1{,}127.$

33  $(1 + x)^{1/3} = 1 + \dfrac{\dfrac{1}{3}}{1!}x + \dfrac{\dfrac{1}{3}\left(-\dfrac{2}{3}\right)}{2!}x^2 + \dfrac{\dfrac{1}{3}\left(-\dfrac{2}{3}\right)\left(-\dfrac{5}{3}\right)}{3!}x^3 + \ldots$

$$= 1 + \frac{1}{3}x - \frac{1}{9}x^2 + \frac{5}{81}x^3 - \ldots$$

37  $(1 + 0.08)^{1/2} = 1 + \dfrac{\dfrac{1}{2}}{1!}(0.08) + \dfrac{\dfrac{1}{2}\left(-\dfrac{1}{2}\right)}{2!}(0.08)^2$

$$+ \frac{\dfrac{1}{2}\left(-\dfrac{1}{2}\right)\left(-\dfrac{3}{2}\right)}{3!}(0.08)^3$$

$$= 1 + \frac{1}{2}(0.08) - \frac{1}{8}(0.0064) + \frac{1}{16}(0.000512) - \ldots$$

$$= 1 + 0.04 - 0.0008 + 0.000032 - \ldots$$

$$\approx 1.04$$

# (∘)

## Important algebraic statements

*The following laws, statements, forms, and formulas are important to your understanding of algebra, and you should recognize their significance. The page references in the margin will direct you to the places in the text where this material is introduced.*

### Addition and subtraction

page 15 A.1 $(a + b) + c = a + (b + c)$      *(Associative law for addition)*

page 15 A.2 $a + b = b + a$      *(Commutative law for addition)*

page 15 A.3 $a + 0 = a = 0 + a$      *(Additive identity law)*

page 15 A.4 *For each real number a, there corresponds a unique real number $-a$, called negative a or the additive inverse of a, such that $a + (-a) = 0 = (-a) + a$*

                         *(Additive inverse law)*

### Multiplication and division

page 19 M.1 $(ab)c = a(bc)$      *(Associative law for multiplication)*

page 19 M.2 $ab = ba$      *(Commutative law for multiplication)*

page 19 M.3 $a \cdot 1 = a = 1 \cdot a$      *(Multiplicative identity law)*

page 19 M.4 *For each nonzero real number a, there corresponds a unique real number $\dfrac{1}{a}$, called the multiplicative inverse or the reciprocal of a, such that $a \cdot \left(\dfrac{1}{a}\right) = 1 = \left(\dfrac{1}{a}\right) \cdot a$*

                         *(Multiplicative inverse law)*

### Order properties of real numbers

page 26 0.1 *The sum of two positive numbers is positive.*

page 26 0.2 *The product of two positive numbers is positive.*

page 27 0.3 *Given $a, b \in R$, exactly one of the following is true: $a < b$, $a = b$, or $a > b$.*      *(Trichotomy law)*

page 27 0.4 *Given $a, b, c \in R$, if $a < b$ and $b < c$, then $a < c$.*

                         *(Transitive law)*

### Exponents

page 33 E.1 $b^m \cdot b^n = b^{m+n}$

page 33 E.2 $(b^m)^n = b^{mn}$

page 33 E.3 $(ab)^n = a^n b^n$

pages 34 and 139 E.4 $\dfrac{b^m}{b^n} = b^{m-n}$

## Binomial products

page 47 BP.1 $(x + a)(x - a) = x^2 - a^2$

page 47 BP.2 $(x + a)^2 = x^2 + 2ax + a^2$

page 47 BP.3 $(x + a)(x + b) = x^2 + (a + b)x + ab$

page 47 BP.4 $(ax + b)(cx + d) = acx^2 + (ad + bc)x + bd$

## Quadratic factoring

page 53 BP.1a $x^2 - a^2 = (x + a)(x - a)$

page 53 BP.2a $x^2 + 2ax + a^2 = (x + a)^2$

page 53 BP.3a $x^2 + (a + b)x + ab = (x + a)(x + b)$

page 53 BP.4a $acx^2 + (ad + bc)x + bd = (ax + b)(cx + d)$

## Fractions

page 62 F.1 $\dfrac{a}{b} = \dfrac{c}{c}$ *if and only if* $ad = bc$ $\quad (b, d \neq 0)$

page 62 F.2 $\dfrac{a}{b} = \dfrac{ac}{bc}$ $\quad (b, c \neq 0)$

page 75 F.3 $\dfrac{a}{b} + \dfrac{c}{b} = \dfrac{a + c}{b}$ $\quad (b \neq 0)$

page 75 F.4 $\dfrac{a}{b} + \dfrac{c}{d} = \dfrac{ad + bc}{bd}$ $\quad (b, d \neq 0)$

page 77 F.5 $\dfrac{a}{b} - \dfrac{c}{d} = \dfrac{a}{b} + \dfrac{-c}{d}$ $\quad (b, d \neq 0)$

page 80 F.6 $\dfrac{a}{b} \cdot \dfrac{c}{d} = \dfrac{ac}{bd}$ $\quad (b, d \neq 0)$

page 81 F.7 $\dfrac{a}{b} \div \dfrac{c}{d} = \dfrac{a}{b} \cdot \dfrac{d}{c}$ $\quad (b, c, d \neq 0)$

## Equivalences

page 102 EQ.1 *The equation* $P = Q$ *is equivalent to the equation* $P + R = Q + R.$

page 102 EQ.2 *The equation* $P = Q$ *is equivalent to the equation* $P \cdot R = Q \cdot R$ *(provided R is never 0).*

## Inequalities

pages 127 and 129 I.1 *If an expression is added to both members of an inequality, the resulting inequality, taken in the same sense, is equivalent to the original one.* $P < Q$ *is equivalent to* $P + R < Q + R.$

*pages 128 and 129* I.2 *If both members of an inequality are multiplied by an expression that is always positive, the resulting inequality, taken in the same sense, is equivalent to the original one. $P < Q$ is equivalent to $P \cdot R < Q \cdot R$ provided R is always positive.*

*page 129* I.3 *If both members of an inequality are multiplied by an expression that is always negative, the resulting inequality, taken in the opposite sense, is equivalent to the original one. $P < Q$ is equivalent to $P \cdot R > Q \cdot R$ provided R is always negative.*

### Radicals

*page 149* RN.1 $\sqrt[n]{b^m} = (\sqrt[n]{b})^m$  $(m, n \in I; n \geq 2)$

*page 149* RN.2 $\sqrt[n]{ab} = \sqrt[n]{a}\,\sqrt[n]{b}$  $(n \in N; n \geq 2)$

*page 150* RN.3 $\sqrt[n]{\dfrac{a}{b}} = \dfrac{\sqrt[n]{a}}{\sqrt[n]{b}}$  $(b \neq 0; n \in N; n \geq 2)$

*page 153* RN.4 $\sqrt[n]{b^m} = \sqrt[cn]{b^{cm}}$  $(m, n, c \in I; c, n \geq 2)$

*page 153* RN.4a $\sqrt[n]{b^m} = \sqrt[n/k]{b^{m/k}}$  $(m, n, k \in I; k, n \geq 2; k\,|\,m, n)$

### Complex numbers

*page 163* CN.1 $(a + bi) + (c + di) = (a + c) + (b + d)i$

*page 165* CN.2 $(a + bi) - (c + di) = (a - c) + (b - d)i$

*page 165* CN.3 $(a + bi)(c + di) = (ac - bd) + (ad + bc)i$

*page 166* CN.4 $\dfrac{a + bi}{c + di} = \dfrac{ac - bd}{c^2 + d^2} + \dfrac{ad + bc}{c^2 + d^2}\, i$

### Quadratic equations

*page 178* QE.1 $ax^2 + bx + c = 0$  $(a \neq 0)$

*page 179* QE.2 $x = \dfrac{-b \pm \sqrt{b^2 - 4ac}}{2a}$

*page 249* QE.3 $x^2 + y^2 = r^2$  $(r > 0)$

*page 251* QE.4 $a^2 x^2 + b^2 y^2 = c^2$

*page 252* QE.5 $a^2 x^2 - b^2 y^2 = c^2$  $(a, b, c > 0)$

*page 253* QE.6 $b^2 y^2 - a^2 x^2 = c^2$  $(a, b, c > 0)$

### Linear equations

*page 221* LEQ.1 $Ax + By + C = 0$  $(A \text{ or } B \neq 0)$  *(Standard form)*

*page 221* LEQ.2 $y = mx + b$  *(Slope-intercept form)*

*page 221* LEQ.3 $y - y_1 = m(x - x_1)$  *(Point-slope form)*

### Distance formula

*page 215* DF.1 $d = \sqrt{(x_2 - x_1)^2 + (y_2 - y_1)^2}$

## Variation

page 257 V.1 $y = kx$        *(x a positive constant)*

page 257 V.2 $y = kx^2$        *(k a positive constant)*

page 258 V.3 $y = kx^n$        *(k a positive constant)*

page 258 V.4 $y = \dfrac{k}{x}$        *(k a positive constant)*

## Logarithms

page 289 LO.1 $b^{\log_b x} = x$    $(x \in R, x > 0)$

page 289 LO.2 $y = \log_b b^y$    $(y \in R)$

page 290 LO.3 $\log_b(x_1 x_2) = \log_b x_1 + \log_b x_2$    $(x_1, x_2 > 0)$

page 290 LO.4 $\log_b \dfrac{x_1}{x_2} = \log_b x_1 - \log_b x_2$    $(x_1, x_2 > 0)$

page 290 LO.5 $\log_b x_1{}^r = r(\log_b x_1)$    $(x_1 > 0)$

## Series and expansions

page 369 AM.1 $1 + 2 + \ldots + k = \dfrac{k(k+1)}{2}$

page 370 AM.2 $S_k = \dfrac{k}{2}[2c + (k-1)d]$

page 370 AM.3 $S_k = \dfrac{k}{2}(s_1 + s_k)$

page 375 GM.1 $S_k = \dfrac{a - ar^k}{1 - r}$    $(r \neq 1)$

page 376 GM.2 $S_k = \dfrac{a - rs_k}{1 - r}$    $(r \neq 1)$.

page 377 GM.3 $a + ar + \ldots + ar^{n-1} + \ldots = \dfrac{a}{1 - r}$

page 384 BI.1 $(x + y)^n = x^n + \dfrac{n}{1!}x^{n-1}y + \dfrac{n(n-1)}{2!}x^{n-2}y^2 +$

$$\dfrac{n(n-1)(n-2)}{3!}x^{n-3}y^3 + \ldots +$$

$$\dfrac{n(n-1)(n-2)\ldots(n-r+2)}{(r-1)!}x^{n-r+1}y^{r-1}$$

$$+ \ldots + y^n$$

page 385 BI.2 $\dfrac{n(n-1)\ldots(n-r+2)}{(r-1)!}x^{n-r+1}y^{r-1}$ .

---

## (∘)
## *Formulas from geometry*

**Square with sides of length s**

(∘)   *Perimeter* = 4*s*

(∘)   *Area* = *s*²

**Rectangle with length b and width a**

(∘)   *Perimeter* = 2*a* + 2*b*

(∘)   *Area* = *ab*

**Parallelogram with base b and altitude h**

(∘)   *Perimeter* = 2*a* + 2*b*

(∘)   *Area* = *bh*

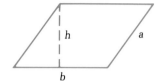

**Trapezoid with parallel sides a and b and altitude h**

(∘)   $Area = \dfrac{h}{2}(a + b)$

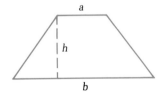

**Triangle with sides of length c and c, base b, and altitude h**

(∘)   *Perimeter* = *a* + *b* + *c*

(∘)   $Area = \dfrac{1}{2}bh$

$= \sqrt{s(s - a)(s - b)(s - c)}$

$where\ s = \dfrac{1}{2}(a + b + c)$

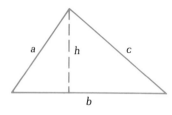

**Right triangle with sides *a* and *b* and hypotenuse *c***

$(\circ)$    $c^2 = a^2 + b^2$

**Circle with radius *r***

$(\circ)$    *Diameter* $d = 2r$

$(\circ)$    *Circumference* $= 2\pi r = \pi d$

$(\circ)$    *Area* $= \pi r^2$

**Rectangular prism with length *a*, width *b*, height *c***

$(\circ)$    *Surface area* $= 2(ab + ac + bc)$

$(\circ)$    *Volume* $= abc$

**Sphere of radius *r***

$(\circ)$    *Surface area* $= 4\pi r^2$

$(\circ)$    *Volume* $= \dfrac{4}{3}\pi r^3$

**Right circular cylinder with radius *r* and height *h***

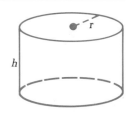

$(\circ)$    *Lateral surface area* $= 2\pi rh$

$(\circ)$    *Volume* $= \pi r^2 h$

| Number | Square | Square root | Number | Square | Square root |
|--------|--------|-------------|--------|--------|-------------|
| 1 | 1 | 1.000 | 51 | 2,601 | 7.141 |
| 2 | 4 | 1.414 | 52 | 2,704 | 7.211 |
| 3 | 9 | 1.732 | 53 | 2,809 | 7.280 |
| 4 | 16 | 2.000 | 54 | 2,916 | 7.348 |
| 5 | 25 | 2.236 | 55 | 3,025 | 7.416 |
| 6 | 36 | 2.449 | 56 | 3,136 | 7.483 |
| 7 | 49 | 2.646 | 57 | 3,249 | 7.550 |
| 8 | 64 | 2.828 | 58 | 3,364 | 7.616 |
| 9 | 81 | 3.000 | 59 | 3,481 | 7.681 |
| 10 | 100 | 3.162 | 60 | 3,600 | 7.746 |
| 11 | 121 | 3.317 | 61 | 3,721 | 7.810 |
| 12 | 144 | 3.464 | 62 | 3,844 | 7.874 |
| 13 | 169 | 3.606 | 63 | 3,969 | 7.937 |
| 14 | 196 | 3.742 | 64 | 4,096 | 8.000 |
| 15 | 225 | 3.873 | 65 | 4,225 | 8.062 |
| 16 | 256 | 4.000 | 66 | 4,356 | 8.124 |
| 17 | 289 | 4.123 | 67 | 4,489 | 8.185 |
| 18 | 324 | 4.243 | 68 | 4,624 | 8.246 |
| 19 | 361 | 4.359 | 69 | 4,761 | 8.307 |
| 20 | 400 | 4.472 | 70 | 4,900 | 8.367 |
| 21 | 441 | 4.583 | 71 | 5,041 | 8.426 |
| 22 | 484 | 4.690 | 72 | 5,184 | 8.485 |
| 23 | 529 | 4.796 | 73 | 5,329 | 8.544 |
| 24 | 576 | 4.899 | 74 | 5,476 | 8.602 |
| 25 | 625 | 5.000 | 75 | 5,625 | 8.660 |
| 26 | 676 | 5.099 | 76 | 5,776 | 8.718 |
| 27 | 729 | 5.196 | 77 | 5,929 | 8.775 |
| 28 | 784 | 5.291 | 78 | 6,084 | 8.832 |
| 29 | 841 | 5.385 | 79 | 6,241 | 8.888 |
| 30 | 900 | 5.477 | 80 | 6,400 | 8.944 |
| 31 | 961 | 5.568 | 81 | 6,561 | 9.000 |
| 32 | 1,024 | 5.657 | 82 | 6,724 | 9.055 |
| 33 | 1,089 | 5.745 | 83 | 6,889 | 9.110 |
| 34 | 1,156 | 5.831 | 84 | 7,056 | 9.165 |
| 35 | 1,225 | 5.916 | 85 | 7,225 | 9.220 |
| 36 | 1,296 | 6.000 | 86 | 7,396 | 9.274 |
| 37 | 1,369 | 6.083 | 87 | 7,569 | 9.327 |
| 38 | 1,444 | 6.164 | 88 | 7,744 | 9.381 |
| 39 | 1,521 | 6.245 | 89 | 7,921 | 9.434 |
| 40 | 1,600 | 6.325 | 90 | 8,100 | 9.487 |
| 41 | 1,681 | 6.403 | 91 | 8,281 | 9.539 |
| 42 | 1,764 | 6.481 | 92 | 8,464 | 9.592 |
| 43 | 1,849 | 6.557 | 93 | 8,649 | 9.644 |
| 44 | 1,936 | 6.633 | 94 | 8,836 | 9.695 |
| 45 | 2,025 | 6.708 | 95 | 9,025 | 9.747 |
| 46 | 2,116 | 6.782 | 96 | 9,216 | 9.798 |
| 47 | 2,209 | 6.856 | 97 | 9,409 | 9.849 |
| 48 | 2,304 | 6.928 | 98 | 9,604 | 9.899 |
| 49 | 2,401 | 7.000 | 99 | 9,801 | 9.950 |
| 50 | 2,500 | 7.071 | 100 | 10,000 | 10.000 |

## (∘)
## *Common*
## *logarithms*

| x | 0 | 1 | 2 | 3 | 4 | 5 | 6 | 7 | 8 | 9 |
|---|---|---|---|---|---|---|---|---|---|---|
| 1.0 | .0000 | .0043 | .0086 | .0128 | .0170 | .0212 | .0253 | .0294 | .0334 | .0374 |
| 1.1 | .0414 | .0453 | .0492 | .0531 | .0569 | .0607 | .0645 | .0682 | .0719 | .0755 |
| 1.2 | .0792 | .0828 | .0864 | .0899 | .0934 | .0969 | .1004 | .1038 | .1072 | .1106 |
| 1.3 | .1139 | .1173 | .1206 | .1239 | .1271 | .1303 | .1335 | .1367 | .1399 | .1430 |
| 1.4 | .1461 | .1492 | .1523 | .1553 | .1584 | .1614 | .1644 | .1673 | .1703 | .1732 |
| 1.5 | .1761 | .1790 | .1818 | .1847 | .1875 | .1903 | .1931 | .1959 | .1987 | .2014 |
| 1.6 | .2041 | .2068 | .2095 | .2122 | .2148 | .2175 | .2201 | .2227 | .2253 | .2279 |
| 1.7 | .2304 | .2330 | .2355 | .2380 | .2405 | .2430 | .2455 | .2480 | .2504 | .2529 |
| 1.8 | .2553 | .2577 | .2601 | .2625 | .2648 | .2672 | .2695 | .2718 | .2742 | .2765 |
| 1.9 | .2788 | .2810 | .2833 | .2856 | .2878 | .2900 | .2923 | .2945 | .2967 | .2989 |
| 2.0 | .3010 | .3032 | .3054 | .3075 | .3096 | .3118 | .3139 | .3160 | .3181 | .3201 |
| 2.1 | .3222 | .3243 | .3263 | .3284 | .3304 | .3324 | .3345 | .3365 | .3385 | .3404 |
| 2.2 | .3424 | .3444 | .3464 | .3483 | .3502 | .3522 | .3541 | .3560 | .3579 | .3598 |
| 2.3 | .3617 | .3636 | .3655 | .3674 | .3692 | .3711 | .3729 | .3747 | .3766 | .3784 |
| 2.4 | .3802 | .3820 | .3838 | .3856 | .3874 | .3892 | .3909 | .3927 | .3945 | .3962 |
| 2.5 | .3979 | .3997 | .4014 | .4031 | .4048 | .4065 | .4082 | .4099 | .4116 | .4133 |
| 2.6 | .4150 | .4166 | .4183 | .4200 | .4216 | .4232 | .4249 | .4265 | .4281 | .4298 |
| 2.7 | .4314 | .4330 | .4346 | .4362 | .4378 | .4393 | .4409 | .4425 | .4440 | .4456 |
| 2.8 | .4472 | .4487 | .4502 | .4518 | .4533 | .4548 | .4564 | .4579 | .4594 | .4609 |
| 2.9 | .4624 | .4639 | .4654 | .4669 | .4683 | .4698 | .4713 | .4728 | .4742 | .4757 |
| 3.0 | .4771 | .4786 | .4800 | .4814 | .4829 | .4843 | .4857 | .4871 | .4886 | .4900 |
| 3.1 | .4914 | .4928 | .4942 | .4955 | .4969 | .4983 | .4997 | .5011 | .5024 | .5038 |
| 3.2 | .5051 | .5065 | .5079 | .5092 | .5105 | .5119 | .5132 | .5145 | .5159 | .5172 |
| 3.3 | .5185 | .5198 | .5211 | .5224 | .5237 | .5250 | .5263 | .5276 | .5289 | .5302 |
| 3.4 | .5315 | .5328 | .5340 | .5353 | .5366 | .5378 | .5391 | .5403 | .5416 | .5428 |
| 3.5 | .5441 | .5453 | .5465 | .5478 | .5490 | .5502 | .5514 | .5527 | .5539 | .5551 |
| 3.6 | .5563 | .5575 | .5587 | .5599 | .5611 | .5623 | .5635 | .5647 | .5658 | .5670 |
| 3.7 | .5682 | .5694 | .5705 | .5717 | .5729 | .5740 | .5752 | .5763 | .5775 | .5786 |
| 3.8 | .5798 | .5809 | .5821 | .5832 | .5843 | .5855 | .5866 | .5877 | .5888 | .5899 |
| 3.9 | .5911 | .5922 | .5933 | .5944 | .5955 | .5966 | .5977 | .5988 | .5999 | .6010 |
| 4.0 | .6021 | .6031 | .6042 | .6053 | .6064 | .6075 | .6085 | .6096 | .6107 | .6117 |
| 4.1 | .6128 | .6138 | .6149 | .6160 | .6170 | .6180 | .6191 | .6201 | .6212 | .6222 |
| 4.2 | .6232 | .6243 | .6253 | .6263 | .6274 | .6284 | .6294 | .6304 | .6314 | .6325 |
| 4.3 | .6335 | .6345 | .6355 | .6365 | .6375 | .6385 | .6395 | .6405 | .6415 | .6425 |
| 4.4 | .6435 | .6444 | .6454 | .6464 | .6474 | .6484 | .6493 | .6503 | .6513 | .6522 |
| 4.5 | .6532 | .6542 | .6551 | .6561 | .6571 | .6580 | .6590 | .6599 | .6609 | .6618 |
| 4.6 | .6628 | .6637 | .6646 | .6656 | .6665 | .6675 | .6684 | .6693 | .6702 | .6712 |
| 4.7 | .6721 | .6730 | .6739 | .6749 | .6758 | .6767 | .6776 | .6785 | .6794 | .6803 |
| 4.8 | .6812 | .6821 | .6830 | .6839 | .6848 | .6857 | .6866 | .6875 | .6884 | .6893 |
| 4.9 | .6902 | .6911 | .6920 | .6928 | .6937 | .6946 | .6955 | .6964 | .6972 | .6981 |
| 5.0 | .6990 | .6998 | .7007 | .7016 | .7024 | .7033 | .7042 | .7050 | .7059 | .7067 |
| 5.1 | .7076 | .7084 | .7093 | .7101 | .7110 | .7118 | .7126 | .7135 | .7143 | .7152 |
| 5.2 | .7160 | .7168 | .7177 | .7185 | .7193 | .7202 | .7210 | .7218 | .7226 | .7235 |
| 5.3 | .7243 | .7251 | .7259 | .7267 | .7275 | .7284 | .7292 | .7300 | .7308 | .7316 |
| 5.4 | .7324 | .7332 | .7340 | .7348 | .7356 | .7364 | .7372 | .7380 | .7388 | .7396 |

| x | 0 | 1 | 2 | 3 | 4 | 5 | 6 | 7 | 8 | 9 |
|---|---|---|---|---|---|---|---|---|---|---|
| 5.5 | .7404 | .7412 | .7419 | .7427 | .7435 | .7443 | .7451 | .7459 | .7466 | .7474 |
| 5.6 | .7482 | .7490 | .7497 | .7505 | .7513 | .7520 | .7528 | .7536 | .7543 | .7551 |
| 5.7 | .7559 | .7566 | .7574 | .7582 | .7589 | .7597 | .7604 | .7612 | .7619 | .7627 |
| 5.8 | .7634 | .7642 | .7649 | .7657 | .7664 | .7672 | .7679 | .7686 | .7694 | .7701 |
| 5.9 | .7709 | .7716 | .7723 | .7731 | .7738 | .7745 | .7752 | .7760 | .7767 | .7774 |
| 6.0 | .7782 | .7789 | .7796 | .7803 | .7810 | .7818 | .7825 | .7832 | .7839 | .7846 |
| 6.1 | .7853 | .7860 | .7868 | .7875 | .7882 | .7889 | .7896 | .7903 | .7910 | .7917 |
| 6.2 | .7924 | .7931 | .7938 | .7945 | .7952 | .7959 | .7966 | .7973 | .7980 | .7987 |
| 6.3 | .7993 | .8000 | .8007 | .8014 | .8021 | .8028 | .8035 | .8041 | .8048 | .8055 |
| 6.4 | .8062 | .8069 | .8075 | .8082 | .8089 | .8096 | .8102 | .8109 | .8116 | .8122 |
| 6.5 | .8129 | .8136 | .8142 | .8149 | .8156 | .8162 | .8169 | .8176 | .8182 | .8189 |
| 6.6 | .8195 | .8202 | .8209 | .8215 | .8222 | .8228 | .8235 | .8241 | .8248 | .8254 |
| 6.7 | .8261 | .8267 | .8274 | .8280 | .8287 | .8293 | .8299 | .8306 | .8312 | .8319 |
| 6.8 | .8325 | .8331 | .8338 | .8344 | .8351 | .8357 | .8363 | .8370 | .8376 | .8382 |
| 6.9 | .8388 | .8395 | .8401 | .8407 | .8414 | .8420 | .8426 | .8432 | .8439 | .8445 |
| 7.0 | .8451 | .8457 | .8463 | .8470 | .8476 | .8482 | .8488 | .8494 | .8500 | .8506 |
| 7.1 | .8513 | .8519 | .8525 | .8531 | .8537 | .8543 | .8549 | .8555 | .8561 | .8567 |
| 7.2 | .8573 | .8579 | .8585 | .8591 | .8597 | .8603 | .8609 | .8615 | .8621 | .8627 |
| 7.3 | .8633 | .8639 | .8645 | .8651 | .8657 | .8663 | .8669 | .8675 | .8681 | .8686 |
| 7.4 | .8692 | .8698 | .8704 | .8710 | .8716 | .8722 | .8727 | .8733 | .8739 | .8745 |
| 7.5 | .8751 | .8756 | .8762 | .8768 | .8774 | .8779 | .8785 | .8791 | .8797 | .8802 |
| 7.6 | .8808 | .8814 | .8820 | .8825 | .8831 | .8837 | .8842 | .8848 | .8854 | .8859 |
| 7.7 | .8865 | .8871 | .8876 | .8882 | .8887 | .8893 | .8899 | .8904 | .8910 | .8915 |
| 7.8 | .8921 | .8927 | .8932 | .8938 | .8943 | .8949 | .8954 | .8960 | .8965 | .8971 |
| 7.9 | .8976 | .8982 | .8987 | .8993 | .8998 | .9004 | .9009 | .9015 | .9020 | .9025 |
| 8.0 | .9031 | .9036 | .9042 | .9047 | .9053 | .9058 | .9063 | .9069 | .9074 | .9079 |
| 8.1 | .9085 | .9090 | .9096 | .9101 | .9106 | .9112 | .9117 | .9122 | .9128 | .9133 |
| 8.2 | .9138 | .9143 | .9149 | .9154 | .9159 | .9165 | .9170 | .9175 | .9180 | .9186 |
| 8.3 | .9191 | .9196 | .9201 | .9206 | .9212 | .9217 | .9222 | .9227 | .9232 | .9238 |
| 8.4 | .9243 | .9248 | .9253 | .9258 | .9263 | .9269 | .9274 | .9279 | .9284 | .9289 |
| 8.5 | .9294 | .9299 | .9304 | .9309 | .9315 | .9320 | .9325 | .9330 | .9335 | .9340 |
| 8.6 | .9345 | .9350 | .9355 | .9360 | .9365 | .9370 | .9375 | .9380 | .9385 | .9390 |
| 8.7 | .9395 | .9400 | .9405 | .9410 | .9415 | .9420 | .9425 | .9430 | .9435 | .9440 |
| 8.8 | .9445 | .9450 | .9455 | .9460 | .9465 | .9469 | .9474 | .9479 | .9484 | .9489 |
| 8.9 | .9494 | .9499 | .9504 | .9509 | .9513 | .9518 | .9523 | .9528 | .9533 | .9538 |
| 9.0 | .9542 | .9547 | .9552 | .9557 | .9562 | .9566 | .9571 | .9576 | .9581 | .9586 |
| 9.1 | .9590 | .9595 | .9600 | .9605 | .9609 | .9614 | .9619 | .9624 | .9628 | .9633 |
| 9.2 | .9638 | .9643 | .9647 | .9652 | .9657 | .9661 | .9666 | .9671 | .9675 | .9680 |
| 9.3 | .9685 | .9689 | .9694 | .9699 | .9703 | .9708 | .9713 | .9717 | .9722 | .9727 |
| 9.4 | .9731 | .9736 | .9741 | .9745 | .9750 | .9754 | .9759 | .9763 | .9768 | .9773 |
| 9.5 | .9777 | .9782 | .9786 | .9791 | .9795 | .9800 | .9805 | .9809 | .9814 | .9818 |
| 9.6 | .9823 | .9827 | .9832 | .9836 | .9841 | .9845 | .9850 | .9854 | .9859 | .9863 |
| 9.7 | .9868 | .9872 | .9877 | .9881 | .9886 | .9890 | .9894 | .9899 | .9903 | .9908 |
| 9.8 | .9912 | .9917 | .9921 | .9926 | .9930 | .9934 | .9939 | .9943 | .9948 | .9952 |
| 9.9 | .9956 | .9961 | .9965 | .9969 | .9974 | .9978 | .9983 | .9987 | .9991 | .9996 |

## ( ∘ )

### Answers to selected problems

Here are the remaining answers to odd-numbered exercise problems. If the odd-numbered answer you are looking for is not here, you will find it in the sample-solution part of the section with which you are concerned.

**(1.1)**   **3** $f$     **5** $a$     **11** $\{2\}$     **13** $\{9, 18, 27, 36,...\}$

**17**   One possible answer is {All positive multiples of 5}.

**19**   {First six natural numbers}

**21**   One of several possible correct answers is {All real numbers whose square is 16}.

**27** Finite     **29** Infinite     **35** $\not\subset$     **37** $\not\subset$     **43** $\notin$

**45** $\notin$     **51** $\subset$     **53** $\in$

**57**   $\{1, 2, 3\}, \{1, 2\}, \{1, 3\}, \{2, 3\}, \{1\}, \{2\}, \{3\}, \emptyset$

**(1.2)**   **3** $\emptyset$     **5** $\{5, 7\}$     **9** $\{1, 2, 3, 4, 5, 6, 7\}$

**11** $\{1, 3, 5, 6, 7\}$     **15** $D$     **19** $A$     **23** $b$     **25** $c$

**29** $c$     **31** $c$     **33** $d$

**(1.3)**   **1**

**5** 4     **9** $\dfrac{1}{2}$     **13** 3     **17** $-2\dfrac{1}{4}$

**21**   $-3$ is greater than or equal to $-4$

**23**   5 is less than or equal to 5     **29** $1 \leq 3$

**33** $a > b$     **37** $a \not> b$

**41**

**43**

**45**

**47**

**51**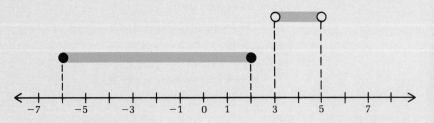

$-3$

**53** 6

**59** $-x$ if $-x \geq 0$ and $x$ if $-x < 0$

**(1.4)**  **1** 14    **3** 22    **7** $-17$    **9** 1    **11** 3    **15** $-7$
**17** 9    **19** 11    **21** $-5$    **23** $-17$    **27** 15
**29** $-1$    **31** $-14$    **33** $-2$    **35** $-2$    **39** 7
**41** $-1$    **43** 0    **47** 0

**(1.5)**  **3** $-12$    **5** 12    **9** 48    **11** $-150$    **13** 9
**17** 10    **19** $-5$    **23** Not defined    **27** $6 \cdot 0 = 0$
**31** $(-11)(8) = -88$

**(1.6)**  **3** $-17$    **7** $-5$    **11** $-6$    **13** 23    **15** 2
**19** 3    **21** $\dfrac{14}{3}$

**(1.7)**  **3** $<$    **5** $>$    **9** $>$    **11** $>$    **13** $<$
**17** $2 < x \leq 3$    **19** $2 < t < 3$
**23** $t$ is less than 6 but is not less than 1
**25** $t$ is greater than 15 but does not exceed 16
**29**

$[-6,-2] \cap (3,5) = \emptyset$

**33** Not an interval

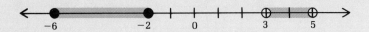

**35**  [1,3]

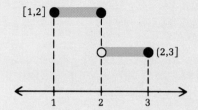

**41**

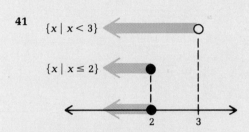

**43**

**(2.1)**  **3**  $8bbb$       **5**  $-xxx$       **9**  $w^1$       **17**  $a^2b^2$

**19**  $9x^4y^2$       **23**  25       **27**  3       **29**  20

**(2.2)**  **3**  Degree 0, coefficient $-8$       **5**  Degree 3, coefficient $-7$

**7**  Degree 2, coefficient 5       **11**  Degree 10, coefficient 2

**15**  Trinomial, degree 3       **19**  Trinomial, degree 54

**23**  Binomial, degree 3       **25**  Trinomial, degree 5

**29**  6, 1, 21       **31**  2, 2, 2       **35**  $P(4) = 3$, $Q(-1) = 2$

**39**  2       **45**  $-9$       **49**  There are none.

**(2.3)**  **3**  $-2st^2$       **7**  $2t^4 + 4$       **11**  $x^2 + 2xy + y^2$

**13**  $x^2 - x$       **17**  $\dfrac{3}{2} - 6yz - 2y^2z^2$       **19**  $3xyz - 2x^2z + 5y$

**23**  $14st^2$       **27**  $4t^3 + t - 2$       **31**  $-x^2 - 2xy - y^2$

**35**  $a^2 - 5ab - b^2 - a$       **37**  $a^3 + 2a^2b - 4ab^2 - b^3$

**39**  $x^2yz + 2z^3 - xyz$       **41**  $a^2 + ab + c^2$

**45** $(3a + 6ab)wx + (5b - b^2)wx^2$

**49** $R(x) = x^5 + x^3 + 1$, $P(-1) = -2$, $Q(-1) = 1$, $R(-1) = -1 = -2 + 1$

**53** $2x^2 + 3x + 2$     **57** $2x - 5y$     **59** $-a - b$

**(2.4)**    **3** $-6w^3$    **5** $6x^2yz$    **7** $30a^3b^2$    **9** $12x^2yz$

**15** $5w^3 - 5w$    **17** $x^3y^2 + 2x^3y^3$

**21** $2x^3 - 6x^2y + 4xy^2$    **23** $2x^{3n} - 3x^{2n} + 4x^n$

**27** $x^2 - 14xy + 49y^2$    **29** $x^2 - 9$    **31** $4x^2 - 25y^2$

**33** $x^2 + x - 6$    **35** $14x^2 + 11x - 15$

**39** $2x^3 + 4x^2 - 5x - 10$

**41** $4x^3 - 8xy + 6x^3y - 12xy^2$

**45** $2y^3 + y^2 - y - 3$    **47** $2x^3 + 3x^2y - xy^2 + 2y^3$

**49** $2t^{5n} + t^{3n} + 2t^{2n} - 15t^n - 5$    **51** $18x^3 - 50x$

**55** $6x^3 - 5x^2y - 17xy^2 + 6y^3$    **59** $-x^4 - 5x^3 - 6x^2 - 8x$

**61** $y^4 - 12y^2 + 36$    **65** $x^2 + (a + b)xy + aby^2$

**(2.5)**    **3** $2y^3$    **5** $3x^2y^4$    **9** $2x^n$

**13** $3t^2 + 6t + 1$    **15** $x^n + 3$

**19** $3x^2 + 4x^3$    **23** $6xz - 8x^2y^3$

**25** $-2y^2 + 4ay^3 + 1$    **29** $5w^2 + w + 1$

**31** $x^2 + y$    **33** $3a$

**(2.6)**    **3** $(x + 3)(x - 3)$    **7** $(3x + 4y)(3x - 4y)$

**9** $(x - 2y)^2$    **13** $(w - 8)(w + 2)$    **15** $2(x - 7)(x - 6)$

**17** $(x + 7y)(x + 2y)$    **19** $(x - 3y)(x + 2y)$

**23** $(7w + 2)(w + 3)$    **25** $2(t + 3)(t - 3)$

**27** $9z(z + 6)$    **31** $2(4a - 3b)(2a + 5b)$    **35** $y^7(y - 1)$

**37** $x(x^2 + 3)^2$    **39** $(x^n - 4)(x^n - 3)$

**43** $(x^3 + 7y^2)(x^3 + 3y^2)$    **47** $(7x + 3)(7x - 7) = 7(7x + 3)(x - 1)$

**51** $(8x + y)(7x - 2y)$    **55** $a = -3$

**(2.7)**    **3** $(2t + 3)(4t^2 - 6t + 9)$    **5** $(3a - b)(9a^2 + 3ab + b^2)$

**7** $[(x - 1) + 1][(x - 1)^2 - (x - 1) + 1] = x(x^2 - 3x + 3)$

**9** $[(x - 1) - 2][(x - 1)^2 + 2(x - 1) + 4] = (x - 3)(x^2 + 3)$

# (∘)

## Answers to selected problems

*Here are the remaining answers to odd-numbered exercise problems. If the odd-numbered answer you are looking for is not here, you will find it in the sample-solution part of the section with which you are concerned.*

**(1.1)**   **3** $f$   **5** $a$   **11** $\{2\}$   **13** $\{9, 18, 27, 36,...\}$

**17** One possible answer is {All positive multiples of 5}.

**19** {First six natural numbers}

**21** One of several possible correct answers is {All real numbers whose square is 16}.

**27** Finite   **29** Infinite   **35** $\not\subset$   **37** $\not\subset$   **43** $\notin$

**45** $\notin$   **51** $\subset$   **53** $\in$

**57** $\{1, 2, 3\}, \{1, 2\}, \{1, 3\}, \{2, 3\}, \{1\}, \{2\}, \{3\}, \emptyset$

**(1.2)**   **3** $\emptyset$   **5** $\{5, 7\}$   **9** $\{1, 2, 3, 4, 5, 6, 7\}$

**11** $\{1, 3, 5, 6, 7\}$   **15** $D$   **19** $A$   **23** $b$   **25** $c$

**29** $c$   **31** $c$   **33** $d$

**(1.3)**   **1**

**5** 4   **9** $\dfrac{1}{2}$   **13** 3   **17** $-2\dfrac{1}{4}$

**21** $-3$ is greater than or equal to $-4$

**23** 5 is less than or equal to 5   **29** $1 \leq 3$

**33** $a > b$   **37** $a \not> b$

**41**

**43**

**45**

**47**

**51**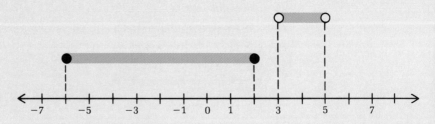

 −3

**53** 6

**59** $-x$ if $-x \geq 0$ and $x$ if $-x < 0$

**(1.4)**　**1** 14　　**3** 22　　**7** −17　　**9** 1　　**11** 3　　**15** −7

　　**17** 9　　**19** 11　　**21** −5　　**23** −17　　**27** 15

　　**29** −1　　**31** −14　　**33** −2　　**35** −2　　**39** 7

　　**41** −1　　**43** 0　　**47** 0

**(1.5)**　**3** −12　　**5** 12　　**9** 48　　**11** −150　　**13** 9

　　**17** 10　　**19** −5　　**23** Not defined　　**27** $6 \cdot 0 = 0$

　　**31** $(-11)(8) = -88$

**(1.6)**　**3** −17　　**7** −5　　**11** −6　　**13** 23　　**15** 2

　　**19** 3　　**21** $\dfrac{14}{3}$

**(1.7)**　**3** <　　**5** >　　**9** >　　**11** >　　**13** <

　　**17** $2 < x \leq 3$　　**19** $2 < t < 3$

　　**23** $t$ is less than 6 but is not less than 1

　　**25** $t$ is greater than 15 but does not exceed 16

　　**29**

$[-6, -2] \cap (3, 5) = \varnothing$

**33** Not an interval

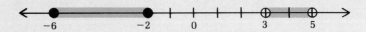

**35** [1,3]

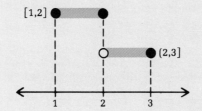

**41**

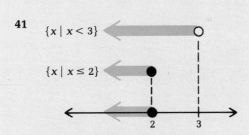

$\{x \mid x < 3\}$

$\{x \mid x \le 2\}$

**43**

**(2.1)**    **3**   $8bbb$     **5**   $-xxx$     **9**   $w^1$     **17**   $a^2 b^2$

      **19**   $9x^4 y^2$     **23**   25     **27**   3     **29**   20

**(2.2)**    **3**   Degree 0, coefficient $-8$     **5**   Degree 3, coefficient $-7$

      **7**   Degree 2, coefficient 5     **11**   Degree 10, coefficient 2

      **15**   Trinomial, degree 3     **19**   Trinomial, degree 54

      **23**   Binomial, degree 3     **25**   Trinomial, degree 5

      **29**   6, 1, 21     **31**   2, 2, 2     **35**   $P(4) = 3, Q(-1) = 2$

      **39**   2     **45**   $-9$     **49**   There are none.

**(2.3)**    **3**   $-2st^2$     **7**   $2t^4 + 4$     **11**   $x^2 + 2xy + y^2$

      **13**   $x^2 - x$     **17**   $\dfrac{3}{2} - 6yz - 2y^2 z^2$     **19**   $3xyz - 2x^2 z + 5y$

      **23**   $14st^2$     **27**   $4t^3 + t - 2$     **31**   $-x^2 - 2xy - y^2$

      **35**   $a^2 - 5ab - b^2 - a$     **37**   $a^3 + 2a^2 b - 4ab^2 - b^3$

      **39**   $x^2 yz + 2z^3 - xyz$     **41**   $a^2 + ab + c^2$

**45** $(3a + 6ab)wx + (5b - b^2)wx^2$

**49** $R(x) = x^5 + x^3 + 1, P(-1) = -2, Q(-1) = 1, R(-1) = -1 = -2 + 1$

**53** $2x^2 + 3x + 2$     **57** $2x - 5y$     **59** $-a - b$

**(2.4)**    **3** $-6w^3$    **5** $6x^2 yz$    **7** $30a^3 b^2$    **9** $12x^2 yz$

**15** $5w^3 - 5w$    **17** $x^3 y^2 + 2x^3 y^3$

**21** $2x^3 - 6x^2 y + 4xy^2$    **23** $2x^{3n} - 3x^{2n} + 4x^n$

**27** $x^2 - 14xy + 49y^2$    **29** $x^2 - 9$    **31** $4x^2 - 25y^2$

**33** $x^2 + x - 6$    **35** $14x^2 + 11x - 15$

**39** $2x^3 + 4x^2 - 5x - 10$

**41** $4x^3 - 8xy + 6x^3 y - 12xy^2$

**45** $2y^3 + y^2 - y - 3$    **47** $2x^3 + 3x^2 y - xy^2 + 2y^3$

**49** $2t^{5n} + t^{3n} + 2t^{2n} - 15t^n - 5$    **51** $18x^3 - 50x$

**55** $6x^3 - 5x^2 y - 17xy^2 + 6y^3$    **59** $-x^4 - 5x^3 - 6x^2 - 8x$

**61** $y^4 - 12y^2 + 36$    **65** $x^2 + (a + b)xy + aby^2$

**(2.5)**    **3** $2y^3$    **5** $3x^2 y^4$    **9** $2x^n$

**13** $3t^2 + 6t + 1$    **15** $x^n + 3$

**19** $3x^2 + 4x^3$    **23** $6xz - 8x^2 y^3$

**25** $-2y^2 + 4ay^3 + 1$    **29** $5w^2 + w + 1$

**31** $x^2 + y$    **33** $3a$

**(2.6)**    **3** $(x + 3)(x - 3)$    **7** $(3x + 4y)(3x - 4y)$

**9** $(x - 2y)^2$    **13** $(w - 8)(w + 2)$    **15** $2(x - 7)(x - 6)$

**17** $(x + 7y)(x + 2y)$    **19** $(x - 3y)(x + 2y)$

**23** $(7w + 2)(w + 3)$    **25** $2(t + 3)(t - 3)$

**27** $9z(z + 6)$    **31** $2(4a - 3b)(2a + 5b)$    **35** $y^7 (y - 1)$

**37** $x(x^2 + 3)^2$    **39** $(x^n - 4)(x^n - 3)$

**43** $(x^3 + 7y^2)(x^3 + 3y^2)$    **47** $(7x + 3)(7x - 7) = 7(7x + 3)(x - 1)$

**51** $(8x + y)(7x - 2y)$    **55** $a = -3$

**(2.7)**    **3** $(2t + 3)(4t^2 - 6t + 9)$    **5** $(3a - b)(9a^2 + 3ab + b^2)$

**7** $[(x - 1) + 1][(x - 1)^2 - (x - 1) + 1] = x(x^2 - 3x + 3)$

**9** $[(x - 1) - 2][(x - 1)^2 + 2(x - 1) + 4] = (x - 3)(x^2 + 3)$

**13** $(5s - 2t)(13s^2 - 32st + 28t^2)$

**17** $(x^3 - 2)(x^6 + 2x^3 + 4)$

**19** $8[x^2 - 3(y + 3)][x^4 + 3x^2(y + 3) + 9(y + 3)^2]$
$= 8(x^2 - 3y - 9)(x^4 + 3x^2 y + 9x^2 + 9y^2 + 54y + 81)$

**23** $2a(3x^2 + a^2)$     **27** $(t^4 - 8)(t^2 - 2)$     **29** $(a - b)(a + c)$

**31** $(x - 3y)(x + s)$     **33** $(3y^2 - x)(x^2 + 4y)$

**37** $(s - 7t + 3t^2)(s - 2t)$     **39** $(a - 3b)(a^2 + 3ab + 9b^2 + 1)$


**(3.1)**    **3** $c, d$     **7** $a, b$     **13** $x - 1$     **15** $-a - 1$

**19** $x \neq 1, x \neq -1$     **23** $x \neq -1, x \neq 3$

**25** $x \neq 1, x \neq -3y, x \neq z$     **27** $x \neq yz$


**(3.2)**    **3** $3 \cdot 5$     **5** $2^2 \cdot 7$     **7** $5 \cdot 7$     **13** $2(x^2 + 1)$

**15** $2 \cdot 3 \cdot 7 \cdot x(x - 2)$     **21** $180$     **23** $180$     **27** $1{,}401{,}400$

**31** $102x^5 y^4 z^3$     **35** $(a + b)^3 (a + 2b)$


**(3.3)**    **3** $\dfrac{2}{3}$     **7** $\dfrac{5}{4}$     **9** $\dfrac{3}{2}$     **13** $\dfrac{3x^3 y}{2}$     $(x, y \neq 0)$

**17** $\dfrac{6stu^2}{3su + 4u^2 - 6}$     $(s, u, \neq 0;\ 3su + 4u^2 \neq 6)$

**19** $\dfrac{y(3x^2 - 4)}{2x^3(x + 3y)}$     $(x, y \neq 0;\ x \neq -3y)$

**23** $x - 2y - 3z$     $(x - 2y + 3z \neq 0)$

**25** $\dfrac{x - 2}{x + 3}$     $(x \neq -3, 2)$     **29** $\dfrac{x - 4y}{5x + y}$     $(2x \neq 3y; 5x \neq -y)$

**31** $\dfrac{1}{x + 1}$     $(x \neq -1, 2y)$     **35** $6$     **39** $x^2 + 2xy + y^2$

**43** $\dfrac{9}{42}, \dfrac{4}{42}$     **45** $\dfrac{15xy}{6x^2 y}, \dfrac{8}{6x^2 y}$     **47** $\dfrac{x - y}{x^2 - y^2}, \dfrac{x + y}{x^2 - y^2}$

**49** $\dfrac{x^2 - xy + y^2}{x^3 + y^3}, \dfrac{x + y}{x^3 + y^3}, \dfrac{x^2 + y^2}{x^3 + y^3}$

**53** $\dfrac{18(x - 1)}{15(x - 1)^2(x - 2)^2}, \dfrac{5x(x - 2)}{15(x - 1)^2(x - 2)^2}$

**55**  $\dfrac{(x+y)^3}{x(x+y)^2(x-y)^3}$ , $\dfrac{x(x-y)}{x(x+y)^2(x-y)^3}$

**(3.4)**  **3**  $\dfrac{1}{24}$  **5**  $\dfrac{5}{x}$  $(x \neq 0)$  **9**  $\dfrac{2x-3}{x+y}$  $(x+y \neq 0)$

**11**  $\dfrac{8-a^2}{ax}$  $(a, x \neq 0)$  **13**  $\dfrac{3y-x}{y^2-x^2}$  $(y \neq \pm x)$

**17**  $\dfrac{5b+9}{b^2+b}$  $(b \neq 0, -1)$

**19**  $\dfrac{-x^2+3x-1}{(x-2)(x-1)}$  $(x \neq 1, 2)$

**23**  $\dfrac{a^2-ab+2b^2}{a^2-4b^2}$  $(a \neq \pm 2b)$

**25**  $\dfrac{a^2+a+2}{a^3-1}$  $(a \neq 1)$

**29**  $\dfrac{4s^2-s+st+t}{(s+3)(s+1)(4s-1)}$  $\left(s \neq -3, -1, \dfrac{1}{4}\right)$

**(3.5)**  **3**  14  **5**  $\dfrac{2}{9}$  **9**  $\dfrac{xy}{6}$  $(x, y \neq 0)$

**13**  $\dfrac{12x}{x-3}$  $(x \neq \pm 3, 0)$

**17**  $\dfrac{(4c-3d)(c+d)}{(2c-d)(2c-3d)}$  $\left(c \neq \dfrac{d}{2}, \pm \dfrac{3d}{2}\right)$

**19**  $\dfrac{s(2s+t)}{2(s+1)}$  $(s \neq 5t, -3t, t, -1)$

**23**  $\dfrac{x^3}{9y^3}$  $(x, y, z \neq 0)$  **27**  $(x-4)(x-3)$  $(x \neq 0, \pm 3)$

**29**  $\dfrac{(x-3)(x+1)}{x-1}$  $(x \neq \pm 1, 2)$

**33**  $\dfrac{x-3}{y+2}$  $(x \neq 2; y \neq -1, -2)$

**35** $\dfrac{x+1}{(x+2)^2}$   $(x \neq -2, -1)$

**37** $\dfrac{(x-6)(y+2)}{x(y-3)}$   $(x \neq 0, -3; y \neq \pm 2, 3)$

**(3.6)**

**3** $y - 6 + \dfrac{5}{y}$   $(y \neq 0)$    **5** $t + 1 - \dfrac{2}{t-4}$   $(t \neq 4)$

**9** $x^2 + 3x + 2 + \dfrac{2}{x+1}$   $(x \neq -1)$

**13** $4z^2 - 1 + \dfrac{2}{z-1}$   $(z \neq 1)$

**17** $P(2) = -1, \dfrac{P(t)}{t-2} = 2t - 1 - \dfrac{1}{t-2}$

**19** $P(-2) = -7, \dfrac{P(x)}{x+2} = x^2 - x + 2 - \dfrac{7}{x+2}$

**25** $P(1) = 5$; so $t - 1$ is not a factor.

**27** $P(-3) = 0$; so $s + 3$ is a factor; $2s^3 + 6s^2 - s - 3 = (2s^2 - 1)(s + 3)$.

**(3.7)**

**3** $\dfrac{7}{3}$    **5** $\dfrac{2}{3}$    **9** $\dfrac{3x^2}{2}$   $(x \neq 0)$    **13** $\dfrac{10}{7}$

**15** $\dfrac{2a-2}{2a+1}$   $\left(a \neq 0, -\dfrac{1}{2}\right)$

**19** $\dfrac{1}{bc}$   $(a, b, c \neq 0; a + b + c \neq 0)$

**21** $\dfrac{xy^2 + xy}{y-1}$   $(y \neq 0, 1)$    **25** $-\dfrac{7}{5}$

**27** $\dfrac{x^2 + 3x + 1}{2x+1}$   $\left(x \neq -1, -\dfrac{1}{2}\right)$

**29** $\dfrac{3(t+2)}{t^2 + 2t + 3}$   $(t \neq -2)$    **33** $\dfrac{s+5}{s+3}$   $(s \neq \pm 2, -3)$

**35** $\dfrac{2(a-b)}{a+b}$   $(a \neq \pm b)$    **37** $\dfrac{5}{8}$

**(4.1)**    **3**   Not a solution      **5**   Solution      **7**   Solution

         **9**   Not a solution      **13**   $-0 + 3 \neq 2 \cdot 0 + 4$

         **15**   $(0 + 3)(0 + 2) \neq 0^2 + 5 \cdot 0 + 5$

         **17**   $(x + 2)^2 - 4x = x^2 + 4$
$$x^2 + 4x + 4 - 4x = x^2 + 4$$
$$x^2 + 4 = x^2 + 4$$

         **21**   $(z + 7)(z + 2) - z^2 - 9z = 14$
$$z^2 + 9z + 14 - z^2 - 9z = 14$$
$$14 = 14$$

         **23**   $3(2 - x) + 6(x - 4) = 4(x - 2) - \dfrac{2x + 20}{2}$
$$6 - 3x + 18x - 24 = 4x - 8 - x - 10$$
$$3x - 18 = 3x - 18$$

**(4.2)**    **3**   $\left\{ \dfrac{8}{5} \right\}$      **5**   $\{2\}$      **9**   $\left\{ -\dfrac{15}{7} \right\}$      **11**   $\left\{ -\dfrac{3}{2} \right\}$

         **15**   $\left\{ -\dfrac{3}{2} \right\}$      **19**   $\{-2\}$      **21**   Empty solution set

         **25**   $\left\{ \dfrac{3}{4} \right\}$

**(4.3)**    **3**   $\dfrac{32 - y}{8}$      **7**   $\dfrac{b}{a}$   $(a \neq 0)$      **9**   $-\dfrac{bc}{a}$   $(a, b \neq 0)$

         **13**   $\dfrac{ab - ay}{b}$   $(a, b \neq 0)$      **15**   $\dfrac{y}{y - 1}$   $(y \neq 0, 1)$

         **17**   $h = \dfrac{A}{b}$      **19**   $r = \dfrac{C}{2\pi}$

         **21**   $c = \dfrac{V}{\pi b^2}$      **25**   $h = \dfrac{A}{2\pi r} - r$

         **27**   $w = \dfrac{V}{lh}$      **29**   $t = \dfrac{v - k}{a}$

**(4.4)**    **1**   24      **3**   John weighs 181 pounds; his dog weighs 19 pounds.

         **5**   7, 8, 9

         **7**   The car is 3 years old.      **9**   \$100 per week

**11** It is impossible; to do so would require $12\frac{1}{2}$ of each kind of coin.

**(4.5)**　　**1** 12 at \$3.95, 8 at \$5.25　　**3** 372　　**5** \$1,400

　　　　　**7** 225 miles　　**9** 11:54 A.M., 19.6 miles

　　　　　**11** 2 hours and 55 minutes

**(4.6)**　　**3** $\{y \mid y > -1\}$

$$\longleftrightarrow \overset{\oplus}{\underset{-1}{\;}} \longrightarrow$$

**5** $\{a \mid a \geq -3\}$

$$\longleftrightarrow \overset{\bullet}{\underset{-3}{\;}} \longrightarrow$$

**7** $\left\{x \mid x > -\dfrac{11}{4}\right\}$

$$\longleftrightarrow \overset{\circ}{\underset{-^{11}/_4}{\;}} \longrightarrow$$

**13** $\left\{w \mid w \geq \dfrac{1}{3}\right\}$

$$\longleftrightarrow \overset{\bullet}{\underset{^1/_3}{\;}} \longrightarrow$$

**17** $[-10, -9] = \{x \mid -10 \leq x \leq -9\}$

$$\longleftrightarrow \overset{\bullet}{\underset{-10}{\;}} \rule{1cm}{0.4pt} \overset{\bullet}{\underset{-9}{\;}} \longrightarrow$$

**19** $\left(0, \dfrac{3}{2}\right] = \left\{x \mid 0 < x \leq \dfrac{3}{2}\right\}$

$$\longleftrightarrow \overset{\oplus}{\underset{0}{\;}} \rule{1cm}{0.4pt} \overset{\bullet}{\underset{^3/_2}{\;}} \longrightarrow$$

**21** $(-4, -1) = \{t \mid -4 < t < -1\}$

$$\longleftrightarrow \overset{\oplus}{\underset{-4}{\;}} \rule{1cm}{0.4pt} \overset{\oplus}{\underset{-1}{\;}} \longrightarrow$$

**23** $\left[\dfrac{4}{3}, 2\right) = \left\{z \mid \dfrac{4}{3} \leq z < 2\right\}$

$$\longleftrightarrow \overset{\circ}{\underset{^4/_3}{\;}} \rule{1cm}{0.4pt} \overset{\oplus}{\underset{2}{\;}} \longrightarrow$$

**27** 75 miles

**3**  $\left\{\dfrac{7}{2}, -\dfrac{1}{2}\right\}$

**5**  $\left\{\dfrac{11}{2}, -\dfrac{1}{2}\right\}$

**7**  $\{1, -2\}$

**9**  $\left\{2, -\dfrac{2}{3}\right\}$

**11**  $(-3, 3)$

**15**  $[-6, 3]$

**17**  $(-16, 26)$

**19**  $\{x \,|\, x \geq 2\} \cup \{x \,|\, x \leq -2\}$

**23**  $\left\{v \,\middle|\, v > \dfrac{1}{2}\right\} \cup \left\{v \,\middle|\, v < -\dfrac{9}{2}\right\}$

**25**  $\left\{s \,\middle|\, s \geq \dfrac{5}{2}\right\} \cup \left\{s \,\middle|\, s \leq \dfrac{1}{2}\right\}$

**27**  $\{-3\}$

**29**  $\left\{\dfrac{5}{2}\right\}$

**(5.1)**  **3** $\dfrac{1}{64}$  **5** $64$  **7** $\dfrac{1}{2}$  **9** $\dfrac{1}{4}$  **13** $\dfrac{1}{8}$  **15** $\dfrac{1}{2}$

**17** $\dfrac{13}{25}$  **19** $\dfrac{1}{160}$  **25** $2 \cdot 3^{-1}$  **27** $2^{-19} \cdot 3^{-7} \cdot 5$

**31** $3^{-1} a^2 c^{-2}$  $(b, c \neq 0)$  **33** $15^{-1} t$  $(s, u \neq 0)$

**35** $5^{-1} x^2 y^{-2}$  $(x, y, z \neq 0)$  **39** $\dfrac{x^3 y}{z^2}$  $(x, z \neq 0)$

**41** $g^{10} h^8 k^{10}$  $(g, h, k \neq 0)$

**45** $2x^2 y^{-1} z^2$  $(x, y, z \neq 0)$

**47** $\dfrac{6}{35} gh^5 k^{-4}$  $(h, k \neq 0)$  **51** $\dfrac{bc + ac + ab}{a^2 b^2 c^2}$  $(a, b, c \neq 0)$

**53** $\dfrac{s^3 t^2 + t}{s^2}$  $(s, t \neq 0)$

**55** $\dfrac{xyz}{yz + 2xz - 3x^2 y^2}$  $(x, y, z \neq 0; \ yz + 2xz - 3x^2 y^2 \neq 0)$

**59** $a^{1-n^2}$  $(a \neq 0)$

**(5.2)**  **3** $-3$  **5** $4$  **9** $-\dfrac{1}{6}$  **11** $\dfrac{2}{3}$

**13** $-2$  **17** $\dfrac{1}{\sqrt[3]{a^2}}$

**19** $\sqrt[3]{125x^4}$  **21** $(\sqrt[3]{a - 2b})^5$  **25** $x^{-2/3}$

**27** $2x^{1/2} y^3$  **31** $w^{1/5}$  **33** $x^{1/3}$  **37** $a^{1/2}$

**41** $x + 4x^{1/2} + 1$  **43** $x^2 - 5x + 2$

**45** $x^{2/3} - x^{1/3} - 6$  **49** $2x^{1/9} + 3y^{1/3}$

**51** $x - 1$  **53** $a^{1/4} + \dfrac{5}{2} a^{-3/4}$  **59** $x^2 y^2$

**(5.3)**  **3** $\sqrt[6]{16} = (\sqrt[6]{2})^4$  **5** $\sqrt[10]{x^3} = (\sqrt[10]{x})^3$  **11** $\dfrac{3}{2}$

**15** $2ab^2$  **19** $\sqrt{18x^2}$  **23** $\sqrt[3]{120}$

**27** $\sqrt{x^3 + 2x^2 y + xy^2}$  **31** $3\sqrt[4]{5}$  **33** $5r\sqrt{x}$

**39** $\sqrt{\dfrac{7}{x}}$  **43** $\sqrt{x}$  **45** $\dfrac{\sqrt{2}}{3}$  **49** $\dfrac{\sqrt[3]{75}}{5}$

**51** $\dfrac{\sqrt{5}}{2x}$  $(x \neq 0)$  **55** $\dfrac{\sqrt[4]{24x}}{2}$  $(x \neq 0)$

**57** $\dfrac{\sqrt{x}}{x}$  $(x \neq 0)$  **59** $\dfrac{\sqrt[3]{20x^2}}{6x}$  $(x \neq 0)$

**(5.4)**  **3** $\sqrt{2}$  **7** $\sqrt[3]{2x^2}$  **9** $\sqrt{4x^2 y}$  **13** $\sqrt[4]{96}$

**17** $\sqrt[6]{135w^7}$  **19** $\sqrt[30]{2^6\, 3^{15}\, x^{25}\, y^{26}}$  **23** $2x\sqrt[4]{5x}$

**25** $3b\sqrt[3]{4a^2}$  **29** $15\sqrt[3]{4}$  **33** $6ab\sqrt[3]{a^2 b^2 c}$

**37** $\dfrac{2}{3}$  **39** $\dfrac{2}{5y}\sqrt[3]{x^2 y^2}$  **41** $\dfrac{2}{7ab^2}\sqrt{7ab}$

**45** $2ab^2\sqrt[3]{9b}$  **47** $\dfrac{\sqrt{x}}{2}$  **49** $\dfrac{\sqrt{18y}}{3y}$

**53** $|x - 1|$  **57** $\dfrac{1}{|3x - 1|}$  $\left(x \neq \dfrac{1}{3}\right)$  **59** $4x^2$

**(5.5)**  **1** $7\sqrt{3}$  **3** $2\sqrt{7}$  **7** $5\sqrt{3}$  **9** $5\sqrt[3]{2}$

**11** $-5\sqrt{2}$  **13** $4\sqrt{5} - 4\sqrt[3]{2}$

**17** $(s - t + 1 - s^2 t)\sqrt[3]{st^2}$

**23** $5 - 2\sqrt{6}$  **25** $3 + 2\sqrt{2}$  **27** $16 - 22\sqrt{3}$

**29** $2x - y$  **33** $6x - \sqrt{x} - 1$  **37** $-\dfrac{3 + \sqrt{3}}{2}$

**39** $-\dfrac{4 + \sqrt{7}}{3}$  **41** $-4 - \sqrt{15}$  **45** $\dfrac{x(\sqrt{x} + 2)}{x - 4}$  $(x \neq 4)$

**47** $\dfrac{a + 2\sqrt{ab} + b}{a - b}$  $(a \neq b)$  **53** $\dfrac{6}{9 - x}$  $(x \neq 9)$

**57** $\dfrac{3(2x\sqrt{x} + x - \sqrt{x})}{(x - 1)(4x - 1)}$  $\left(x \neq 1, \dfrac{1}{4}\right)$

**59** $\dfrac{2(x^2 + 1)\sqrt{x^2 + 2}}{x(x^2 + 2)}$  $(x \neq 0)$

**(5.6)**    **3**   $2 + 3i$     **5**   $12 - 6i\sqrt{3}$     **9**   $5 + 3i$

        **11**   $2 + 0i$     **13**   $-3 + 6i$     **15**   $-1 + 2i$     **19**   $-8 + \dfrac{3}{2}i$

**(5.7)**    **3**   $-3 - 12i$     **5**   $-3 + 2i$     **7**   $10 - 2i$

        **11**   $5 - 2i\sqrt{2}$     **13**   $20 + 2i\sqrt{5}$     **15**   $-5 + 12i$

        **19**   $5 + 0i$     **23**   $0 + \dfrac{5}{3}i$     **25**   $1 - i$

        **27**   $-\dfrac{1}{2} + \dfrac{1}{2}i$     **29**   $-\dfrac{3}{5} - \dfrac{6}{5}i$     **33**   $0 - i$

        **35**   $\dfrac{46}{25} - \dfrac{3}{25}i$     **37**   $3 - 2i$

        **41**   $1, -\dfrac{1}{2} + \dfrac{i\sqrt{3}}{2}, -\dfrac{1}{2} - \dfrac{i\sqrt{3}}{2}, -1, \dfrac{1}{2} + \dfrac{i\sqrt{3}}{2}, \dfrac{1}{2} - \dfrac{i\sqrt{3}}{2}$

**(6.1)**    **3**   $\{0, -6\}$     **7**   $\{4, 6\}$     **9**   $\{-5\}$

        **13**   $\{-5, 3\}$     **15**   $\left\{-\dfrac{1}{2}, 3\right\}$     **17**   $\left\{\dfrac{3}{4}, -\dfrac{5}{2}\right\}$

        **19**   $\left\{-\dfrac{5}{4}, \dfrac{3}{4}\right\}$     **21**   $\{-2, -3\}$     **23**   $\left\{\dfrac{1}{2}, -3\right\}$

        **27**   $\{-8, 3\}$     **31**   $x^2 + 2x - 35 = 0$

        **35**   $x^2 + 5x = 0$     **39**   $6x^2 - 37x + 35 = 0$

**(6.2)**    **3**   $\{-\sqrt{8}, \sqrt{8}\}$     **5**   $\{-\sqrt{6}, \sqrt{6}\}$     **7**   $\{-\sqrt{15}, \sqrt{15}\}$

        **13**   $\{4, -2\}$     **15**   $\left\{\dfrac{1}{2}, -\dfrac{3}{2}\right\}$     **17**   $\left\{\dfrac{7}{3}, -1\right\}$

        **19**   $\left\{\dfrac{4}{7}, -\dfrac{10}{7}\right\}$     **23**   $\dfrac{25}{4}, \left(y - \dfrac{5}{2}\right)^2$     **25**   $\dfrac{4}{9}, \left(a - \dfrac{2}{3}\right)^2$

        **29**   $5, \left(t + \dfrac{5}{2}\right)^2$     **33**   $\{-3, -6\}$

        **35**   $\left\{\dfrac{-9 + \sqrt{13}}{2}, \dfrac{-9 - \sqrt{13}}{2}\right\}$

**37** $\left\{\dfrac{-1+\sqrt{17}}{2}, \dfrac{-1-\sqrt{17}}{2}\right\}$    **43** $\left\{-1, -\dfrac{1}{2}\right\}$

**47** $\left\{\dfrac{4}{3}\right\}$    **49** $\left\{\dfrac{a}{2}(-1+\sqrt{13}), \dfrac{a}{2}(-1-\sqrt{13})\right\}$

**(6.3)**    **3** 2 real roots    **7** 2 real roots    **9** 2 real roots    **13** $\{4\}$

**15** $\{2+i, 2-i\}$    **17** $\left\{-\dfrac{3}{4}, 1\right\}$

**23** $\left\{\dfrac{-1+\sqrt{1+3a^2}}{2a}, \dfrac{-1-\sqrt{1+3a^2}}{2a}\right\}$    $(a\neq 0)$. If $a=0$, the only solution is $z=0$.

**27** $k \geq -16$    **29** $k \leq \dfrac{9}{16}$    **31** $k = \dfrac{25}{8}$    **33** $k = \dfrac{9}{16}$

**37** $k > \dfrac{16}{9}$    **41** Sum $= 3$, product $= -7$

**43** Sum $= \dfrac{3}{2}$, product $= \dfrac{5}{2}$    **45** Sum $= -\dfrac{27}{5}$, product $= -27$

**(6.4)**    **3** $\{3, -3, i\sqrt{2}, -i\sqrt{2}\}$    **5** $\{3, -3, 1, -1\}$    **7** $\{-2, 2\}$
**9** $\{\sqrt[5]{2}, 1\}$    **15** $\{27\}$    **17** $\{0, 1, -3\}$

**(6.5)**    **3** $\{4\}$    **5** $\{4\}$    **9** $\{2, 5\}$    **13** No solution
**15** $\{12\}$    **17** $\{1, 7\}$

**(6.6)**    **3** $[-1, 0]$

**5** $\{x \mid x \leq 2\} \cup \{x \mid x \geq 3\}$

**7** $[1, 2]$

**9** $R =$ set of real numbers

**13** $\left[-\dfrac{5}{2}, -1\right)$

**15**  $\{x \mid x > 2\} \cup \{x \mid x < 1\}$

**17**  $\{x \mid x < 3, x \neq 0\}$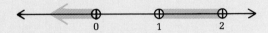

**21**  $\{x \mid x < 0\} \cup \{x \mid 1 < x < 2\}$

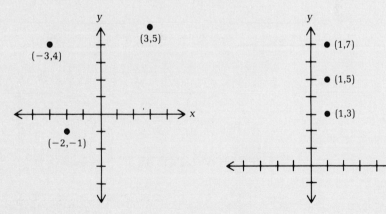

**(6.7)**  **3**  10 and 12

**7**  There are two possible answers, 8 feet by 12 feet or 6 feet by 16 feet.

**11**  25¢    **13**  5% and 6%    **17**  50 mph    **21**  6 hours

**(7.1)**  **3**                                            **5**

**9**

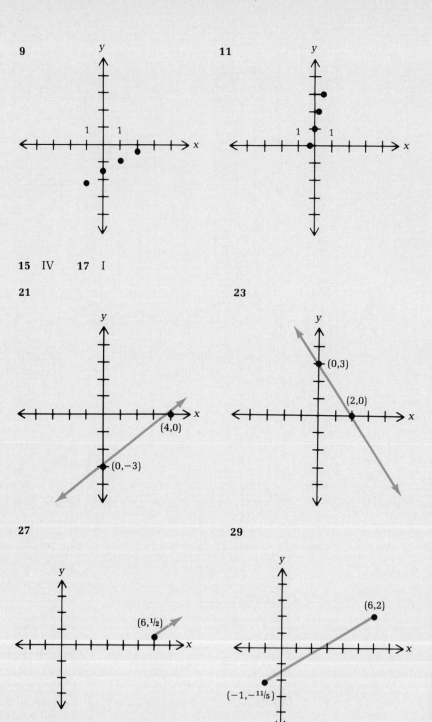

**11**

**15** IV    **17** I

**21**

**23**

(0,3)

(2,0)

**27**

**29**

(6,¹/₂)

(6,2)

(−1,−¹¹/₅)

**3** Domain = {1}, range = {1, 2, 3}, not a function

**5** Domain = {1, 2, 3}, range = {1, 3, 4}, one-one function

**7** Domain = {0, 1}, range = {2, 3}, one-one function

**9** Domain = {1}, range = {2}, one-one function

**13** $\dfrac{1}{2}$   **17** 16   **21** $-\dfrac{1}{1+h}$

**25** Domain = $R$, one-one function

**27** Domain = $\{x \mid x \in R, x \neq 0\}$, one-one function

**29** Domain = $R$, not a function

**31** Domain = $R$, function, not one-one

**33** Domain = $[0, 1]$, one-one function

(7.3)   **3** $\sqrt{65}$   **5** $\sqrt{34}$   **7** 3   **9** $\sqrt{74}$

**13** Not a right triangle   **15** Right triangle

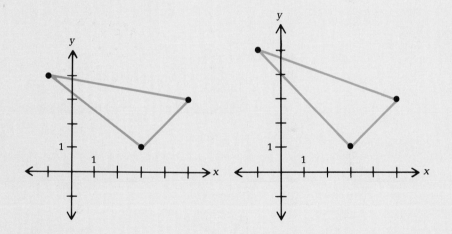

**19** $2x + 3y = 13$

(7.4)   **3** 2   **5** $-\dfrac{2}{3}$   **7** $\dfrac{3}{2}$

**11** $y = -\dfrac{9}{8}x + \dfrac{3}{2}$, slope $= -\dfrac{9}{8}$, intercept $= \dfrac{3}{2}$

**13** $y = -2x - 5$, slope $= -2$, intercept $= -5$

**17** $2x + y - 1 = 0$    **19** $x + 0y - 4 = 0$

**23** $230x - y + 3 = 0$    **25** $x - y - 8 = 0$

**29** $3x - 2y - 6 = 0$    **33** $11x + 8y + 4 = 0$

**35** $x + y - 6 = 0$

**39** Slope of first line $= -\dfrac{1}{2}$, slope of second line $= 2$; hence the lines are perpendicular.

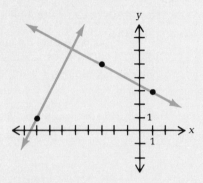

**41** $3x - 5y - 8 = 0$

**45** Parallel line, $x = 5$; perpendicular line, $y = 4$

**47** Parallel line, $y - 1 = -\dfrac{7}{3}(x - 1)$; perpendicular line, $y - 1 = \dfrac{3}{7}(x - 1)$

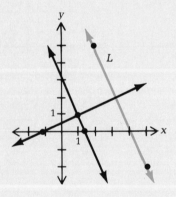

**49** Slope of sides is 1, −1, 4.

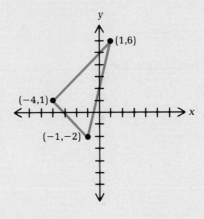

**53** Slope of sides 1 and 3, −1; sides 2 and 4, 1.

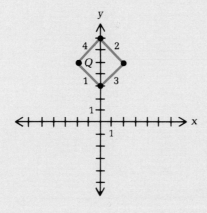

**55** $k = -28$ or $k = -2$

**3**

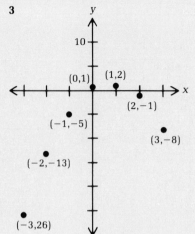

**7**

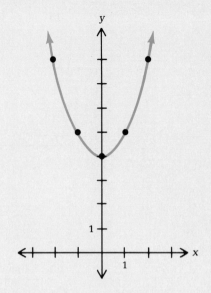

**9**

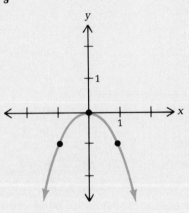

**11**

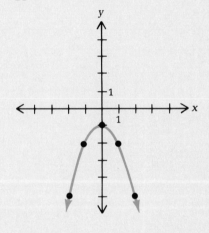

**13**

**15**

**17**

**19**

**23**

**25**

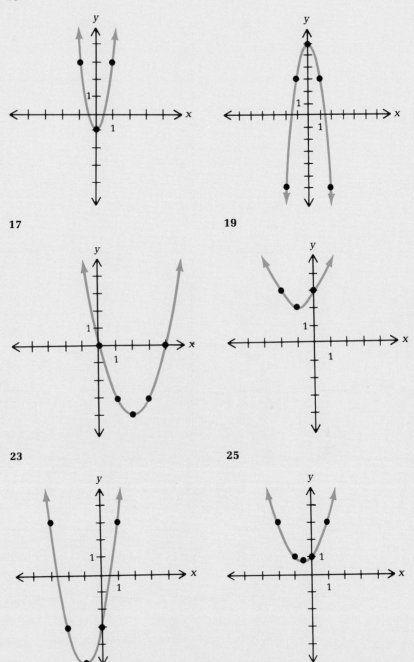

**27**

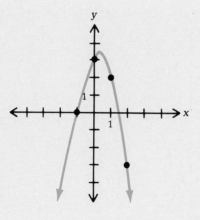

**29**

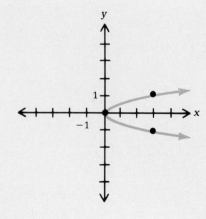

**33**

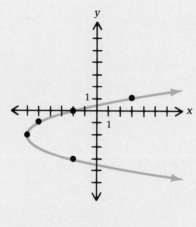

**35**

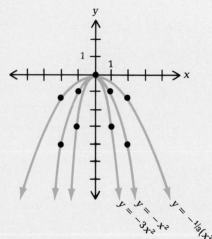

$y = -3x^2$   $y = -x^2$   $y = -\frac{1}{3}(x^2)$

(7.6)   **3**   $x$-intercepts $0$, $-9$; $y$-intercept $0$; turning point $\left(-\dfrac{9}{2}, \dfrac{81}{4}\right)$

**5**   $x$-intercepts $-3$, $2$; $y$-intercept $6$; turning point $\left(-\dfrac{1}{2}, \dfrac{25}{4}\right)$

**7**   $x$-intercepts, none; $y$-intercept $2$; turning point $\left(\dfrac{1}{4}, \dfrac{15}{8}\right)$

**9** x-intercepts $-\dfrac{1}{2}$, $-\dfrac{2}{3}$; y-intercept 1; turning point $\left(-\dfrac{7}{12}, -\dfrac{1}{48}\right)$

**13**

**15**

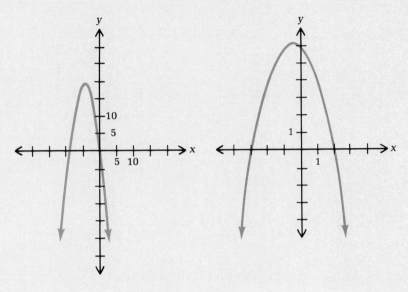

**17**

**19**

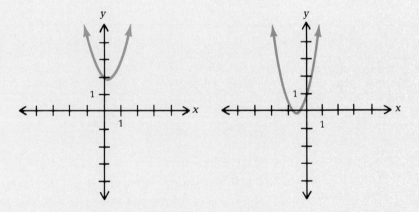

**23** $x$-intercept $-4$; $y$-intercepts $\dfrac{-3 \pm \sqrt{41}}{4}$; turning point $\left( -\dfrac{41}{8}, -\dfrac{3}{4} \right)$

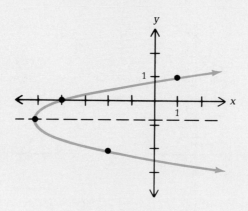

**27** $\dfrac{9}{2}$ and $-\dfrac{9}{2}$     **29**  30 feet by 30 feet

**33**  He should build a circular pool whose radius is $\dfrac{100}{\pi}$.

**(7.7)**  **3**  Circle; intercepts $\pm\dfrac{2}{3}\sqrt{3}$ ; $\left\{ x \left| -\dfrac{2}{3}\sqrt{3} \leq x \leq \dfrac{2}{3}\sqrt{3} \right. \right\}$

**5**  Ellipse; $x$-intercepts $\pm\sqrt{3}$, $y$-intercepts $\pm 3$; $\{x | -\sqrt{3} \leq x \leq \sqrt{3}\}$

**9**  Ellipse; $x$-intercepts $\pm\sqrt{6}$, $y$-intercepts $\pm 3$; $\{x | -\sqrt{6} \leq x \leq \sqrt{6}\}$

**13**  Hyperbola; $y$-intercepts $\pm\dfrac{1}{5}\sqrt{15}$ ; domain = all real numbers

**15**  Hyperbola; $x$-intercepts $\pm 3$; domain $\{x | x \geq 3\} \cup \{x | x \leq -3\}$

**17**  Parabola; intercepts 0; domain = all real numbers

**19**  Hyperbola; $y$-intercepts $\pm 1$; domain = all real numbers

**23**

**25**

**29**

**33**

**35**

**37**

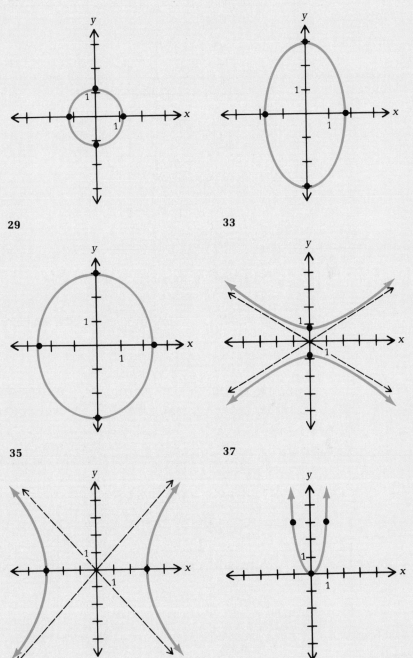

**39**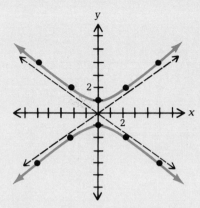

**(7.8)**  **3**  $C = kr$  $(k > 0)$    **5**  $S = kH$  $(k > 0)$

**7**  $H = \dfrac{k}{d^2}$  $(k > 0)$    **9**  $V = khr^2$  $(k > 0)$

**13**  $y = \dfrac{12}{x}$    **15**  $W = \dfrac{dt^2}{18}$    **19**  40 miles    **21**  95.2 pounds


**(8.1)**  **3**  0.0239    **5**  5,000,000,000,000    **7**  0.31    **9**  12,230

**13**  $(2.5)10^7$    **15**  $(3.1)10^{-3}$    **19**  $(1.4001)10^1$

**23**  $2 \cdot 10^9$    **27**  $(1.956)10^{13}$    **29**  $(9.741)10^{11}$


**(8.2)**  **3**

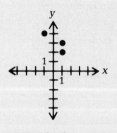

**(a)** Graph of $S$

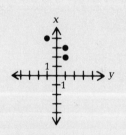

**(b)** Interchange $x$ and $y$

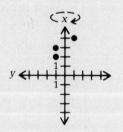

**(c)** Reflect

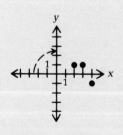

**(d)** Rotate 90° for graph of $S^{-1}$

**5**

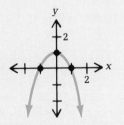

**(a)** Graph of $S$

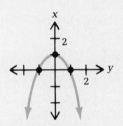

**(b)** Interchange
$x$ and $y$

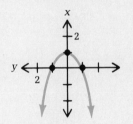

**(c)** Reflect

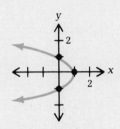

**(d)** Rotate 90° for
graph of $S^{-1}$

**7**

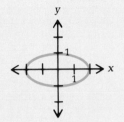

**(a)** Graph of $S$

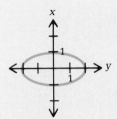

**(b)** Interchange
$x$ and $y$

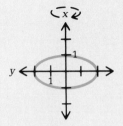

**(c)** Reflect

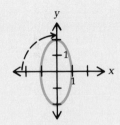

**(d)** Rotate 90° for
graph of $S^{-1}$

**9**

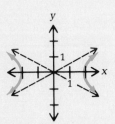

**(a)** Graph of $S$

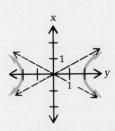

**(b)** Interchange
$x$ and $y$

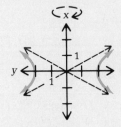

**(c)** Reflect

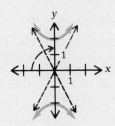

**(d)** Rotate 90° for
graph of $S^{-1}$

**13**  $f^{-1}(x) = 4x - 6$     **15**  $f^{-1}(x) = \dfrac{10 - 7x}{4}$

**17**  $f^{-1}(x) = \dfrac{1}{x} - 1 \qquad (x \neq 0)$

**21**  The function $h$ is one-one. Its inverse is the function $h^{-1}(x) = x^2 - 1$ with domain $\{x \mid x \geq 0\}$.

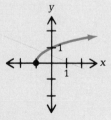

**(a)** Graph of $h$

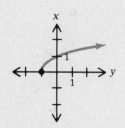

**(b)** Interchange $x$ and $y$

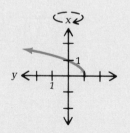

**(c)** Reflect

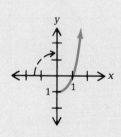

**(d)** Rotate 90° for graph of $h^{-1}$

**23**  The function $f$ is not one-one.

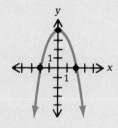

**(a)** Graph of $f$

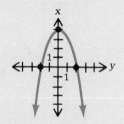

**(b)** Interchange $x$ and $y$

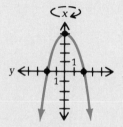

**(c)** Reflect

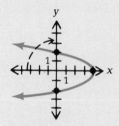

**(d)** Rotate 90° for graph of $f^{-1}$

**27**  This function is not one-one since, for example, $f(-1) = f(1)$.

**29**  One-one; $g^{-1}(x) = \sqrt{\dfrac{x + 9}{2}}$

**31**  One-one; $f^{-1}(x) = x^2 - 1$, with domain $\{x \mid x \leq 0\}$

**35**  The time it takes for the object to drop $x$ feet is represented by $d^{-1}(x) = \dfrac{1}{4}\sqrt{x}$. The domain of $d^{-1}$ is $[0,200]$; its range is $\left[0, \dfrac{5}{2}\sqrt{2}\,\right]$.

**39**  $V^{-1}(x) = \sqrt[3]{\dfrac{3x}{4\pi}}$ represents the radius of a sphere that has volume $x$; domain = range = $\{x \mid x > 0\}$.

**41**  $C^{-1}(x) = \dfrac{9}{5}(x + 32)$ expresses $x$ Celsius as a Fahrenheit temperature.

**(8.3)**  **3** 7   **5** 9   **7** $\dfrac{1}{10}$

**13** $\left(-3, \dfrac{1}{125}\right), \left(-1, \dfrac{1}{5}\right), (2, 25)$

**15** $(-3, 8), (-2, 4), \left(4, \dfrac{1}{16}\right)$   **19** $\left(-2, \dfrac{1}{4}\right), \left(-3, \dfrac{1}{8}\right), (5, 32)$

**23**   **25**   **27**

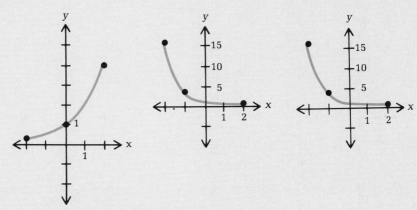

**33** $\dfrac{1}{4}$ milligram at 60 seconds, $\dfrac{1}{8}$ milligram at 90 seconds

**37** 30.0 inches at sea level, 19.2 inches at 2 miles

**(8.4)**  **3** $\log_3 1 = 0$   **5** $\log_4 \dfrac{1}{16} = -2$   **7** $\log_4 2 = \dfrac{1}{2}$

**9** $\log_{2/3} \dfrac{9}{4} = -2$   **13** $4^{1/2} = 2$   **15** $\left(\dfrac{1}{6}\right)^{-2} = 36$

**17** $10^3 = 1,000$   **19** $\left(\dfrac{1}{8}\right)^{-4/3} = 16$   **23** $1$

**25** $-4$   **27** $-2$   **29** $2$   **33** $4, -2, 1$

**35** $-\dfrac{1}{2}, -\dfrac{3}{2}, 2$   **37** $49, \dfrac{1}{7}, 1$   **39** $2$   **43** $\dfrac{1}{27}$

**47**                                         **49**

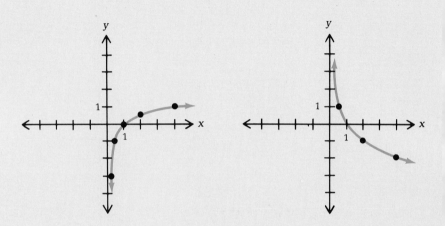

**(8.5)**   **3** $1 + 2 \log_4 x + \log_4(x+1) - \dfrac{1}{2}\log_4(x-2)$

**5** $\dfrac{1}{2} - \dfrac{1}{2}\log_{16}(y-1)$   **9** $\log_5 \dfrac{(x+5)^3}{5x}$

**11** $35$   **15** $5$   **19** $1.1461$   **21** $-0.5441$

**25** $1.3980$   **29** $30$

**33**   Note that $b^{2(\log_{b^2} x)} = (b^2)^{\log_{b^2} x} = x$. Since $\log_b x$ is the unique number $y$ such that $b^y = x$, it follows that $\log_b x = 2 \log_{b^2} x$.

**(8.6)**   **3** $0.9345$   **5** $1.5441$   **9** $0.0414 - 3$   **11** $2 \cdot 10^0 = 2$

**15** $(9.05)10^5 = 905,000$   **17** $(7.53)10^{-2} = 0.0753$

**19** $(1.02)10^{-1} = 0.102$   **23** $1.9161$   **25** $3.7591$

**27** $0.4739 - 1$   **29** $0.0913 - 4$   **33** $(8.862)10^2 = 886.2$

**35** $(4.525)10^4 = 45,250$   **39** $(2.398)10^{-4} = 0.0002398$

**43** $3.8991$   **45** $7.7582$   **47** $0.7934 - 6$   **49** $0.5160 - 3$

(8.7)   **3**   0.7143; actual answer 0.7143       **5**   (1.740)10$^6$       **7**   2.55

**13**   0.8670       **15**   (1.673)10$^5$       **17**   324       **19**   21.37

**23**   2.574       **27**   8.22       **29**   11.8

**33**   (2.607)10$^{11}$ cubic miles       **35**   2.429       **39**   39.37

**41**   2.216 seconds       **43**   1.872 square inches       **45**   $1,724


(8.8)   **1**   1.16       **5**   −0.38       **7**   1.23       **11**   1.47       **15**   4.644

**17**   1.792       **19**   0.5       **21**   $2,226, 8.66 years

**25**   11.55 seconds       **29**   1.6 miles


(9.1)   **3**   Dependent system

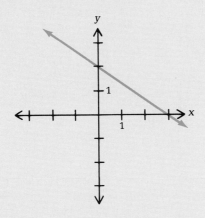

**5**                                        **7**

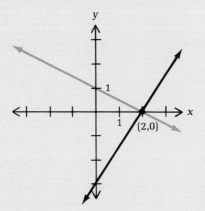                         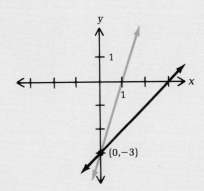

**11** $\left\{\left(\dfrac{1}{3},-\dfrac{1}{2}\right)\right\}$    **13** $\left\{\left(\dfrac{13}{33},-\dfrac{29}{11}\right)\right\}$    **17** $\left\{\left(\dfrac{1}{2},-1\right)\right\}$

**19** $\left\{\left(\dfrac{27}{5},-\dfrac{14}{5}\right)\right\}$    **21** 14 and 9

**25** He wants \$4,500 invested at 4% and \$8,000 at 7%.

**27** The plumber earned \$8 per hour and his helper \$5.

**29** 80 pounds of the deluxe mixture and 65 pounds of the economy blend

**33** Original pump 6 hours, borrowed pump 14 hours

**(9.2)**    **3** 2      **5** 0      **7** $-\dfrac{1}{8}$      **11** $\left\{\left(\dfrac{1}{3},-\dfrac{1}{2}\right)\right\}$

**13** $\left\{\left(\dfrac{13}{33},-\dfrac{29}{11}\right)\right\}$      **17** $\left\{\left(\dfrac{1}{2},-1\right)\right\}$

**19** $\left\{\left(\dfrac{27}{5},-\dfrac{14}{5}\right)\right\}$      **23** Dependent system

**25** $\left\{\left(-\dfrac{12}{7},-\dfrac{22}{7}\right)\right\}$      **29** Dependent system

**(9.3)**    **3** $\{(-4, 0, 9)\}$      **5** $\left\{\left(-\dfrac{79}{5},\dfrac{61}{10},-\dfrac{22}{5}\right)\right\}$      **7** $\{(2, -3, 1)\}$

**11** Inconsistent    **13** 25, 12, and 8

**17** \$60 for a three-speed, \$85 for a five-speed, and \$108 for a ten-speed

**19** 20 at \$10, 35 at \$6, 15 at \$4

**21** $y = -x^2 + 3x + 5$

**(9.4)**    **3** 0      **5** 20      **9** $\{(2, 3, 0)\}$      **11** $\{(1, -3, -3)\}$

**15** Inconsistent system    **17** $\left\{\left(-\dfrac{3}{7},\dfrac{12}{7},-\dfrac{1}{7}\right)\right\}$    **19** $\left\{\left(\dfrac{1}{2},\dfrac{1}{2},\dfrac{3}{2}\right)\right\}$

**3** {(1,3), (3,−1)}          **5** {(2,0), (0,3)}

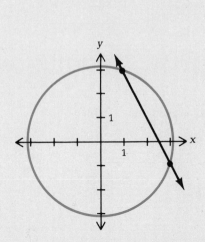

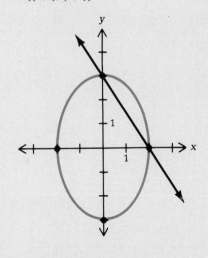

**7** {(1, 4)}          **9** {(−1 + 2i, −3 + 4i), (−1 − 2i, (−3 − 4i)}

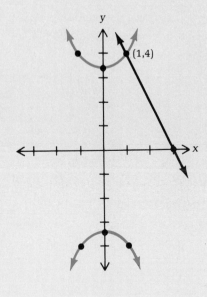

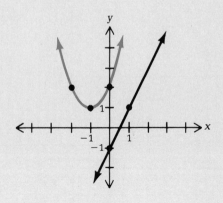

**13** {(2, i), (2, −i), (−2, i), (−2, −i)}

**15** {($\sqrt{2}$, $\sqrt{3}$), ($\sqrt{2}$, −$\sqrt{3}$), (−$\sqrt{2}$, $\sqrt{3}$), (−$\sqrt{2}$, −$\sqrt{3}$)}

**19** $\{(4, 1)\}$      **21** $\{(3, 1), (-3, -1), (i, -3i), (-i, 3i)\}$

**23** $\left\{\left(1, \dfrac{1}{2}\right), \left(-1, -\dfrac{1}{2}\right)\right\}$

**27** $\left\{(3, 1), (-3, -1), \left(2, \dfrac{3}{2}\right), \left(-2, -\dfrac{3}{2}\right)\right\}$      **29** $1 + 3i$

**31** $3 - 2i$      **35** 10 and 6      **37** 4 inches and 6 inches

**39** Car 50 mph, climbing rate 2 mph

**(10.1)**    **3** $s(n) = 1 \ (n = 1, 2, 3, 4, 5)$      **5** $s(n) = \dfrac{n}{n + 1} \ (n = 1, 2, 3, 4)$

**9** $\dfrac{1}{5}, \dfrac{1}{7}, \dfrac{1}{9}, \dfrac{1}{11}, \dfrac{1}{13}$      **11** $-1, 1, -1, 1$

**15** $1, \dfrac{1}{4}, \dfrac{1}{9}, \dfrac{1}{16}, \dfrac{1}{25}$      **17** $5, 8, 11, 14, 17$

**19** $\dfrac{1}{2}, \dfrac{4}{3}, \dfrac{9}{4}, \dfrac{16}{5}, \dfrac{25}{6}$

**23** First sequence 3, 5, 7, 9, 11; second sequence 3, 5, 7, 9, 35

**25** First sequence 1, 4, 9; second sequence 1, 4, 11

**29** One possible answer is $s_n = n^2$;   $s_n = n^2 + (n - 1)(n - 2)(n - 3)(n - 4)$.

**(10.2)**    **3** $5, 2; s_n = 5 + (n - 1)(-3)$      **5** $\dfrac{5}{2}, \dfrac{7}{2}; s_n = \dfrac{5}{2} + (n - 1)(1)$

**7** $-3, -\dfrac{11}{2}; s_n = 2 + (n - 1)\left(-\dfrac{5}{2}\right)$      **11** $-32$      **15** 46

**19** $-16$      **21** 90      **23** 64      **27** 2,450      **29** $4,850

**31** 8 years

**(10.3)**    **3** $1, \dfrac{1}{3}, \dfrac{1}{9}; s_n = 9\left(\dfrac{1}{3}\right)^{n-1} = 3^{3-n}$

**5** $8, -32, 256; s_n = \dfrac{1}{2}(-4)^{n-1}$      **9** 4      **11** $-3^6$

**13** $9,930      **17** $\dfrac{11}{2}$      **19** 90      **23** $-2,457$

**27** 5,460 letters     **31** $\dfrac{125}{6}$     **33** No sum     **35** 27

**37** $\dfrac{27}{5}$     **41** 52 inches     **43** \$320

**47** $\dfrac{16}{99}$     **49** $\dfrac{3}{11}$     **51** $\dfrac{4,577}{1,110}$

**(10.4)**   **3** $243w^5 - 810w^4 + 1,080w^3 - 720w^2 + 220w - 32$

**5** $64x^6 - 64x^5 y + \dfrac{80}{3}x^4 y^2 - \dfrac{160}{27} x^3 y^3 + \dfrac{20}{27} x^2 y^4$

$- \dfrac{4}{81} xy^5 + \dfrac{1}{729} y^6$

**7** $\dfrac{1}{27} a^3 + \dfrac{1}{3} a^2 b^2 + ab^4 + b^6$

**11** $x^4 - 8x^{23} y + \dfrac{92}{3} x^{22} y^2 - \dfrac{2,024}{27} x^{21} y^3 + \dfrac{2,552}{27} x^{20} y^4$

**13** $8,192a^{13} + 53,248a^{12} b + 159,744a^{11} b^2 + 292,864a^{10} b^3 + 366,080a^9 b^4$

**15** $h^{12} - 36h^{11} + 594h^{10} - 5940h^9 + 40,095h^8$

**19** $17,856a^{30}$     **21** $-\dfrac{819}{2} t^{25}$     **25** 0.83     **27** 1.20

**31** 11,046     **35** $1 - x + x^2 - x^3$     **39** 1.02     **41** 0.95

## (∘)
## *Index*

Abscissa, 202
Absolute inequality, 126
Absolute value, 11, 133–36
Addition, 14
Additive identity law, 15
Additive inverse law, 15
Algebraic expression, 36
Arithmetic, fundamental theorem
   of, 20
Arithmetic sequence:
   finite, 368
   infinite, 368
Arithmetic series, 369
Associative law:
   for addition, 15
   for multiplication, 19
Asymptotes, 252
Axis:
   horizontal, 201
   vertical, 201
   $x$-, 201
   $y$-, 201
Axis of symmetry, 231

Base, 32
Base $e$, 311
Binomial:
   defined, 36
   expansion, 383
   series, 383
   theorem, 384

Cancellation law, 20
Cartesian plane, 202, 267
Cartesian system, 202
Characteristic, 296
Circle, 249
Clearing factor, 91
Closed interval, 27
Coefficient, 36
Common logarithms, 294
Common multiple, 67
Complete factorization, 67
Completing the square, 174
Complex fractions, 90
Complex numbers, 162
Conditional equation, 100
Conditional inequality, 126
Conic sections, 254
Conjugate, 158
Constant:
   of proportionality, 257
   of variation, 257
   term, 36, 37
Coordinate, 202

Coordinate axes, 202
Cramer's rule, 331–33

Degree, 37
Denominator, least common, 71
Dependent system, 318
Determinant:
   second-order, 330
   third-order, 346
Difference, 15
Difference between successive
   terms, 369
Direct variation, 257
Discriminant, 179
Disjoint sets, 8
Distance, 214
Distance formula, 215
Distributive law, 22
Dividend, 20
Division, 20
Divisor, 20
Domain, 207
Double negative law, 16

Element, 2
Elimination of variables, 320
Ellipse, 250
Empty set, 2
Endpoints, 27
Equal, 3
Equation:
   conditional, 100
   defined, 99
   defining the relation, 208
   exponential, 310
   first-degree, 103
   graph of, 202
   linear, 203
   logarithmic, 292
   member of, 99
   quadratic, 169
   second-degree, 169
   systems, 317
Equivalent:
   defined, 100
   inequalities, 127
Exponent, 32
Exponential:
   decay function, 279
   equation, 310
   function with base $b$, 277
   growth function, 278
Expression, algebraic, 36
Extraction of roots, 173
Extraneous root, 185

Factor:
  clearing, 91
  defined, 20
Factoring:
  defined, 50
  solution by, 170
Factorization:
  complete, 67
  proper, 67
Finite:
  arithmetic sequence, 368
  sequence, 364
  set, 2
First-degree equation, 103
First term, 369
Formula, quadratic, 179
Fractions:
  complex, 90
  defined, 61
  fundamental principle of, 62
Function:
  defined, 209
  exponential decay, 279
  exponential growth, 278
  exponential with base $b$, 277
  logarithmic with base $b$, 284
  one-one, 210
Fundamental principle of fractions, 62
Fundamental theorem of arithmetic, 20

General term, 365
Geometric sequence, 374
Geometric series, 375
Graph:
  defined, 10
  of $(a,b)$, 202
  of an equation, 202
  sign, 185, 189
Graphing, 9
Greater than, 10

Horizontal axis, 201
Hyperbola, 252

Identity, 100
Imaginary number, 163
Inconsistent system, 318
Index, 148
Inequalities, equivalent, 127
Inequality:
  absolute, 126
  conditional, 126
  defined, 126
  member of, 126

Inequality, *cont.*
  solution of, 126
Infinite:
  arithmetic sequence, 368
  sequence, 365
  set, 2
Integer, 2
Intercept:
  defined, 220
  $x$-, 220, 239
  $y$-, 220, 239
Intercepts, 239
Interpolation, linear, 296
Intersection:
  defined, 7
  of intervals, 27–8
Interval:
  closed, 27
  open, 27
Inverse, multiplicative, 19
Inverse variation, 258
Irrational number, 2

Joint variation, 258

Law:
  additive identity, 15
  additive inverse, 15
  cancellation, 20
  distributive, 22
  double negative, 16
  multiplicative identity, 19
  multiplicative inverse, 19
  trichotomy, 27
Law for addition:
  associative, 15
  commutative, 15
Law for multiplication:
  associative, 19
  commutative, 19
Least common denominator (LCD), 71
Least common multiple (LCM), 67
Less than, 10
Like terms, 36
Line, number, 9
Linear equation:
  defined, 203
  point-slope form, 221
  slope-intercept form, 220
  standard form, 220
Linear interpolation, 296
Logarithm:
  common, 294
  natural, 311

Logarithmic equation, 292
Logarithmic function with base $b$, 284
Lowest terms, 69

Mantissa, 296
Member:
  of equation, 99
  of inequality, 126
  of set, 2
Method of substitution, 319
Monomial, 36
Multiple, least common, 67
Multiplication, 19
Multiplicative identity law, 19
Multiplicative inverse, 19
Multiplicative inverse law, 19

Natural logarithm, 311
Natural number, 2
Negative number, 10
Non-negative number, 10
Notation:
  exponential, 145
  radical, 145
  scientific, 263
  set-builder, 7
$n$th root, 144
$n$th term, 365
Null set, 2
Number:
  complex, 162
  imaginary, 113
  irrational, 2
  natural, 2
  negative, 10
  non-negative, 10
  positive, 9
  prime, 20
  rational, 2
  real, 3
Number line, 9
Numeral, 2

One-one function, 210
Open interval, 27
Opposite sense, 126
Ordered pair, 201
Ordered triple, 337
Order of radical, 148
Ordinate, 202
Origin, 9, 201

Point-slope form (linear equation), 221

Polynomial:
    defined, 36
    prime, 67
    quadratic, 53
Positive number, 9
Power, 32
Prime:
    number, 20
    polynomial, 67
Product:
    defined, 19
    of polynomials, 46
Proper factorization, 67
Proportionality, constant of, 257

Quadrants, 202
Quadratic:
    equation, 169
    formula, 179
    polynomial, 53
Quotient, 23

Radical:
    notation, 145
    sign, 148
Radical, order of, 148
Radicand, 148
Range, 207
Ratio, 374
Rationalizing the denominator, 150
Rational number, 2
Real number, 3
Reciprocal, 19
Rectangular coordinate system, 202
Reflection, 231
Relation:
    defined, 207
    equation defines the, 208
Remainder:
    defined, 85
    theorem, 87
Replacement set, 99
r factorial, 383
Root:
    cube, 144
    extraneous, 185
    nth, 144
    of equation, 100
    square, 144

Same sense, 126
Scientific notation, 263
Second-degree equation, 169
Second-order determinant, 330
Sense:
    same, 126
    opposite, 126
Sequence:
    arithmetic, 368
    defined, 364
    finite, 364
    finite arithmetic, 368
    geometric, 374
    infinite, 365
    infinite arithmetic, 368
Series:
    arithmetic, 369
    binomial, 383
    geometric, 375
Set:
    defined, 2
    finite, 2
    infinite, 2
    replacement, 99
    solution, 100, 127
Set-builder notation, 7
Set intersection, 7
Set union, 6
Sets, disjoint, 8
Sign, radical, 148
Sign graph, 189
Slope, 218
Slope-intercept form (linear equation), 220
Solution:
    by factoring, 170
    defined, 100
    of inequality, 126
Solution set, 100, 127
Solve:
    an equation, 100
    an inequality, 127
Square root, 144
Standard form (linear equation), 220
Subset, 3
Substitution, method of, 319
Subtraction, 16
Sum, 14
System:
    Cartesian, 202

System, cont.
    dependent, 318
    inconsistent, 318
    rectangular coordinate, 202
System of equations, defined, 317

Term:
    constant, 36, 37
    defined, 36
    first, 369
    general, 365
    nth, 365
Terms:
    difference between successive, 369
    like, 36
    lowest, 69
    of a sequence, 364
    unlike, 36
Theorem, remainder, 87
Third-order determinant, 346
Trichotomy law, 27
Trinomial, 36
Turning point, 231

Union:
    defined, 6
    of intervals, 27–8
Unknown, 99
Unlike terms, 36

Variable, 7
Variables, elimination of, 320
Variation:
    constant of, 257
    direct, 257
    inverse, 258
    joint, 258
Vertical axis, 201

x-axis, 201
x-intercept, 220

y-axis, 201
y-intercept, 220